V. M. Akulin   N. V. Karlov

Intense Resonant Interactions in Quantum Electronics

Translated by
O.N.Tselikova and V.S.Potapchouk

With 102 Figures

Springer-Verlag
Berlin Heidelberg New York
London Paris Tokyo
Hong Kong Barcelona
Budapest

V. M. Akulin

Moscow Institute
of Physics and Technology
Dolgoprudny, Institutskij per. d. 9
141700 Moscow, USSR

N. V. Karlov

Moscow Institute
of Physics and Technology
Dolgoprudny, Institutskij per. d. 9
141700 Moscow, USSR

Translators

O. N. Tselikova

Moscow Institute
of Physics and Technology
Dolgoprudny, Institutskij per. d. 9
141700 Moscow, USSR

V. S. Potapchouk

Moscow Institute
of Physics and Technology
Dolgoprudny, Institutskij per. d. 9
141700 Moscow, USSR

Editors

Elliott H. Lieb

Jadwin Hall
Princeton University
P.O. Box 708
Princeton, NJ 08544-0708, USA

Wolf Beiglböck

Institut für Angewandte Mathematik
Universität Heidelberg
Im Neuenheimer Feld 294
W-6900 Heidelberg 1, FRG

Tullio Regge

Istituto di Fisica Teorica
Università di Torino
C. so M. d'Azeglio, 46
I-10125 Torino, Italy

Robert P. Geroch

Enrico Fermi Institute
University of Chicago, 5640 Ellis Ave.
Chicago, IL 60637, USA

Walter Thirring

Institut für Theoretische Physik der
Universität Wien, Boltzmanngasse 5
A-1090 Wien, Austria

Title of the original Russian edition:
Intensivnye resonansnye vsaimodeistviya v kvantovoi elektronike
© Nauka, Moscow 1987

ISBN-13: 978-3-642-64757-4   e-ISBN-13: 978-3-642-61241-1
DOI: 10.1007/978-3-642-61241-1

Typesetting: Data conversion by Springer-Verlag
56/3140 – 5 4 3 2 1 0 – Printed on acid-free paper

Preface to the English Edition

This book is a course of lectures given for senior students at the Moscow Institute of Physics and Technology. For those who have graduated in the USSR this information should be sufficient to give an idea of the level and the manner in which the subject matter is presented. On the other hand, readers outside of this country may never have heard about this well-known Soviet institution, and so we would like to say few words about it now.

The Moscow Institute of Physics and Technology (MFTI or Fiztekh) was founded in 1947 as the result of a special directive of Stalin in order to supply the space and nuclear program with highly educated experts. The best scientists in the country were involved in the foundation process. They invented an effective and flexible educational system that includes basic education according to an university program followed by specialization at leading scientific centers. Being organized initially as a department of Moscow State University, MFTI recently separated into an independent institution. In the sixties it lost its mainly top secret and military character and became the most prestigious place in the country for an education in physics. The political changes of the last few years have opened it to contacts with other countries.

The course of lectures comprising this book is dedicated to the subject of the intense resonant interaction of laser radiation with matter and contains a significant part of the Ph. D. exam program in quantum electronics. It should also be of interest to postgraduate students and those postdoctoral scientists who are just starting to deal with lasers and their applications and would like to become acquainted with the subject in more detail. The main purpose of the book is to give readers a self-consistent and up-to-date view of the subject, with an emphasis on the quantum mechanical background of elementary processes. But we have also kept in mind the necessity of introducing important terms, mathematical methods and ideas from adjacent fields in the context of a physical description.

The four years which have passed since the Russian edition of this book appeared have not affected contemporary interest in the main concepts. The development of quantum electronics has just added some new physical information, which can still be considered in the same conceptual framework. For those readers who would like to have recent information, we give an additional list of references, which is also included in case some of the books and articles we had suggested to Russian readers are not available to English-speaking readers. This additional list, like the primary one, does not pretend to be comprehensive or complete – it just gives the papers which we feel are similar in level and manner of presentation to this book.

We hope that our English-speaking readers will find this book interesting.

Moscow, April 1991   *V. M. Akulin*
                                                      N. V. Karlov

Preface to the Russian Edition

This book concerns the physical basis and applications of intense resonant interactions of laser light with matter. It contains lectures given to the students of the Moscow Institute of Physics and Technology in the frame of a general course of quantum electronics. This course starts with the fundamentals of laser physics followed by a description of the operation principles, the construction principles and the properties of the best-known lasers. We consider that a course on the fundamentals of quantum electronics should be immediately followed by a course of lectures on intense resonant interactions between laser light and matter, since the physics of these elementary interactions is in close proximity to, and sometimes coincides with, the physics of the processes that bring about laser generation.

The book is meant for students who are going to take up theoretical or experimental work in the domain of intense resonant interactions. We should particularly stress the peculiarities of narration that, to our mind, make the material presented here interesting for both experimentalists and theoreticians. For convenience, we have not hesitated to use modern mathematical methods, trying, nevertheless, to put them into a clear physical form. Preserving the rigor of presentation we have nonetheless tried to avoid unnecessary details and awkward calculations. As often as possible we call the attention of the students to similarities in certain interaction processes, stressing at the same time their peculiarities. We also hope that the form of presentation of the material makes it possible for students who do not aim at a detailed consideration of the mathematics of the processes considered to clearly understand their physical meaning. To this end we start each investigation with a qualitative description of the phenomenon under study.

As a rule, the resonant interactions of laser light with matter are basically quantum phenomena. We have tried to stress this important fact by the order of presentation of the material. The lectures start by considering phenomena whose quantum nature is more evident. We mean here the coherent processes of the resonant interaction of light with two-level quantum systems: optical nutation, photon echo, and self-induced transparency among them. All of these occur in a single isolated two-level quantum system. A further complication lies in considering an ensemble of such systems interacting or not interacting among themselves or with their surroundings. It is in these system ensembles that the phenomena of coherent damping and transverse and longitudinal relaxation manifest themselves. We next come to an investigation of these phenomena, together with the saturation processes of homogeneously and inhomogeneously broadened lines. Any variations in the internal state of the particles of the ensemble in a resonant laser field may lead to changes in the statistical properties of the ensemble. This is the nature of the effects of light pressure in the laser fields of travelling and standing waves and of the light-induced drift taken up in the next block of lectures entitled Laser Gas Kinetics.

Under the conditions of a resonant laser field of high intensity it is essential that the internal structure is complex, the particle being excited is multilevel and the quantum transitions are multiphoton transitions. We then proceed to describe the multiphoton processes and the excitation dynamics of spectrally complex multilevel systems. Many phenomena known in quantum electronics as nonlinear parametric effects have a pronounced quantum-mechanical origin easily interpreted in terms of multilevel-system dynamics. This circumstance has also found its place in our description of the multitude of processes possible in an ensemble of multilevel systems; we have chosen to dwell only upon the processes of collisional relaxation and on cascade processes of molecular excitation by a laser field.

Speaking of applications, we think it worthwhile to consider the resonant processes of laser excitation of atoms, their selective photoionization included, and the molecular excitation at both vibrational and electronic transitions in a field of intense laser light. We pay great attention to the selective dissociation of molecules by IR-laser light, the combined influence of infrared and ultraviolet resonant radiation of molecules, the problem of selective laser photochemistry, resonant laser photochemistry and resonant processes near a surface. From the point of view of experimental realization it is these processes that are the source of intense resonant interactions of laser light with matter. They have many aspects but the most essential of each of them is, as a rule, one of the resonant quantum phenomena we have already studied, and we pay special attention to this.

We are not going to give the reader all the details of the experimental applications of the processes considered in the book, nor their theoretical descriptions, nor the technicalities of the respective installations. All those interested in the above problems may consult the list of recommended literature which, although far from complete, nevertheless permits a deeper insight into the matter. We should emphasize here that the works on this list are methodologically similar to the problems posed in our book and correspond didactically to the manner of presentation chosen by the authors. Such a list should be concise; otherwise it would not be useful to the reader. Obviously, the works chosen according to this principle cannot cover the latest developments in the corresponding branch of science; neither do they touch upon its history.

In addition, we have not intended to introduce the reader to the problem, taken up traditionally in courses of general and theoretical physics, of the theory of oscillations and of atomic and molecular spectroscopy. We recommend, if necessary, turning to the following well-known books[*]: L. Landau, Eu. Lifshitz "Quantum Mechanics"; Eu. Lifshitz, L. Pitaievsky "Physical Kinetics"; V. Silin "Introduction to the Kinetic Theory of Gases"; Yu. Rumer, M. Ryvkin "Thermodynamics, Statistical Physics and Kinetics"; A. Andronov, A. Vitt, S. Haikin "The Theory of Oscillations"; I. Sobelman "Introduction to the Theory of Atomic Spectra"; M. Elyashevitch "Atomic and Molecular Spectroscopy"; G. Herzberg "The Vibrational and Rotational Spectra of Polyatomic Molecules". Only once have we strayed from this rule by including a lecture on molecular spectra. The subject matter of this lecture is not directly concerned with intense resonant interactions but is nevertheless necessary to the study of the laser-induced processes in molecules in the form we have previously chosen. But the reader who comes across the problem of polyatomic molecules for the first time will find that, along with the general texts mentioned above, Hehts's paper in the list of references is of much use.

To help the reader with the mathematics of the present course we would suggest the book "Mathematical Physics" by Ch. Mathews.

We also recommend the following books as guides to the fundamentals of quantum electronics: N. Karlov "Lectures on Quantum Electronics", O. Svelto "Laser Principles".

The authors thank V. Bagratashvilli, A. Giardini-Guidoni, V. Kovalev, B. Lukjanchuk, G. U. Rich, Ch. Treanor, L. Shelepin and A. Shalagin for the material they presented and for discussions of problems of IR-laser photochemistry, the inversion of the wave front, laser thermochemistry, collisional relaxation of anharmonic oscillators and the light-induced drift. Without these discussions it would have been impossible to include these problems in the book.

---

[*] The corresponding bibliography is given at the end of the book.

The authors appreciate the students and postgraduates who, by voicing their opinions on the manner of presentation of the material, have contributed to its improvement.

The authors are grateful to our reviewers, the Chair of Optics and Spectroscopy of the Physics Department of Moscow State University, and to A. Dykhne for a number of essential suggestions gratefully accepted at the final stage of preparation of the manuscript for publication.

The authors express their profound gratitude to Alexander Prokhorov. Under his guidance and together with him they worked over the problems included in this book.

And, finally, the authors are grateful to the Moscow Institute of Physics and Technology, from which they graduated and where they now work.

Contents

Linear and nonlinear coupling between oscillators, linear and nonlinear polarizabilities. Harmonics and undertones. Cubic nonlinearity. Compound level schemes, multilevel transitions in them. Intermediate resonances. IRS, the Stokes and anti-Stokes components. CARS. Spontaneous processes, probability of spontaneous decay and spontaneous scattering. Spontaneous processes in the presence of a strong field.

Selective two-step ionization of atoms. Spectroscopic condition. Probability of excitation and ionization. Resonant transfer of excitation energy, resonant recharge. Laser isotope separation in atomic vapor, conditions of its effectiveness. Autoionization, methods of artificial autoionization. Multiphoton ionization of atoms. Example of three-photon ionization of metastable helium atoms. Parametric generation.

The Born-Oppenheimer parameter. Hierarchy of spectra. Normal vibrations. Anharmonicity. Rotations. K-degeneracy, hyperfine splitting. Vibrational-rotational interactions. Three-dimensional oscillator, anharmonic removal of degeneracy. The Coriolis splitting. Hybrid states. The Fermi resonance. Dipole transitions. IR-active modes. Rotational selection rules. P-, Q- and R-branches. Hot bands. Stochastization of vibrations.

Simple-oscillator model. Anharmonicity and overcoming it by means of a strong field. Polyatomic molecules. Model of the excited mode damped by the interaction with a thermostat. Classical and quantum approaches. Large polyatomic molecules. Longitudinal and transverse relaxations. Fermi resonances, soft modes. Small polyatomic molecules. Number of atoms. Three stages of excitation. Excitation of quasicontinuum. Kinetic coefficients. Phase-volume average. Single degree of freedom of three-fold degenerate mode excitation. Spectral dependencies.

Lower levels. Multiphoton resonances. Fraction of excited molecules. Two-photon Q-branch of a spherical top. Saturation. Large molecules. RRKM model. Small polyatomic molecules, multidimensional motion in a potential well, stochasticity, probability of dissociation. Experimental studies. Examples of CF_3I- and SF_6-molecules. Fermi resonances. Spectral dependencies of kinetic coefficients. Red shift. Sharp resonant structure of the spectrum of lower levels.

Vibrational exchange in collisions of simple oscillators. Vibrational heating in the absence and in the presence of vibrational-translational relaxation. V-V'-exchange, preferable heating of a softer oscillator. Radiational-collisional cascade, vibrational temperature greatly exceeding translational one, pulsed and continuous irradiation regimes. Collisions of anharmonic oscillators, possibility of population inversion. Polyatomic molecules, bottleneck broadening. Photon-assisted collisions, the Rabi and Weisskopf frequencies.

Diatomic molecule. Franck-Condon principle. Polyatomic molecules, generalization of Franck-Condon principle. Term-to-term transition at vibrational motion. Repulsion of "intersecting" terms. Landau-Zener transition. Intramolecular conversion. Photochemistry. Single- and multiphoton excitation. IR–UV irradiation, possibility of selective breaking of bonds. Change of distribution functions of reactants under the action of radiation. Chemical-kinetic equation. IR-laser irradiation.

Laser energy supply, its rate and resonant properties. Ambiguity, hysteresis, catastrophes. Back loops, periodic regimes. Stochastic attractors. Laser thermochemical pyrolysis and synthesis. Nontriviality of the products of laser thermochemistry. Thermochemistry on phase boundary surface, oxidation and combustion of metals in laser radiation field. Spatial instability. Cold wall thermochemistry. Physical adsorption. Van der Waals forces. Polarizational interaction of molecules in resonant laser field. Deepening the adsorption potential, hampering the diffusion through radiation-transparent structures with a developed surface.

Polarization and susceptibility. Self-influence at collisionless multiphoton excitation of molecules. Shift of dispersion curve. Pulse shortening. Self-focusing, self-defocusing, waveguide modes. Wave front inversion in four-wave interaction. Inverting mirror. Dynamic holography. Delayed self-influence at thermochemical interactions.

Lecture One

Introduction

We shall begin with defining the subject of our lectures. The well-known ability of quantum electronics to concentrate the energy of electromagnetic radiation in extremely small spatial, temporal and spectral intervals, limited only by the uncertainty principles, makes laser light interactions with matter and their applications the problem of interest. Of considerable interest are the intense interactions, including the irreversible ones, which result in macroscopically significant changes in the state of the exposed object.

The set of possible changes is rather diverse. Laser radiation may change the velocity and the coordinates of the irradiated particles, i.e., their velocity and coordinate distribution functions. It may also influence the structure of the irradiated particles, leading to ionization, dissociation, chemical rebuilding, isomerization, polymerization, etc. It may cause phase transitions such as evaporation, condensation, melting, crystallization, amorphization, quenching and so on. In case all these and analogous changes are produced by monochromatic light, and should the result depend upon frequency, we are dealing with the intense resonant interaction of laser light with matter.

Among all possible interactions, the intense resonant ones are of special interest since they utilize a whole set of unique properties of laser light most fully and directly.

The effects of the interaction of laser light with matter may be of the threshold type or may manifest smooth intensity or flux dependence. In either case the effect will bring about appreciable changes in the sample only if the population distribution under the action of laser light deviates noticeably from the initial or equilibrium distribution. At some time the distribution may turn into a stationary one or come back to equilibrium but with another characteristic temperature; i.e., in any case the redistribution of populations should occur once again. So, the redistribution of populations among the states is the fundamental cause of any effect of the resonant interaction of laser light with matter. The resonant action of laser light is intense only if the population redistribution caused by the irradiation is considerable and becomes macroscopically apparent in the modification of the properties or the behavior of the system. This is the border line between the intense resonant interactions and the linear spectroscopy, including laser one, where the influence of radiation on the substance is small.

Thus, the subject of our lectures is the resonant action of laser light on matter, leading to the significant redistribution of populations among the energy levels, and the subsequent changes in the properties and structure of the irradiated substance. We shall also discuss possible applications.

The redistribution of the populations between two states coupled via a radiative transition occurs differently under the conditions of stationary or essentially pulsed irradiation.

In the former case the redistribution of populations becomes noticeable near the saturation point, when the probability of the radiative occupation of a state becomes equal to the probability of the inverse process of the state decay determined by the relaxation. The particular mechanism of how the state is decaying, in a radiative or a nonradiative way, is unimportant. The competition of the occupation and decay rates leads to a steady-state distribution of populations that differs from the equilibrium one. In the limit of high intensities the initial and final states are equally populated. Taking into account degeneracies g_1 and g_2 of the corresponding states, we obtain the following relation for populations n_1 and n_2:

$$n_1/g_1 = n_2/g_2 \,. \tag{1.1}$$

It is known that the rate of the radiative occupation may be evaluated as $I\sigma/h\nu$, where I is the light intensity; σ, the absorption cross section; and $h\nu$, the energy of the corresponding quantum. If the state lifetime is τ, its decay rate will be τ^{-1}. Then in stationary conditions, i.e. when the irradiation is continuous or in a steady state for a time much longer than τ, the saturation may be achieved at $I\sigma/h\nu > \tau^{-1}$. Assuming $g_1 = g_2$, we have the following estimate for the saturation intensity:

$$I_{\text{sat}} \approx h\nu/\sigma\tau \,. \tag{1.2}$$

In the visible and near-IR spectral regions at a typical value of $\sigma \approx 10^{-18}$ cm^2 and lifetimes not exceeding 10^{-4}–10^{-3} s, the range of the possible values of I_{sat} is quite broad: from several dozen watts to several dozen kilowatts per cm^2.

The saturation in the continuous regime is often either unattainable because of the absence of sufficiently powerful continuous-wave lasers or undesirable due to severe overheating of the sample.

In the pulsed regime the saturation is much easier to achieve and the sample is not exposed to overheating. But the saturation in the pulsed regime differs from that of the continuous one. The irradiation regime should be treated as pulsed if the pulse duration does not exceed the lifetime of the excited level, i.e., $\tau_{\text{pulse}} < \tau$. This lifetime is determined by the so-called *longitudinal*, or energy relaxation time. The longitudinal relaxation time essentially exceeds the *transverse*, or phase, relaxation time, often by several orders of magnitude. The longitudinal relaxation time is usually denoted by T_1 and the transverse one, by T_2. At $\tau_{\text{pulse}} > T_2$ the phase relations are broken and the interaction of laser light with matter is essentially incoherent. If, further, $\tau_{\text{pulse}} < T_1$, the saturation has a pulsed character in the sense that the relaxation processes may be neglected during the irradiation pulse. Thus, at $T_2 < \tau_{\text{pulse}} < T_1$, the saturation may be achieved if the integral of the occupation rate of the excited level taken across the pulse duration approaches unity. In other words, in the incoherent pulsed regime the saturation is achieved at the energy density of radiation,

$$F_{\text{sat}} \approx \int_{\tau_{\text{pulse}}} I(t)dt = h\nu/\sigma. \tag{1.3}$$

In the visible and near-IR regions of the spectrum at the typical value $\sigma = 10^{-18}$ cm^2, the value of F_{sat} lies within the interval between several hundredths and several tenths of a joule per cm^2. These radiation-energy densities are readily attainable.

The phenomenological rate equations used for estimates (1.2) and (1.3) do not allow for the coherent effects of the interaction of radiation with matter that become apparent at times less than the phase relaxation time of the level being excited. In this time scale the systems are hamiltonian and should be described by the Schrödinger equation. To take account of the interaction with the field, the additional terms modifying the initial Schrödinger equation of the quantum system considered should be inserted into the Hamiltonian.

Provided the resonant approximation of the perturbation theory is made when considering the resonant electromagnetic field, and any relaxation processes as well as the spontaneous decay of levels are neglected, the interaction of the quantum system with the field is known to be reduced to the coherent oscillations of the probability of findind a particle in a state. The frequency of these oscillations (the so-called Rabi frequency) is determined by the electromagnetic-wave field amplitude $\mathcal{E}_0$ and by the matrix element of the dipole moment operator corresponding to the transition between the considered levels μ_{21}:

$$\Omega_0 = \mu_{21}\mathcal{E}_0/\hbar. \tag{1.4}$$

For allowed transition $\mu_{21} = 1$ D $= 10$ CGSE. The laser light intensity of 10^6 W/cm^2 corresponds to $\mathcal{E}_0 = 10^2$ CGSE. Then $\Omega_0 = 10^{11}$ Hz.

In accordance with the above we can say that the coherent interaction may be treated as intense if, during the pulse, a noticeable fraction of particles reaches the excited level. Then we may rewrite the requirement for the interaction to be intense as $\Omega_0\tau_{\text{pulse}} \geq 1$. This is equivalent to the condition

$$\mathcal{E}_0 \geq \hbar/\mu_{21}\tau_{\text{pulse}}, \tag{1.5}$$

where τ_{pulse} should be shorter than all the relaxation times. The corresponding intensity is

$$I_0 = c\mathcal{E}_0^2/4\pi \geq c\hbar^2/4\pi\mu_{21}^2\tau_{\text{pulse}}^2, \tag{1.6}$$

where c is the velocity of light. For allowed transitions the last relation yields the value of several dozen kW/cm^2 when illuminating the sample by nanosecond pulses.

Summing up the above discussion, we conclude that one should distinguish between the coherent and incoherent interactions and also between the continuous and pulsed regimes. The estimates (1.2), (1.3) and (1.5) give the order-of-magnitude values of the intensity, the energy density and the field strength of laser light that correspond to the essentially intense resonant interaction of this radiation with matter.

The following discussion will be devoted to the basic physical principles of the resonant interactions of intense laser light with matter. Before geting started, we believe it reasonable to remind the reader that quantum electronics usually uses a semiclassical approach. In this approach the radiation field is treated classically while the particles of the substance are considered as quantum objects. In particular, this approach assumes that the classical oscillating electromagnetic field $E(t) = \mathcal{E}_0 \cos \omega t$ induces an oscillating dipole moment in a quantum particle that is responsible for many aspects of the interparticle interaction as well as their interaction with the surroundings and the radiation field.

We shall start with the analysis of coherent interactions not complicated by any relaxation processes. The relaxation processes are rather slow compared to the period of the particle oscillations, if the quality of the resonant processes is not too low. In other words, the optical frequency of the transition usually essentially exceeds the relaxation rate, $\omega \tau \gg 1$. Thus, we can always choose a sufficiently short time interval where the relaxation processes are unessential. It is this time scale in which the interaction of resonant radiation with a quantum system leads to coherent effects. In case the coherent interaction is weak and the process itself is much longer, one should take into consideration the irreversible relaxation effects. After discussing the coherent effects we shall go on with the analysis of the incoherent resonant interactions of intense laser light with matter.

Some possible applications of the intense resonant interaction of laser radiation with matter will be discussed later, either during consideration of pertinent phenomena or individually, as for the selective ionization of atoms, the dissociation of molecules, laser photochemistry and selective heterogeneous processes.

Lecture Two

Coherent Interaction

Resonant interaction. Relative populations and polarization. Excitation in constant-amplitude field. Dynamic field broadening. Nutations. Pulsed excitation. Slow-varying-amplitude method. Pulse area, $\frac{\pi}{2}$- and π-pulses. Coherent damping.

It is convenient to consider the coherent effects of the interaction of laser light with matter using the wavefunction technique and the Schrödinger equation.

Let a two-level particle interact in a dipole way with the electric component of the radiation field

$$\mathcal{E}(t) = \mathcal{E} \cos \omega t = \frac{1}{2}\mathcal{E}\left(e^{j\omega t} + e^{-j\omega t}\right). \tag{2.1}$$

Then the interaction energy may be represented as

$$V = -\boldsymbol{\mu}\boldsymbol{E}(t), \tag{2.2}$$

where $\boldsymbol{\mu}$ is the dipole moment of the transition. In the following discussion, for simplicity, we shall consider vectors $\boldsymbol{\mu}$ and $\boldsymbol{E}$ to be parallel if the contrary is not stated.

Let us denote the energy levels of the particle by E_2 and E_1, E_2 being the upper level; i.e.,

$$E_2 - E_1 = \hbar\omega_{21} > 0. \tag{2.3}$$

We denote the detuning of the transition frequency ω_{21} from that of radiation ω by

$$\Delta = \omega_{21} - \omega \tag{2.4}$$

and assume that the following relation is always satisfied

$$\Delta \ll \omega_{21}, \omega. \tag{2.5}$$

Condition (2.5) is the requirement for the interaction to be resonant. It holds as long as the perturbation of particle eigenfrequencies remains sufficiently small, i.e., as long as the interaction energy does not become too great:

$$|V| = \mu\mathcal{E} \ll \hbar\omega_{21}. \tag{2.6}$$

When the interaction energy exceeds the interlevel distance, condition (2.6) is violated and the definition of the resonant interaction should be specified more precisely. For allowed transitions, i.e., at $\mu \approx 10^{-18}$ CGSE (1 Debye), the noticeable deviations from condition (2.6) take place for infrared light at intensities of $10^{11} - 10^{13}$ W/cm^2 and for visible and ultraviolet light, at $10^{13} - 10^{15}$ W/cm^2.

Generally speaking, one should also bear in mind that for the interaction to remain resonant the radiation field strength should not exceed or even approach the so-called atomic field strength $\mathcal{E}_{at}$;

$$\mathcal{E} \ll \mathcal{E}_{at}. \tag{2.7}$$

The atomic field strength $\mathcal{E}_{at} = e/a_0^2 = 5 \cdot 10^9$ V/cm ($e = 4.18 \cdot 10^{-10}$ CGSE is the electronic charge, $a_0 = 0.53 \cdot 10^{-8}$ cm, the radius of the first Bohr orbit in a hydrogen atom) corresponds to the intensity of linearly polarized light of the order of $3.5 \cdot 10^{16}$ W/cm^2. Thus condition (2.6) is, as a rule, more rigorous.

In case conditions (2.5)–(2.7) are valid the Schrödinger equation for the system considered may be solved in the so-called resonant or rotating-wave approximation.

Let the total energy of the particle and the field without the interaction term be described by the Hamiltonian $\widehat{H}_0$. Then for the Ψ-function of the particle in the radiation field we have

$$j\hbar \frac{\partial \Psi}{\partial t} = (\widehat{H}_0 + \widehat{V})\Psi, \tag{2.8}$$

where $\widehat{V}$ is the operator of interaction between the particle and the field corresponding to the interaction energy (2.2).

For a two-level system in the resonant approximation (2.8) may be reduced to two linear equations for the wavefunction amplitudes of the first and the second particle states. Here we take into account only the terms which lead to the solutions with small resonant denominator of the type Δ^{-1}.

Now we shall obtain the resonant approximation equations.

Let the Ψ-function we are seeking for be the superposition of initial wavefunctions with time-dependent coefficients,

$$\Psi = a_1(t)\Phi_1 + a_2(t)\Phi_2, \tag{2.9}$$

where

$$\begin{aligned} \Phi_1 &= \phi_1(x, y, z) \exp(-jE_1 t/\hbar), \\ \Phi_2 &= \phi_2(x, y, z) \exp(-jE_2 t/\hbar), \end{aligned} \tag{2.10}$$

and the spatial parts of wavefunctions $\phi_1(x, y, z)$ and $\phi_2(x, y, z)$ are the solutions of stationary Schrödinger equations:

$$\widehat{H}_0 \phi_1 = E_1 \phi_1, \quad \widehat{H}_0 \phi_2 = E_2 \phi_2. \tag{2.11}$$

The substitution of (2.9) in (2.8) along with (2.10) and (2.11) gives the following equation:

$$\begin{aligned} j\hbar\phi_1 \exp(-jE_1 t/\hbar)\dot{a}_1 &+ j\hbar\phi_2 \exp(-jE_2 t/\hbar)\dot{a}_2 \\ &= a_1 \widehat{V}\phi_1 \exp(-jE_1 t/\hbar) + a_2 \widehat{V}\phi_2 \exp(-jE_2 t/\hbar). \end{aligned} \tag{2.12}$$

The multiplication of the obtained equation by ϕ_1^* or by ϕ_2^* with the following integration over all the spatial coordinates results in

$$
\begin{aligned}
j\hbar\dot{a}_1 &= a_2 \exp(-j\omega_{21}t) \int_{-\infty}^{\infty} \phi_1^* \widehat{V} \phi_2 dx\, dy\, dz\,, \\
j\hbar\dot{a}_2 &= a_2 \exp(j\omega_{21}t) \int_{-\infty}^{\infty} \phi_2^* \widehat{V} \phi_1 dx\, dy\, dz\,,
\end{aligned}
\tag{2.13}
$$

Here we use the orthonormality of wavefunctions ϕ_1 and ϕ_2, the symmetry and definition (2.3) of the transition frequency.

Since (2.1) and (2.2) give

$$
\widehat{V} = -\frac{1}{2}\mathcal{E}\left(e^{j\omega t} + e^{-j\omega t}\right)\widehat{\mu}\,,
\tag{2.14}
$$

where $\widehat{\mu}$ is the transition dipole moment, we get

$$
\begin{aligned}
j\hbar\dot{a}_1 &= -\frac{1}{2}a_2[\exp(j\omega t) + \exp(-j\omega t)]\exp(-j\omega_{21}t)\mathcal{E}\mu_{12}\,, \\
j\hbar\dot{a}_2 &= -\frac{1}{2}a_1[\exp(j\omega t) + \exp(-j\omega t)]\exp(j\omega_{21}t)\mathcal{E}\mu_{12}\,.
\end{aligned}
\tag{2.15}
$$

In (2.15) we have introduced the definition of the transition dipole moment,

$$
\mu_{12} = \int_{-\infty}^{\infty} \phi_1^* \widehat{\mu}\phi_2 dx\, dy\, dz = \int_{-\infty}^{\infty} \phi_2^* \widehat{\mu}\phi_1 dx\, dy\, dz = \widehat{\mu}_{21}.
\tag{2.16}
$$

Now we can simplify (2.15), taking into account condition (2.5) for the interaction to be resonant and omitting the rapidly oscillating terms with $\exp(\pm 2j\omega t)$. Then

$$
\begin{aligned}
j\hbar\dot{a}_1 &= -\frac{1}{2}\mu_{12}\mathcal{E}a_2 \exp[-j(\omega_{21} - \omega)t]\,, \\
j\hbar\dot{a}_2 &= -\frac{1}{2}\mu_{21}\mathcal{E}a_1 \exp[j(\omega_{21} - \omega)t]\,.
\end{aligned}
\tag{2.17}
$$

Now we represent coefficients a_1 and a_2 slowly varying in time in the symmetric form:

$$
a_1 = \psi_1 \exp\left(-j\frac{\omega_{21} - \omega}{2}t\right)\,, \qquad a_2 = \psi_2 \exp\left(j\frac{\omega_{21} - \omega}{2}t\right)\,.
\tag{2.18}
$$

Then from (2.17) we obtain the following equations for the amplitudes of their oscillations:

$$
\begin{aligned}
j\hbar\dot{\psi}_1 &= -\frac{1}{2}(\hbar\omega_{21} - \hbar\omega)\psi_1 - \frac{1}{2}\mu_{21}\mathcal{E}\psi_2\,, \\
j\hbar\dot{\psi}_2 &= \frac{1}{2}(\hbar\omega_{21} - \hbar\omega)\psi_2 - \frac{1}{2}\mu_{21}\mathcal{E}\psi_1\,.
\end{aligned}
\tag{2.19}
$$

The time evolution of the ψ_1- and ψ_2-functions with respect to which the resonant approximation equations (2.19) are written determines the dynamics of the sought-for Ψ-function of the considered two-level particle (2.9) under the action of field $E(t)$ (see (2.1)). So let us turn to the physical sense of functions ψ_1 and ψ_2. To this end we shall represent (2.9) in the form

$$\Psi = \Psi_1 + \Psi_2 \tag{2.20}$$

and write down the combinations $\Psi_1\Psi_1^*$, $\Psi_1\Psi_2^*$, $\Psi_2\Psi_1^*$, $\Psi_2\Psi_2^*$ pairwise. From (2.20), (2.18), (2.10) and (2.9) it follows that

$$
\begin{aligned}
\Psi_1\Psi_1^* &= \psi_1\psi_1^*\phi_1\phi_1^*, \\
\Psi_1\Psi_2^* &= \psi_1\psi_2^*\phi_1\phi_2^* \exp(j\omega t), \\
\Psi_2\Psi_1^* &= \psi_2\psi_1^*\phi_2\phi_1^* \exp(-j\omega t), \\
\Psi_2\Psi_2^* &= \psi_2\psi_2^*\phi_2\phi_2^*.
\end{aligned}
\tag{2.21}
$$

Having integrated the first and the last of equations (2.21) over the whole space, we get the probability of detecting the particle on levels 1 and 2, respectively. We denote these probabilities by ρ_{11} and ρ_{22}. Then, due to the orthonormality of the stationary-state wavefunctions ψ_1 and ψ_2, we have

$$\rho_{11} = \psi_1\psi_1^* \int \phi_1\phi_1^* dx\, dy\, dz = \psi_1\psi_1^*, \tag{2.22}$$

$$\rho_{22} = \psi_2\psi_2^* \int \phi_2\phi_2^* dx\, dy\, dz = \psi_2\psi_2^*. \tag{2.23}$$

Thus, the squared modulus of functions ψ_1 and ψ_2 determines the population dynamics of states 1 and 2 when the considered particle is exposed to the radiation field. Now let us find the dipole moment of this particle. According to the quantum-mechanical definition of a mean value, the dipole moment of the particle with wavefunction Ψ is given by

$$\langle \mu \rangle = \int \Psi\hat{\mu}\Psi^* dx\, dy\, dz. \tag{2.24}$$

Then, taking into account (2.20) and (2.21), we have

$$
\langle \mu \rangle = \psi_1\psi_1^* \int \phi_1\hat{\mu}\phi_1^* dx\, dy\, dz + \psi_1\psi_2^* \int \phi_1\hat{\mu}\phi_2^* dx\, dy\, dz \\
+ \psi_2\psi_1^* \int \phi_2\hat{\mu}\phi_1^* dx\, dy\, dz + \psi_2\psi_2^* \int \phi_2\hat{\mu}\phi_2^* dx\, dy\, dz. \tag{2.25}
$$

Since the dipole moment operator $\hat{\mu}$ is the odd function of the coordinate, the first and the last integrals vanish. The second and the third ones represent the matrix elements $\mu_{12} = \mu_{21}$ of the dipole moment operator (see (2.16)). The result is

$$\langle \mu \rangle = \mu_{21} \left(\psi_1\psi_2^* e^{j\omega t} + \psi_2\psi_1^* e^{-j\omega t} \right). \tag{2.26}$$

Thus, the off-diagonal elements $\rho_{12} = \psi_1\psi_2^*$ and $\rho_{21} = \psi_2\psi_1^*$ ($\rho_{12} = \rho_{21}^*$), composed from functions ψ_1 and ψ_2, determine the magnitude and the dynamics of the polarization of the particle considered, under the action of the radiation field.

Now let us go back to the resonant-approximation equations (2.19). We consider the resonant medium exposed to the action of the constant amplitude field turned on at $t = 0$. Taking the initial conditions $\psi_1(t = 0) = 1$ and $\psi_2(t = 0) = 0$, let's try the solution of the linear differential equations (2.19) in the form

$$\psi_1 = \cos \Omega t + B \sin \Omega t, \quad \psi_2 = C \sin \Omega t. \tag{2.27}$$

Now substituting (2.27) into (2.19) and equating the coefficients, multiplying terms $\sin \Omega t$ and $\cos \Omega t$, we obtain four new equations, one of which follows from the other three and is, therefore, of no interest to us. The remaining three equations,

$$-j\hbar\Omega = \frac{1}{2}B(\hbar\omega - \hbar\omega_{21}) - \frac{1}{2}C\mathcal{E}_0\mu_{21},$$

$$j\hbar\Omega B = \frac{1}{2}(\hbar\omega - \hbar\omega_{21}), \tag{2.28}$$

$$j\hbar\Omega C = -\frac{1}{2}\mathcal{E}_0\mu_{12},$$

allow us to determine the three unknown constants Ω, B, C:

$$\Omega = \frac{1}{2\hbar}[(\hbar\omega - \hbar\omega_{21})^2 + (\mu_{21}\mathcal{E}_0)^2]^{1/2}, \tag{2.29}$$

$$B = \frac{1}{2j\hbar\Omega}(\hbar\omega - \hbar\omega_{21}), \tag{2.30}$$

$$C = -\frac{\mu_{21}\mathcal{E}_0}{2j\hbar\Omega}. \tag{2.31}$$

Thus,

$$\psi_1 = \cos \Omega t + j\frac{\omega_{21} - \omega}{2\Omega} \sin \Omega t, \tag{2.32}$$

$$\psi_2 = j\frac{\mu_{21}\mathcal{E}_0}{4\hbar\Omega} \sin \Omega t. \tag{2.33}$$

The wave functions of the states, which are stationary in the absence of an external perturbation, are seen to oscillate in time, the frequency of these oscillations being Ω.

Now let us find out the probability of the particle being found on level 2.

$$\rho_{22} = \psi_2\psi_2^* = \frac{\mu_{21}^2\mathcal{E}_0^2}{8\hbar^2\Omega^2}(1 - \cos 2\Omega t). \tag{2.34}$$

This probability includes the term oscillating with double frequency, 2Ω. For convenience we rewrite expression (2.29) for frequency Ω in the form

$$\Omega = \frac{(\Delta^2 + \Omega_0^2)^{1/2}}{2}, \tag{2.35}$$

where $\Delta = \omega_{21} - \omega$, and

$$\Omega_0 = \mu_{21}\mathcal{E}_0/\hbar \tag{2.36}$$

is the Rabi frequency (see (1.4)). Then the occupation probability of the second level comprises

$$\rho_{22} = \frac{1}{2}\frac{\Omega_0^2}{\Delta^2 + \Omega_0^2}(1 - \cos 2\Omega t), \qquad (2.37)$$

which clearly shows the main characteristic features of the coherent process of the upper level occupation under the action of the field close to the resonant one. The occupation probability equals the relative population of level 2; i.e., it may be treated as the upper level population per one particle. Then, if N is the density of the resonant particles, the level 2 population density is

$$N_2 = N\rho_{22}. \qquad (2.38)$$

Thus the upper level population of the resonant transition which is proportional to the radiation intensity is seen to oscillate between zero and a certain maximum value, with the frequency being determined by both the detuning and the Rabi frequency. The spectral dependence of the population has the Lorentz shape, i.e., it corresponds to the case of homogeneous broadening with the width determined by the Rabi frequency.

Let us consider the most clearly extreme cases, namely, the limit of large detuning or small radiation field strength,

$$\rho_{22} = \Omega_0^2(1 - \cos \Delta t)/2\Delta^2, \quad \Delta^2 \gg \Omega_0^2; \qquad (2.39)$$

and the opposite extreme of small detuning or large radiation intensity

$$\rho_{22} = \Omega_0^2(1 - \cos \Omega_0 t)/2, \quad \Delta^2 \ll \Omega_0^2; \qquad (2.40)$$

The meaning of the last formulae is clear. At intense irradiation with the field strength sufficiently high for the Rabi frequency to considerably exceed the detuning of the field frequency of the exact resonant value, the two-level quantum system oscillates between the upper and lower levels with the Rabi frequency. If the field strength is small, so that the corresponding Rabi frequency is much less than the detuning, the probability of detecting the particle in the upper state never reaches unity, the frequency of the probability oscillation being equal to the detuning.

If the particle is precisely in resonance with the external electromagnetic field, it necessarily reaches the upper level, even at a small field strength, but in this case the process will take much more time due to the slowness of the Rabi oscillations. One should keep in mind that this time should not exceed the interval during which the relaxation processes may be neglected.

Let us consider the spectral dependence (2.37) in more detail. In the above discussion we emphasized that if the field is detuned from resonance, i.e., $\Delta \neq 0$, the field strength being small, $\Delta^2 \gg \Omega_0^2$, the probability of detecting the particle on the upper level never reaches unity. However, increasing the field strength and, hence, the Ω_0-value, the influence of detuning may be overcome. As we already know, at $\Omega_0 \gg \Delta$ the population of the upper level rapidly oscillates between zero and unity. Hence, the field has the effect of broadening the transition.

The question of field broadening of the transition is extremely important. Here we discuss only the dynamic broadening in a strong field which, despite detuning, results in the periodic transitions, the transition probability approaching unity. The spectral dependence (2.37) $\frac{\Omega_0^2}{\Delta^2 + \Omega_0^2}$ of the population probability amplitude is Lorentzian – the resonant denominator along with the squared detuning contains the squared Rabi frequency which, therefore, plays the part of homogeneous broadening. Hence, in the absence of the relaxation processes, we can take into account the dynamic field broadening $\mu_{21}\mathcal{E}_0/\hbar$, which considerably changes the resonant properties of the quantum system at the high field strength of an electromagnetic wave.

If the transition line is broadened by Ω_0, the corresponding maximum change in energy amounts to $\hbar\Omega_0$. The only interaction capable of changing the energy is, for the problem in question, the electro-dipole interaction with energy $\mu_{21}\mathcal{E}_0$. If we equate these energies to each other we obtain, as would be expected, the Rabi frequency value (2.36). Note, by the way, that the same result may be obtained from the dimension considerations. The only parameter of the problem with the frequency dimension is the combination $\mu_{21}\mathcal{E}_0/\hbar$.

Now let us turn to the polarization of the medium in the case considered. The appearance of polarization is caused by the mixing of states 1 and 2 by the radiation field, i.e., by the appearance of the non-zero off-diagonal elements ρ_{21} and ρ_{12}.

Using (2.32) and (2.33) along with definitions (2.4) and (2.36), we may derive that

$$\rho_{12} = \frac{\Omega_0}{4\Omega}\left(\frac{\Delta}{\Omega}\sin^2\Omega t - j\sin 2\Omega t\right). \tag{2.41}$$

Then, in accordance with (2.26), the quantum average dipole moment of a particle is

$$\langle\mu\rangle = \frac{\mu_{21}\Omega_0\Delta}{\Omega_0^2 + \Delta^2}(1 - \cos 2\Omega t)\cos\omega t + \frac{\mu_{21}\Omega_0}{(\Omega_0^2 + \Delta^2)^{1/2}}\sin 2\Omega t \sin\omega t. \tag{2.42}$$

As one should expect, the field-induced dipole moment oscillates at the field frequency, its magnitude being determined by the same resonant denominator $\Omega_0^2 + \Delta^2$ as the upper level population value. The medium polarization, i.e., the dipole moment per unit volume equals $N\langle\mu\rangle$.

The field-induced medium polarization a the backward effect on the field. It is well known that the oscillating dipole emits radiation on the frequency of oscillations. The strength of the radiated field is proportional to the second derivative of the dipole moment with respect to time. That is clear. A dipole moment may be represented as the product of a certain effective charge by the distance between two such charges of different signs. Then the second time derivative of the dipole moment is proportional to the acceleration of the charges in their relative motion, the electromagnetic field being irradiated only by an accelerated charge.

If the particle is in resonance and the Rabi frequency, Ω_0, is small as compared to the frequency of light, $\Omega_0 \ll \omega$, then from (2.42) it is obvious that

$$\ddot{P} \propto N\mu_{21}\omega^2 \sin \Omega_0 t \sin \omega t. \tag{2.43}$$

Hence, field $\mathcal{E}_0 \cos \omega t$ acquires an additional coherent component, which is sinusoidally modulated in amplitude and shifted in phase by $\pi/2$. Then the resulting field is of the form:

$$\mathcal{E}(t) = \mathcal{E}_0(\cos \omega t + m \sin \Omega_0 t \sin \omega t), \tag{2.44}$$

where the modulation depth m is proportional to $\omega^2 N\mu_{21}$, and the modulation frequency $\Omega_0 = \mu_{21}\mathcal{E}_0/\hbar$ is the Rabi frequency. The slow (as compared to the oscillation frequency of light) periodical change of the resulting field strength for the given polarizing field strength in the process of the resonant interaction of radiation with matter is referred to as the phenomenon of light nutations. The measurement of the nutation frequency is one of the most accurate methods to determine the matrix elements μ_{21} of the dipole moment operator of the resonant transition being studied. The modulation damping after the external field has been switched off is indicative of the damping of the dipole moment induced by this field and may be used to measure the typical damping time and to study relaxation processes. The short powerful resonant shock after which the damping nutations may be observed is most convenient. However, a pulsed excitation requires special analysis.

We have considered the coherent beats in the resonant medium caused by the action of the constant-amplitude field $\mathcal{E}(t) = \mathcal{E}_0 \cos \omega t$. In general, any other dependence $\mathcal{E}(t)$ does not allow (2.19) to be solved. However, if the system is precisely in resonance , i.e., $\omega = \omega_{21}$, the solution may be found for an arbitrary form of the high-frequency field $\mathcal{E}(t)$ envelope, given it is slow in comparison with time ω^{-1}.

Let field $\mathcal{E}(t)$ be written in the form

$$\mathcal{E}(t) = \mathcal{E}_0(t) \cos \omega t, \tag{2.45}$$

where the slowness of the amplitude $\mathcal{E}(t)$ change is determined by the condition

$$|\dot{\mathcal{E}}_0(t)| \ll \omega |\mathcal{E}_0(t)|. \tag{2.46}$$

In the considered case of strict resonance Eqs. (2.19) may be rewritten in a rather symmetric form:

$$j\hbar\frac{d\psi_1}{dt} = -\frac{\mu_{21}\mathcal{E}_0(t)}{2}\psi_2, \quad j\hbar\frac{d\psi_2}{dt} = -\frac{\mu_{21}\mathcal{E}_0(t)}{2}\psi_1. \tag{2.47}$$

Note that in the case of pulsed action, the exact-resonance conditions are fulfilled and equations (2.47) are valid if the product of detuning Δ by the pulse duration τ_{pulse} is much less than unity, $\Delta\tau_{\text{pulse}} \ll 1$.

Let us introduce the dimensionless time θ by the relation

$$\frac{\mu_{21}\mathcal{E}_0(t)}{2\hbar}dt = \frac{d\theta}{2}. \tag{2.48}$$

Then (2.47) gives a system of two equations:

$$j\frac{d\psi_1}{d\theta/2} = -\Psi_2, \quad j\frac{d\psi_2}{d\theta/2} = -\Psi_1. \tag{2.49}$$

The latter system is equivalent to the equations of the undamped harmonic oscillator type. For example, for ψ_2 we have

$$\frac{d^2}{d(\theta/2)^2}\psi_2 + \psi_2 = 0. \tag{2.50}$$

The solution of such equations is well known. Under the chosen initial conditions $(\psi_1(0) = 1, /; \psi_2(0) = 0)$ we get

$$\psi_1 = \cos(\theta/2), \quad \psi_2 = j\sin(\theta/2). \tag{2.51}$$

If we turn back to nonscaled time, using (2.48) we obtain

$$\psi_1 = \cos\int_0^t \frac{\mu_{21}\mathcal{E}_0(t)}{2\hbar}dt, \quad \psi_2 = j\sin\int_0^t \frac{\mu_{21}\mathcal{E}_0(t)}{2\hbar}dt. \tag{2.52}$$

These formulae allow us to obtain the relative populations of the upper and lower levels:

$$\rho_{11} = \psi_1\psi_1^* = \cos^2\int_0^t \frac{\mu_{21}\mathcal{E}_0(t)}{2\hbar}dt = \frac{1}{2}\left(1 + \cos\int_0^t \frac{\mu_{21}\mathcal{E}_0(t)}{\hbar}dt\right), \tag{2.53}$$

$$\rho_{22} = \psi_2\psi_2^* = \sin^2\int_0^t \frac{\mu_{21}\mathcal{E}_0(t)}{2\hbar}dt = \frac{1}{2}\left(1 - \cos\int_0^t \frac{\mu_{21}\mathcal{E}_0(t)}{\hbar}dt\right). \tag{2.54}$$

To find out the polarization (see(2.26)) we have to know the off-diagonal element

$$\rho_{12} = \psi_1\psi_2^* = \frac{1}{2}j\sin\int_0^t \frac{\mu_{21}\mathcal{E}_0(t)}{\hbar}dt, \tag{2.55}$$

which leads to

$$P = N\langle\mu\rangle = N\mu_{21}\sin\omega t\sin\int_0^t \frac{\mu_{21}\mathcal{E}_0(t)}{\hbar}dt. \tag{2.56}$$

From the relations obtained, the population dynamics and the polarization are seen to be determined by the value of the variable

$$\theta(t) = \int_0^t \frac{\mu_{21}\mathcal{E}_0(t)}{\hbar}dt. \tag{2.57}$$

In practice, in the presence of the relaxation processes for the interaction of the radiation with the considered medium to be coherent, the irradiation should be momentary, i.e., essentially pulsed. Hence, the envelope $\mathcal{E}(t)$ in (2.31) should correspond to the pulse of the finite duration, which is considerably longer than the high-frequency oscillation period but shorter than the relaxation time. The quantity

$$\theta = \int_0^\infty \frac{\mu_{21}\mathcal{E}_0(t)}{\hbar} dt \qquad (2.58)$$

is referred to as the area under the pulse envelope or, in short, the pulse area. The statement that the behavior of the system exposed to a resonant field is determined by the integral of the field strength envelope of the past radiation over time multiplied by the ratio of the transition matrix element to Planck's constant may be considered as the most general formulation of the so-called area theorem.

The comparison of (2.56) and (2.54) with (2.42) and (2.40), respectively, shows that for zero detunings the latter are merely instances of the equations just obtained. Indeed, if the field strength is constant, the effective area of the radiation acting during time t may be calculated by merely multiplying the Rabi frequency Ω_0 by this time interval t. The area theorem is valid for the product $\Omega_0 t$. In particular, at $\Omega_0 t = \pi$, the population distribution is inverted: $\rho_{11} = 0$ and $\rho_{22} = 1$. In the case of pulsed irradiation this occurs at

$$\int_0^t \frac{\mu_{21}\mathcal{E}_0(t)}{\hbar} dt = \pi .$$

The radiation pulse with the total area of π, $\theta = \pi$, is called the π-pulse. After the π-pulse has elapsed, the population distribution is inverted, the polarization remaining zero. At $\theta = \pi/2$ (the $\pi/2$-pulse), the pulse equalizes the populations, the polarization reaching a maximum value. It is this case that is most suitable for the measurement of the polarization relaxation time after the radiation has been turned off.

However, the damping of the nutations is also conceivable in the presence of the field due to the so-called coherent damping effect. Indeed, the results obtained above may be applied directly only to the ensemble of identical particles. In case the external field interacts with the ensemble of systems differing in the transition frequencies and/or in the matrix element values, an additional analysis should be carried out, if we are interested in the ensemble averaged characteristics.

Consider for a start the mean population and the mean polarization for the case of the spread in frequencies only[*]. Let the detuning distribution $f(\Delta)$ define the probability $f(\Delta)d\Delta$ for the difference $\hbar\omega_{21} - \hbar\omega$ to be found in interval $\hbar(\Delta \pm d\Delta/2)$. First of all let us determine the probability of the population of level 2 by using formula (2.37) for ρ_{22} along with (2.35). Then the ensemble average may be represented by

$$\langle\rho_{22}\rangle = \frac{1}{2} \int_{-\infty}^\infty \frac{\Omega_0^2}{\Delta^2 + \Omega_0^2} \left(1 - \cos\left[(\Delta^2 + \Omega_0^2)^{1/2}t\right]\right) f(\Delta)d\Delta . \qquad (2.59)$$

[*] The spread in frequencies corresponds to the so-called inhomogeneous broadening to be discussed in Lecture 6.

There are two simple extremes. In the first the width of distribution $f(\Delta)$ is much greater than $\Omega_0 = \mu_{21}\mathcal{E}_0/\hbar$ and in the second, much less. First consider the simplest case, the first of the two. Here $f(\Delta)$ is being replaced by $f(0)$ so that the integration in (2.59) may be accomplished. The first term in (2.59) is an exact integrand and the integral of the second term may be reduced in the longtime limit, $\Omega_0 t \gg 1$, to the tabulated integral

$$\int_{-\infty}^{\infty} \cos z^2 dz = (\pi/2)^{1/2}$$

As a result, we obtain the following asymptotic expression:

$$\langle \rho_{22} \rangle = \frac{\pi}{2}\Omega_0 f(0) - \sqrt{\frac{\pi}{2}}\Omega_0 \frac{\cos \Omega_0 t}{(\Omega_0 t)^{1/2}} f(0). \tag{2.60}$$

In the time scale longer than the Rabi oscillation period the spread in the transition eigenfrequency values causes the gradual decrease of the amplitude of the upper-level population oscillations, even in the coherent interaction regime, the final population of the upper level tending to $\pi\mu_{21}\mathcal{E}_0 f(0)/2\hbar$. This is the so-called coherent damping of the Rabi oscillations.

The opposite extreme of distribution $f(\Delta)$, narrow as compared to Ω_0, may also be treated only at large times. At a narrow distribution, $\Omega_0^2/(\Omega_0^2 + \Delta^2) \approx 1$. It is convenient to consider the difference $\delta\rho = \rho_{11} - \rho_{22} = 1 - 2\rho_{22}$. After averaging over the ensemble this difference becomes equal to

$$\langle \delta\rho \rangle = \int_{-\infty}^{\infty} f(\Delta) \cos[(\Delta^2 + \Omega_0^2)^{1/2}t]d\Delta. \tag{2.61}$$

Only small values of Δ are significant in this case. Therefore, the square root in (2.61) may be expanded in a series, all the terms except the first one being omitted.

$$\langle \delta\rho \rangle = \cos \Omega_0 t \int_{-\infty}^{\infty} f(\Delta) \cos \frac{\Delta^2}{2\Omega_0}t d\Delta. \tag{2.62}$$

The analysis of the integral in (2.62) shows that the oscillation amplitude decreases with t. In the particular case of normal distribution of detunings we have

$$f(\Delta) = \pi^{-1/2}\Delta^{-1} \exp\left(-\Delta^2/\Delta_0^2\right), \tag{2.63}$$

where Δ_0 is the effective distribution width which, in the given case, is much less than Ω_0. Then the integral may be evaluated and

$$\langle \delta\rho \rangle = \left[1 + \left(\frac{\Delta_0^2}{2\Omega_0}t\right)^2\right]^{-1/4} \cos\left[\frac{1}{2}\arctan\left(\frac{\Delta_0^2}{2\Omega_0}t\right)\right] \cos \Omega_0 t. \tag{2.64}$$

At large t the population difference keeps on oscillating while the oscillation amplitude decreases as $t^{-1/2}$. Hence, even a relatively small spread in detunings of the levels involved in the interaction with radiation brings about the coherent damping of the Rabi oscillations.

The coherent damping of the Rabi oscillations somewhat resembles the saturation effect in the incoherent interaction process.

Now let us pay attention to the damping of the polarization. The expression for the dipole moment is given by (2.42). Consider the case of the symmetric distribution function $f(\Delta)$. Then, when integrating over the interval between $-\infty$ and $+\infty$, the first term of (2.42) vanishes and

$$\langle \mu \rangle = \mu_{21} \Omega_0 \sin \omega t \int_{-\infty}^{\infty} f(\Delta) \frac{\sin[(\Delta^2 + \Omega_0^2)^{1/2} t]}{(\Delta^2 + \Omega_0^2)^{1/2}} d\Delta. \qquad (2.65)$$

In the case of a broad distribution of detunings,

$$\langle \mu \rangle = \mu_{21} \Omega_0 \sin \omega t f(0) \int_{-\infty}^{\infty} \frac{\sin[(1+x^2)^{1/2} \Omega_0 t]}{(1+x^2)^{1/2}} dx, \qquad (2.66)$$

where $x = \Delta/\Omega_0$. The substitution $x = \sinh \phi$ reduces the integral from (2.66) to

$$\int_{-\infty}^{\infty} \sin(\Omega_0 t \cosh \phi) d\phi,$$

This, in accordance with one of the integral representations of Bessel functions, gives the zero-order Bessel function of $\Omega_0 t$, $\pi J_0(\Omega_0 t)$. As a result we have

$$\langle \mu \rangle = \pi \mu_{21} \Omega_0 f(0) J_0(\Omega_0 t) \sin \omega t. \qquad (2.67)$$

The polarization decreases with time as the Bessel function $J_0(\Omega_0 t)$.

Therefore, in the absence of any irreversible relaxation processes the statistics of the spread in the transition frequency values causes the damping of the coherent oscillations of the ensemble averaged population and polarization of two-level quantum particles.

By analogy, the spread in the values of the transition operator matrix elements (the case of degenerate levels) for the systems tuned in resonance also brings about the damping of oscillations. Let us consider, for instance, an ensemble of two-level resonant systems differing in the orientation of the transition dipole moment μ_{21}. Then the interaction energy that determines the pulse area may be written in the form $\mathcal{E}_0(t) \mu_{21} \cos \phi$, where ϕ is the angle included between vectors $\boldsymbol{E}(t)$ and $\mu(\mathbf{t})$. The case of the uniform distribution in directions (spatial degeneracy) is described in the spherical coordinate system by the function

$$f(\theta, \phi) = \frac{\sin \phi}{2\pi^2}, \qquad (2.68)$$

where ϕ is the polar angle and θ is the latitude angle, the polar axis coinciding with $\boldsymbol{E}(t)$. Then using (2.53) and (2.54) we can write the ensemble-averaged difference of populations in the form

$$\langle \delta \rho \rangle = \frac{1}{2\pi^2} \int_0^{2\pi} d\theta \int_0^{\pi} d\phi \cos \left(\int dt \frac{\mathcal{E}_0(t)\mu_{21}}{\hbar} \cos \phi \right) \sin \phi. \qquad (2.69)$$

The substitution $\cos\phi = z$ reduces the polar-angle integral entering this equation to the tabulated one. Then the ensemble-averaged difference of populations,

$$\langle\delta\rho\rangle = \frac{2}{\pi}\left(\int\frac{\mathcal{E}_0(t)\mu_{21}}{\hbar}dt\right)^{-1}\sin\int\frac{\mathcal{E}_0(t)\mu_{21}}{\hbar}dt = \frac{2}{\pi}\frac{\sin\theta(t)}{\theta(t)}, \qquad (2.70)$$

(where $\theta(t)$ is given by (2.57)) damps while oscillating as the area under the pulse increases.

In the above-mentioned case of isotropic distribution we can write the expression for the ensemble-averaged polarization using formula (2.56)

$$\langle P\rangle = \frac{N\mu_{21}}{\pi}\int_0^\pi d\phi\cos\phi\sin\left(\int dt\frac{\mu_{21}\mathcal{E}_0(t)}{\hbar}\cos\phi\right)\sin\phi\sin\omega t. \qquad (2.71)$$

The previously used substitution reduces the integral over ϕ to

$$\int_{-1}^1 z\sin az\, dz = 2(\sin a - a\cos a)/a^2.$$

As a result we have

$$\langle P\rangle = \pi^{-1}N\mu_{21}\sin\omega t\, 2\theta^{-2}(\sin\theta - \theta\cos\theta). \qquad (2.72)$$

In the limit of large values of $\int\frac{\mu_{21}\mathcal{E}(t)}{\hbar}dt$, the polarization damps with the area under the pulse in the same way as the population difference.

The process of coherent damping of the population difference (inversion) and of the polarization may be essential in practice for the interaction of radiation with the system of particles with degeneracy or with detuning. The interaction of laser light with the vibrational states of molecules in gases is an example.

Lecture Three

Photon Echo and Self-Induced Transparency Phenomena

Polarization produced by two short resonant pulses. Reversible unphasing. Recollection in phase. Dependence on pulse area. Duration of echo signal. Wave equation. Sine-Gordon equation. Automodel solution. 2π-pulse. Propagation velocity. Spectroscopic applications.

In the previous lecture we have considered the population dynamics and the polarization of a two-level medium caused by the coherent interaction with laser light. The results obtained are essential for the analysis of a number of important phenomena including the photon echo and self-induced transparency.

We shall start with the photon echo. For this purpose we shall consider the polarization of a resonant medium produced by two laser pulses separated in time. Let the medium be characterized by the distribution $f(\Delta)$ of frequency detunings. The width of this function will be denoted by the symbol Δ_0. We shall only pay attention to the case of a rather strong field,

$$\mu_{21}\mathcal{E}_0/\hbar \gg \Delta_0, \qquad (3.1)$$

and a fairly short pulse,

$$\Delta_0 \ll \tau_1^{-1}, \qquad (3.2)$$

where τ_1 is the duration of the first pulse. In the frame of these assumptions all the particles of the medium during the radiation pulse may be considered resonant. By the time the first pulse is over, $t = \tau_1$; then, irrespective of the detuning value Δ, we have for each particle

$$\psi_1(\Delta, t = \tau_1) = \cos(\theta_1/2)\,, \quad \psi_2(\Delta, t = \tau_1) = j\sin(\theta_1/2)\,, \qquad (3.3)$$

where according to (2.57) we denote

$$\theta_1 = \theta(\tau_1) = \int_0^{\tau_1} \frac{\mu_{21}\mathcal{E}_0(t)}{\hbar}\,dt\,.$$

Let $\tau \gg \tau_1$ be the time between the termination of the first pulse and the moment the second pulse has been turned on. Then at $\mathcal{E} = 0$ the initial equations (2.19) lead to

$$\psi_1(\Delta, t = \tau + \tau_1) = \exp(j\tau\Delta/2)\cos(\theta_1/2)\,,$$
$$\psi_2(\Delta, t = \tau + \tau_1) = j\exp(-j\tau\Delta/2)\sin(\theta_1/2)\,. \qquad (3.4)$$

At the moment when $t = \tau_1 + \tau$, the second pulse lasting τ_2 is being turned on. It meets the requirements similar to (3.1) and (3.2) which were imposed on the first pulse. The modification of wavefunctions caused by the second pulse may be found using (2.19), if we assume that the system is precisely in resonance and adopt (3.4) for the initial conditions. Then the general solution may be written in the form

$$\begin{aligned} \psi_1 &= A\cos(\theta_2/2) + B\sin(\theta_2), \\ \psi_2 &= C\cos(\theta_2/2) + D\sin(\theta_2), \end{aligned} \tag{3.5}$$

where

$$\theta_2 = \int_0^{\tau_2} \frac{\mu_{21}\mathcal{E}_0(t)}{\hbar} dt.$$

Constants A, B, C, D are determined by the values of ψ_1, ψ_2, $d\psi_1/dt$, $d\psi_2/dt$ at $t = \tau_1 + \tau$. From (3.4) using (2.47) we find

$$\begin{aligned} A &= \psi_1(\Delta, t = \tau + \tau_1), & B &= -jC, \\ C &= \psi_2(\Delta, t = \tau + \tau_1), & D &= -jA. \end{aligned} \tag{3.6}$$

Finally, we obtain the values of wavefunctions ψ_1 and ψ_2 at the moment of the second pulse termination

$$\begin{aligned} \psi_1 &= \cos\frac{\theta_1}{2}\cos\frac{\theta_2}{2}\exp\left(j\frac{\Delta}{2}\tau\right) - \sin\frac{\theta_1}{2}\sin\frac{\theta_2}{2}\exp\left(-j\frac{\Delta}{2}\tau\right), \\ \psi_2 &= j\sin\frac{\theta_1}{2}\cos\frac{\theta_2}{2}\exp\left(-j\frac{\Delta}{2}\tau\right) + j\cos\frac{\theta_1}{2}\sin\frac{\theta_2}{2}\exp\left(j\frac{\Delta}{2}\tau\right). \end{aligned} \tag{3.7}$$

At time T after the second pulse termination, in accordance with (2.19) and by analogy with (3.4) functions ψ_1 and ψ_2 acquire the values

$$\begin{aligned} \psi_1 &= \cos\frac{\theta_1}{2}\cos\frac{\theta_2}{2}\exp\left[j\frac{\Delta}{2}(T+\tau)\right] \\ &\quad - \sin\frac{\theta_1}{2}\sin\frac{\theta_2}{2}\exp\left[-j\frac{\Delta}{2}(T-\tau)\right], \\ \psi_2 &= j\sin\frac{\theta_1}{2}\cos\frac{\theta_2}{2}\exp\left[-j\frac{\Delta}{2}(T+\tau)\right] \\ &\quad - j\cos\frac{\theta_1}{2}\sin\frac{\theta_2}{2}\exp\left[j\frac{\Delta}{2}(T-\tau)\right]. \end{aligned} \tag{3.8}$$

The expressions obtained allow us to determine the medium polarization in accordance with the well-known procedures (see (2.26)). When evaluating $\psi_2\psi_1^* = (\psi_1\psi_2^*)^*$ one should not forget that the terms containing $T + \tau$ in the exponent oscillate very fast. If $\Delta_0 T, \Delta_0 \tau \gg 1$ they vanish after averaging over the frequency detuning distribution at large values of $T + \tau$. Then we have to keep only one term

$$\psi_2\psi_1^* = -j\cos\frac{\theta_1}{2}\sin\frac{\theta_1}{2}\sin^2\frac{\theta_2}{2}\exp[-j\Delta(T-\tau)]. \tag{3.9}$$

After substituting in (2.26) this gives

$$\langle \mu \rangle = -\mu_{21} \sin \theta_1 \sin^2 \frac{\theta_2}{2} \sin[\Delta(T - \tau) + \omega t]. \qquad (3.10)$$

The value of the induced dipole moment is seen to be determined by the area under both the first and the second pulses. Besides, we can conclude that the spread in frequency detunings Δ brings about a spread in the phase of the oscillations of the dipoles comprising the macroscopic medium polarization. The average over the frequency detuning distribution always gives the resulting zero dipole moment except for the evident case of $T = \tau$ (see (3.10)), when all the microdipoles have the same phase shift despite the difference in Δ.

Indeed, the averaging of (3.10) over, say, the normal distribution of the detunings

$$f(\Delta) = \pi^{-1/2} \Delta_0^{-1} \exp\left(-\Delta^2/\Delta_0^2\right) \qquad (3.11)$$

leads to

$$\langle \mu \rangle = -\mu_{21} \sin \omega t \sin \theta_1 \sin^2 \frac{\theta_2}{2} \exp\left[-\frac{\Delta_0^2}{4}(T - \tau)^2\right]. \qquad (3.12)$$

This expression corresponds to a peak of duration of $1/\Delta_0$ at $T = \tau$, i.e., exactly at the time after the second pulse that equals the time span between the first and the second pulses. This oscillating dipole moment emits electromagnetic waves. The moment at which all the microdipoles comprising the resonant medium macrodipole are re-collected in phase is characterized by an intense burst of radiation. That is why this effect is referred to as the photon echo.

Note that the echo signal reaches its maximum at $\theta_1 = \pi/2$ and $\theta_2 = \pi$, i.e., in case the first pulse is the $\pi/2$-pulse,

$$\int_0^{T_1} \frac{\mu_{21} \mathcal{E}_0(t)}{\hbar} dt = \frac{\pi}{2}, \qquad (3.13)$$

while the second one is the π-pulse,

$$\int_0^{T_2} \frac{\mu_{21} \mathcal{E}_0(t)}{\hbar} dt = \pi. \qquad (3.14)$$

In this case the magnitude of the echo signal is proportional to

$$S \sim \omega^2 \mu_{21} \exp\left[-\frac{\Delta_0^2}{4}(T - \tau)^2\right]. \qquad (3.15)$$

If the area under the pulses is small, i.e., $\theta_1, \theta_2 \ll 1$, the magnitude of the echo is small, too:

$$S = \frac{1}{4}\omega^2 \mu_{21} \theta_1 \theta_2^2 \exp\left[-\frac{\Delta_0^2}{4}(T - \tau)^2\right]. \qquad (3.16)$$

We stress again that the photon echo phenomenon depends dramatically on the area under the first and the second pulses, while the shape of the echo signal does not depend on the shape of these pulses if the latter are short (see (3.21)). In addition, we note that at small values of τ_1 and τ_2 the study of the shape of the echo signal allows the determination of the width of the microdipole distribution in frequency detunings. Figure 3.1 illustrates the disposition of the photon echo pulses.

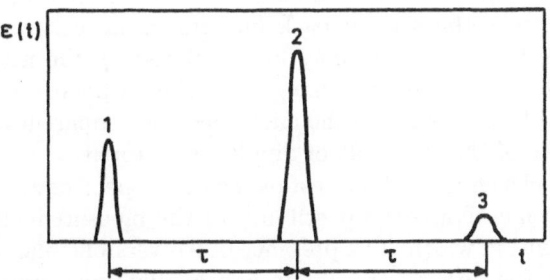

Fig. 3.1. Photon echo: $1 - \frac{\pi}{2}$ pulse, $2 - \pi$-pulse, $3 -$ echo pulse

The photon echo effect was first observed as the spin echo by E. Hahn as far back as 1950 in his experiments on NMR. In fact, it was the first coherent effect of the intense resonant interaction of radiation with matter ever discovered. Thanks to its striking experimental manifestation, this effect attracted the attention of physicists to the possibility of investigating a wide range of phenomena taking place in the coherent interaction of intense resonant radiation with matter. It is noteworthy that the coherent effects caused by the action of a light wave on a substance present an optical analogue of such well-known radio-spectroscopic phenomena as, for instance, spin echo and nutations. The study and application of these coherent interaction effects in optics became possible only after the advent of lasers. The photon echo effect proper was observed for the first time in a ruby using a ruby laser.

So, the photon echo effect occurs under successive irradiation of a medium by two light pulses, the frequency of the light wave coinciding with that of the medium absorbing transition. For the effect to be observed a medium should be characterized by a certain spread in the transition frequencies, i.e., the transition should be inhomogeneously broadened. In a gaseous medium inhomogeneous broadening is usually caused by the Doppler shift. If the first exciting pulse is a $\pi/2$-pulse, it equalizes the populations of the upper and lower levels, thus producing a maximum polarization of the medium. If the pulse is short, the spread in the frequency detunings may be neglected when considering the process of polarization. However, the spread in eigenfrequencies leading to a distribution of frequency detunings with an effective width of Δ_0 in the absence of a field brings about complete unphasing in the time interval of the order of Δ_0^{-1}. In the time interval τ, the relative phase shift for the particles with their eigenfrequencies separated from the distribution center by the distance Δ equals $\Delta\tau$ and is different for the particles with different

Δ. Such an unphasing of dipole oscillations brings about the loss of coherence in the particle ensemble, thus terminating the process of coherent dipole radiation of the whole system in a time that is small against the longitudinal and the transverse relaxation times.

However, the polarization decay is caused by the distribution of the eigenfrequencies of dipoles comprising an ensemble, not by the relaxation processes, and, hence, in principle it is reversible. So, the speculative time inversion at the moment τ brings the system back into its initial coherent state at the moment 2τ. The time inversion may be simulated by the action of a short π-pulse. It inverts the system and changes the sign of polarization. The radiation pulse caused by the photon echo effect becomes apparent at $t = 2\tau$ when the coherent state of the ensemble of dipoles is restored.

The photon echo effect is of interest owing to its spectroscopic applications. We have already mentioned the possibility of the measurement of the eigenfrequency distribution width. The presence of irreversible relaxation processes which develop at the rate T_2^{-1} results in irreversible unphasing with the time constant T_2. This, in turn, leads to an exponential decrease in the magnitude of the echo signal which is proportional to $\exp(-2\tau/T_2)$. This effect may be used as a convenient technique for measuring the transverse relaxation time T_2. The study of polarization dependencies of the echo signal, i.e., of how the polarization and the intensity of the echo signal depend on the polarization of the exciting pulses and on the angle between them, allow the identification of the type of resonant transition responsible for echo formation.

Concluding the consideration of the photon echo we would like to mention that for an allowed transition at $\mu_{21} \approx 0.3$ D and for a pulse duration of 1 ns, the conditions for the π-pulse will be met at the radiation intensity of about 10^4 W/cm^2.

The photon echo has been observed in doped luminescent crystals and in molecular gases.

In optics practically all the media may be considered extensive. So, the radiation propagation effects are essential. The radiation caused by a change of the field-induced medium polarization should be taken into consideration in the wave equation that describes the field propagation. Let us put down this equation for the radiation propagating in the z-direction

$$\frac{\partial^2}{\partial z^2}\mathcal{E} - \frac{1}{c^2}\frac{\partial^2}{\partial t^2}\mathcal{E} = \frac{4\pi}{c^2}\ddot{P}, \tag{3.17}$$

where c is the light velocity. Field $E(t, z)$, in contrast to (2.45), is represented as

$$\mathcal{E}(t, z) = \mathcal{E}_0(t, z)\cos(\omega t - kz). \tag{3.18}$$

One should bear in mind that the condition for the envelope of $\mathcal{E}(t, z)$ to be slow in contrast to (2.46) must be written as

$$|\partial\mathcal{E}_0/\partial t| \ll \omega|\mathcal{E}_0|, \quad |\partial\mathcal{E}_0/\partial t| \ll \frac{|\mathcal{E}_0|}{\lambda}, \tag{3.19}$$

where $\lambda = 2\pi c/\omega$ is the wavelength. The polarization in the right-hand side of (3.17) will then (see (2.56)) have the form

$$P = N\mu_{21} \sin\left(\int_{-\infty}^{t} \frac{\mu_{21}\mathcal{E}_0(t,z)}{\hbar} dt\right) \sin(\omega t - kz). \qquad (3.20)$$

Using (3.18) and (3.20) and taking into account (3.19) we can find the terms of (3.17). Neglecting all the terms of order higher than $\left(\frac{\partial \mathcal{E}}{\partial z} \frac{1}{k\mathcal{E}}\right)^2$ we get

$$\frac{\partial^2 \mathcal{E}}{\partial z^2} = -k^2 \mathcal{E}_0(z,t) \cos(\omega t - kz) + 2k\frac{d}{dz}\mathcal{E}_0(z,t) \sin(\omega t - kz), \qquad (3.21)$$

$$\frac{\partial^2 \mathcal{E}}{\partial t^2} = -\omega^2 \mathcal{E}_0(z,t) \cos(\omega t - kz) - 2\omega\frac{d}{dt}\mathcal{E}_0(z,t) \sin(\omega t - kz), \qquad (3.22)$$

$$\frac{\partial^2 P}{\partial t^2} = -\omega^2 N\mu_{21} \sin\theta \sin(\omega t - kz) + 2N\mu_{21}\omega\dot\theta \cos\theta \cos(\omega t - kz), \qquad (3.23)$$

where in contrast to (2.57) we denote the area under the pulse $\theta(z,t)$ by

$$\theta = \theta(z,t) = \frac{\mu_{21}}{\hbar} \int_{-\infty}^{t} \mathcal{E}_0(z,t)dt. \qquad (3.24)$$

Equation (3.17) must be valid both for the cosinusoidal and sinusoidal components.

For the terms prior to $\cos(\omega t - kz)$ we obtain

$$\left(k^2 - \frac{\omega^2}{c^2} + \frac{8\pi}{c^2}N\omega\frac{\mu_{21}^2}{\hbar}\cos\theta\right)\mathcal{E}_0(z,t) = 0, \qquad (3.25)$$

where we used $\dot\theta \equiv \partial\theta/\partial t = \mu_{21}\mathcal{E}(z,t)/\hbar$. The modulus of the last parenthesized term does not exceed

$$\frac{8\pi}{c^2}N\hbar\omega\mu_{21}^2\frac{\mathcal{E}_0^2}{\hbar^2\mathcal{E}_0^2} = 2\frac{N\hbar\omega}{\mathcal{E}_0^2/4\pi}\Omega^2\omega\omega^2 k^2,$$

where $\Omega = \mu_{21}\mathcal{E}_0/\hbar$ is the Rabi frequency, $N\hbar\omega/(\mathcal{E}_0^2/4\pi)$ being the ratio of the energy accumulated by the particles in the field to the field energy itself. Since we have $\Omega \ll \omega$, this term that is responsible only for a small dispersive influence of the medium may be neglected. Then we get

$$(-k^2 + \omega^2/c^2)\mathcal{E}_0(z,t) = 0, \qquad (3.26)$$

which leads to the usual relation between the wave number and the frequency

$$k = \omega/c. \qquad (3.27)$$

The equation for the coefficients multiplying $\sin(\omega t - kz)$ is of the form

$$\left(\frac{1}{c}\frac{\partial}{\partial t} + \frac{\partial}{\partial z}\right)\mathcal{E}_0(z,t) = -2\pi\frac{\omega}{c}N\mu_{21}\sin\theta(z,t). \qquad (3.28)$$

But in accordance with (3.24)

$$\mathcal{E}_0(z,t) = \frac{\hbar}{\mu_{21}} \frac{\partial}{\partial t} \theta(z,t) \,, \tag{3.29}$$

and for $\theta(z,t)$ we have

$$\left(\frac{1}{c}\frac{\partial}{\partial t} + \frac{\partial}{\partial z}\right) \frac{\partial}{\partial t} \theta(z,t) = -\frac{2\pi N \omega \mu_{21}^2}{c\hbar} \sin \theta(z,t) \,. \tag{3.30}$$

Thus, if we introduce a new variable,

$$x = ct + z \,, \tag{3.31}$$

we obtain

$$\frac{\partial^2}{\partial x \partial t} \theta(x,t) = -\frac{2\pi N \omega \mu_{21}^2}{c\hbar} \sin \theta(x,t) \,. \tag{3.32}$$

This equation may be rewritten more concisely

$$\theta''_{x,t} = \kappa \sin \theta \,, \tag{3.33}$$

where we denote

$$\kappa = -2\pi N \hbar \omega \mu_{21}^2 / c\hbar^2 \,. \tag{3.34}$$

A complete solution of this equation, which is usually referred to as the sin-Gordon equation, is rather complex, and is the subject of up-to-date studies in mathematical physics. In a fairly general case the solution of a sin-Gordon equation may be represented as a set of isolated pulses propagating in the z-direction, their velocity dependent on the interpulse distances (the so-called soliton solutions).

The case of a single pulse or a single soliton is most simple. We try the solution in the form

$$\theta(z,t) = \theta(z - vt) \,, \tag{3.35}$$

i.e., in the form of a pulse with an unknown but invariant shape

$$\mathcal{E}_0(z,t) = \mathcal{E}_0(z - vt) = \frac{\hbar}{\mu_{21}} \dot{\theta}(z - vt) \,, \tag{3.36}$$

which propagates in the z-direction with velocity v (the so-called automodel solution). Let us perform the replacement of variables that is quite natural in the case considered:

$$\phi = z - vt \,. \tag{3.37}$$

Then

$$\mathcal{E}(z,t) = -v\frac{\hbar}{\mu_{21}} \frac{\partial}{\partial \phi} \theta(\phi) \,. \tag{3.38}$$

Substituting (3.38) in (3.28) and taking into account that $\partial \phi / \partial z = 1$ and $\partial \phi / \partial t = -v$, for $\theta(\phi)$ we obtain the physical pendulum equation,

$$\frac{\partial^2 \theta}{\partial \phi^2} = \frac{2\pi N \omega \mu_{21}^2}{v(c - v)\hbar} \sin \theta \,, \tag{3.39}$$

where θ is the angle of the deviation from the equilibrium position. Denoting the square of the "frequency" of the oscillations of this pendulum about the equilibrium point by ν^2,

$$\nu^2 = \frac{2\pi N \omega \mu_{21}^2}{v(c-v)\hbar},$$ (3.40)

we obtain in the most concise form,

$$\theta''_\phi(\phi) = \nu^2 \sin\theta,$$ (3.41)

which dramatically differs from the sin-Gordon equation (3.33). This equation is much simpler due to the fact that (3.41) describes the automodel solution of (3.35).

Equation (3.41) has the solution

$$\theta = 4 \arctan e^{\lambda\phi},$$ (3.42)

which leads to a decreasing field at $\phi \to \pm\infty$. Then

$$\theta' = 4\lambda e^{\lambda\phi}/(1 + e^{2\lambda\phi}),$$ (3.43)

It is seen that $\theta' = 0$ at $\phi \to \pm\infty$. This implies that $\mathcal{E} \to 0$ at $\phi \to \pm\infty$ (see (3.38)). Further differentiation of this expression after a number of transformations leads to

$$\theta''_\phi \phi = \lambda^2 \sin(4 \arctan e^{\lambda\phi}),$$ (3.44)

which, allowing for (3.42), satisfies (3.41) at $\lambda = \nu$.

The envelope of the obtained automodel pulse has a symmetric bell shape (Fig. 3.2). In accordance with (3.43) and (3.38),

$$\mathcal{E}(\phi) = 8\nu v \frac{\hbar}{\mu_{21}} \text{sech}\,\phi.$$ (3.45)

Substituting the value of ν from (3.40) and using the notation (3.37) we finally get

$$\mathcal{E}(z - vt) = 8 \sqrt{\frac{2\pi v N \hbar \omega}{c - v}} \text{sech} \left[\frac{\mu_{21}}{\hbar}(z - vt) \sqrt{\frac{2\pi N \hbar \omega}{v(c-v)}} \right].$$ (3.46)

We emphasize that irrespective of the initial shape of the 2π-pulse entering the resonant medium, it acquires the universal shape (3.46) while propagating.

Indeed, solution (3.42) describes the 2π-pulse, since $\theta(+\infty) - \theta(-\infty) = 2\pi$. Hence, it is the 2π-pulse that propagates in the medium without damping and, finally, acquires the hyperbolic secant shape. The propagation velocity depends on the pulse intensity.

The propagation of radiation pulses without damping results in the effect called self-induced transparency.

Now let us find the propagation velocity of the 2π-pulse. First, we bind together the pulse duration τ_{pulse} and the propagation velocity v. As the automodel solution (3.43) predicts, the maximum pulse amplitude is attained

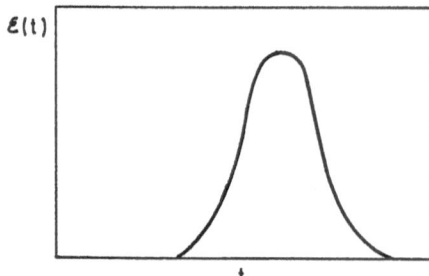

Fig. 3.2. 2π pulse of self-indused transparency

at $\phi = 0$ and comprises $\theta'_\phi(0) = 2\nu$. Let us define the duration of the self-induced transparency pulse as an interval on the ϕ-axis where θ'_ϕ exceeds half its maximum value. Denoting the boundaries of this interval by $\phi_{1/2}$ we can determine it by solving the equation $\theta'_\phi(\phi_{1/2}) = \theta'_\phi(0)/2$. Then (see (3.43)) $\phi_{1/2} = (\mathrm{Arch}2)/\nu$, and the pulse duration is given by

$$\Delta\phi = 2\phi_{1/2}(2\mathrm{Arch}2)/\nu\,.$$

But according to (3.37) $\phi = z - vt$. At a fixed point z the modulus of the interval considered equals $v\Delta t$. Hence,

$$\tau_{\mathrm{pulse}} = (2\mathrm{Arch}2)/\nu v\,. \tag{3.47}$$

Raising (3.47) to the second power and substituting the value of ν^2 from (3.40) we obtain the linear equation with respect to v, its solution being

$$v = c\left[1 + \frac{2\pi}{(2\mathrm{Arch}2)^2}N\hbar\omega\frac{\mu_{21}^2}{\hbar^2}\tau_{\mathrm{pulse}}^2\right]^{-1}. \tag{3.48}$$

For compactness we omit factor $2/(2\mathrm{Arch}2)^2 \approx 1$. Let us recall that according to the meaning of the term 2π-pulse, the following relation for the τ_{pulse} is valid

$$\tau_{\mathrm{pulse}}\frac{\mu_{21}}{\hbar}\mathcal{E}_m \approx 2\pi\,, \tag{3.49}$$

where $\mathcal{E}_m$ is the field amplitude at the pulse maximum. In other words, the order-of-magnitude value of the duration is determined by the period of the Rabi oscillations corresponding to the maximum field in the pulse. Allowing for (3.49) we may write

$$v \approx c[1 + (2\pi)^2 N\hbar\omega/\mathcal{E}_m^2]^{-1}\,. \tag{3.50}$$

If the energy of the phonons which may be absorbed by the medium unit volume is less than the energy of the field enclosed in the same volume, $\frac{N\hbar\omega}{\mathcal{E}_m/8\pi} \ll 1$, the propagation velocity is close to that of light and its relative delay is

$$(c - v)/c \approx (2\pi)^2 N\hbar\omega/\mathcal{E}_m^2\,. \tag{3.51}$$

The delay, though increasing with the resonant particle density and the transition frequency, decreases with the growth of the resonant field intensity. At $N = 3 \cdot 10^{19}$ cm^{-3}, $\hbar\omega = 5 \cdot 10^{-19}$ J, $\mathcal{E}_m = 3 \cdot 10^4$ W/cm ($I_m = 1$ MW/cm^2) the value of $(c-v)/c$ is about $5 \cdot 10^{-7}$. Note that the particle density of $3 \cdot 10^{19}$ cm^{-3} corresponds not only to the density of gases at normal conditions but also to the concentration of the resonant impurity ions in a solid lattice (e.g., chromium in corundum, i.e., ruby).

So, the 2π-pulse in a resonant medium propagates with retardation, but without any damping or change in shape. After it has passed, the medium remains in its ground state.

Briefly, the essence of the phenomenon is as follows. When pulsed resonant radiation propagates in a medium, the particles of the medium are exposed to the action of the field; not simultaneously, but as the pulse passes the region in space which they occupy. The front of the pulse promotes the particles to an upper level. The region of the medium covered by the pulse becomes polarized and accumulates energy from the radiation that has already passed. If the pulse area is sufficiently large, then its front already effectively excites the particles. The rest of the pulse then propagates in the medium of the excited particles. Under the action of this "tail", these particles radiate coherently, passing the energy over to the field. If the area under the pulse equals 2π, such a transfer of energy from the front to the tail of the pulse does not influence its shape. After the pulse has passed, the particles from the region the radiation has already left return to their initial state. The process of absorption and re-radiation of energy requires some finite time, which leads, as a consequence, to a delay of propagation of the resulting pulse of a fixed form.

The absorption and re-radiation are only balanced if the area under the pulse is a multiple of 2π. Indeed, let us consider in more detail the change in the energy flow for the case of a short pulse propagation. Let us multiply (3.28) by $\mu_{21}\mathcal{E}(z,t)\hbar$ and, allowing for (3.24), integrate it over time in the limits from $-\infty$ to $+\infty$. After integration the first term vanishes, since it may be reduced to $\int_{-\infty}^{\infty} \mathcal{E}d\mathcal{E} = 0$ according to the condition of the zero field at infinity. The second term is

$$\frac{\mu_{21}}{2} \frac{\partial}{\partial z} \int_{-\infty}^{\infty} \mathcal{E}_0{}^2 dt \,,$$

and the right-hand side may be reduced to

$$-\frac{2\pi}{c}\omega N \mu_{21} \left(1 - \cos \int_{-\infty}^{\infty} \frac{\mu_{21}}{\hbar}\mathcal{E}_0(z,t)dt\right) .$$

As a result we obtain

$$\frac{\partial}{\partial z} \int_{-\infty}^{\infty} I dt = -N\hbar\omega \left(1 - \cos \int_{-\infty}^{\infty} \frac{\mu_{21}\mathcal{E}_0(z,t)}{\hbar} dt\right)$$

where we have made allowance for $I = c\mathcal{E}^2/4\pi$.

Thus, the change in the energy flow proportional to $N\hbar\omega$ is seen to be determined by the total area under the pulse that turns to zero at $\theta = m2\pi$. It should be emphasized again that among all the possible $m2\pi$-pulses only the 2π-pulse may be considered as an automodel (since (3.41) admits a single solution of the form of (3.42) with the derivative vanishing at $t \leftarrow \pm\infty$). The 2π-pulse is the only self-induced transparency pulse of a strictly determined and invariant shape. In contrast, all other $m2\pi$-pulses break into separate solitons while propagating in the medium, i.e. the 4π-pulse will break into two 2π-pulses, the 6π-pulse, into three 2π-pulses, etc.

Somewhat special is the so-called 0π-pulse. The area under it equals zero (Fig. 3.3). Such a pulse may be formed only if in the course of its duration the field amplitude passes the zero value, while its phase jumps up by π. In fact, such a 0π-pulse is equivalent to several closely spaced pulses. The evolution of such an intense pulse propagating in a resonant medium is rather complicated and its analytic study is fairly cumbersome.

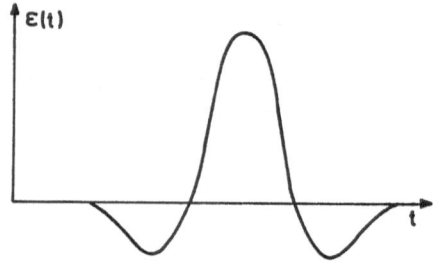

Fig. 3.3. 0π pulse

Summing up the above analysis we can see that the phenomenon of self-induced transparency occurs when a monochromatic radiation pulse passes through a resonant absorbing medium. The pulse duration should be much shorter than the relaxation times of the field-induced polarization in the medium. The essence of this phenomenon is that for a 2π-pulse the absorption disappears and the medium becomes transparent.

The effect of self-induced transparency has been experimentally observed using a ruby laser in a ruby crystal and also in a great number of molecular gases like SF_6, BCl_3, etc., all of which have vibrational-rotational transitions in the work-regions of CO_2-lasers. The study of this effect allows us to estimate the values of the dipole moment of the absorption transition. The artificial violation of the coherence condition $\tau_{\text{pulse}} \ll T_2$ makes it possible to estimate the irreversible relaxation time T_2 when observing the changes in the character of the 2π-pulse propagation. Even though these parameters of the resonant medium admit more reliable measurement with the photon echo and the nutation effects, self-induced transparency, being most simple experimentally, permits the carrying out of preliminary measurements. The results obtained in such experiments may be used to specify the approximate regions of duration and intensity of the radiation pulses to be $\pi/2$-, π- and 2π-pulses with respect to the transition under consideration.

As an illustration we offer the photo in Fig. 3.4 of a self-induced transparency pulse of frequency 941.16 cm^{-1} propagating in a cuvette with BCl$_3$. The cuvette length is 3 m; the gas pressure, 0.1 T; the radiation pulse duration, 100 ns; and the intensity, 25 kW/cm^2.

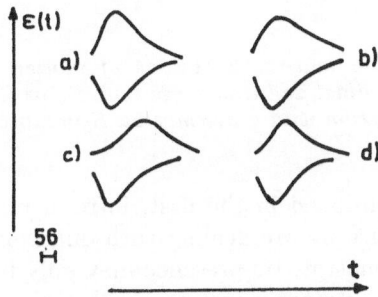

Fig. 3.4. Set of pulses of resonant radiation propagating in a long cuvette with BCl$_3$ gas. Input pulses are plotted with maxima upward, and corresponding output pulses with maxima downward. The pulses intensity increases from a) to d). Curve c) corresponds to 2π-pulse. One can see the symmetry of self-indused transparency pulses

Having mentioned the similar origins of the photon echo, optical nutation and self-induced transparency phenomena, we should emphasize the importance of observing and studying the coherent effects for the investigation of manifold degenerate vibrational-rotational transitions in polyatomic molecules as well as of the band-band and intraband transitions in semiconductors. It is also noteworthy that self-induced transparency may play an important role in laser photochemistry processes, especially in the process of the selective IR dissociation of molecules.

Lecture Four

Relaxation Processes

Density matrix. Relaxation matrix. Equations of motion of the density matrix. Bloch's equations. Longitudinal and transverse relaxations. Relaxation in a gas. Relaxation due to the interaction with a thermostat. Spontaneous radiation.

As we have already mentioned in the first, introductory, lecture one should not forget that in practice we are dealing with quantum systems surrounded by a medium, or a thermostat. Its presence may only be ignored if the interaction of laser light with matter is instantaneous, otherwise, it is to be taken into account. This implies that we should be able to describe open quantum systems. In other words, we are interested in the behavior of a small open subsystem which, in fact, is part of a large closed system. Below we give the main idea that underlies the description of such a subsystem.

To begin with, let us assume that the subsystem considered does not interact with the other parts of a large system previously referred to as a thermostat. Then it may be described using a basis of quantum-mechanical states which do not include the quantum numbers of these other parts of the system. For example, this may be a set of energy eigenstates of the subsystem. After the interaction has been turned on, it affects the motion of both the subsystem and the thermostat. The internal motion of the subsystem brings about the force acting on the thermostat, and vice versa: quite a complex thermostat motion causes an involved force, resembling a random one, to act from the thermostat upon the subsystem.

Such a random force called the Langevin force will cause both a time drift of the subsystem energy levels and transitions between them. The thermostat will behave similarly. The changes in the subsystem and in the thermostat provoked by the casual force, however, are small, since they are not correlated with the respective motions of the subsystem and the thermostat. Thus, first-order perturbation theory in the interaction between the subsystem and the thermostat does not imply a modification of the equation of motion and, hence, the motion itself. This contrasts with the second order, where we take into consideration the influence of those small motions of the thermostat which are themselves caused by the motions of subsystem. In other words, the second order provides for the self-action of the subsystem via the thermostat. Such an influence is resonant, inphase and, thus, essential. It brings about the viscous-friction-type effects due to the appearance of extra relaxation terms in the equations of motion, corresponding to the longitudinal and transverse relaxations. At the same time the self-action of the thermostat via the subsystem is usually neglected owing to the relative smallness of the latter.

In this lecture we shall consider the main concepts relevant to the term *density matrix*. We shall also show how the equations of motion look in the simplest case of a two-level system when the relaxation is present, and also demonstrate how these equations may be written in Bloch's form. Then we shall consider the implementation of this general derivation technique in two particular cases: the case of the interaction of a small subsystem with a large thermostat, and the case of a gaseous phase, where the translational degrees of freedom of the colliding particles act as a thermostat.

For the description of a closed system two equivalent approaches can be used. One uses Schrödinger equations for Ψ-functions, while the other is based on the equations of motion of the density matrix. In fact, both can be used as a basis for the description of an open system. However, they are far from being equivalent in complexity. When the description of an open system is based on the Schrödinger equation, the reduction of the problem to a system of linear differential equations for Ψ-functions turns out to be impossible. But if we use the second approach based on the density matrix formalism, the reduction may be performed quite successfully.

For a closed system the density matrix is introduced so that its elements satisfy the equation

$$\rho_{ij} = \Psi_i \Psi_j^* \,, \tag{4.1}$$

where wavefunction $\Psi_i = \psi_i \exp(-jE_i t/\hbar)$ describes the system state of energy E_i. In this case the quantum-mechanical average of operator $\hat{A}$ corresponding to the physical variable A is the trace of the product of matrix $\hat{\rho}$ by operator $\hat{A}$. The trace is denoted by

$$\langle A \rangle = \mathrm{Sp}(\hat{\rho}\hat{A}) = \sum_{m,n} \rho_{mn} A_{mn} \,. \tag{4.2}$$

In the case of an open system the latter relation (4.2) remains unchanged even though the elements of the density matrix are to be averaged over all the possible states of the external (surrounding) system.

In the simplest case of a two-level quantum system interacting with the electric component $\mathcal{E}(t)$ of the radiation field and with the surrounding medium thermostat the equations of motion of the density matrix are

$$j\hbar\frac{\partial \rho_{11}}{\partial t} = -\mu_{21}\mathcal{E}(t)(\rho_{21} - \rho_{12}) - j\hbar(R_{1111}\rho_{11} + R_{1122}\rho_{22}) \,,$$

$$j\hbar\frac{\partial \rho_{12}}{\partial t} = -\mu_{21}\mathcal{E}(t)(\rho_{22} - \rho_{11}) + (E_1 - E_2)\rho_{12} - j\hbar R_{1212}\rho_{12} \,,$$

$$j\hbar\frac{\partial \rho_{21}}{\partial t} = -\mu_{21}\mathcal{E}(t)(\rho_{11} - \rho_{22}) + (E_2 - E_1)\rho_{21} - j\hbar R_{2121}\rho_{21} \,,$$

$$j\hbar\frac{\partial \rho_{22}}{\partial t} = -\mu_{21}\mathcal{E}(t)(\rho_{12} - \rho_{21}) - j\hbar(R_{2222}\rho_{22} + R_{2211}\rho_{11}) \,.$$

$$(4.3)$$

Thus written, the terms containing the relaxation matrix elements R_{ijkl} correspond to the interaction with the thermostat. The elements of the relaxation matrix satisfy the following relations

$$R_{1111} = -R_{2211}, \quad R_{2222} = -R_{1122}, \quad R_{1212} = R_{2121},$$

$$R_{1111} = R_{2222} \exp\left(-\frac{E_2 - E_1}{kT}\right). \tag{4.4}$$

In the absence of an external field, equations (4.3) along with (4.4) bring about the relaxation of the system to the steady state, where

$$\rho_{12} = \rho_{21} = 0, \quad \rho_{11} + \rho_{22} = \text{const}, \quad R_{1111}\rho_{11} + R_{1122}\rho_{22} = 0. \tag{4.5}$$

Besides, the third equation of (4.5) allowing for (4.4) entails the Boltzmann relation between the populations of the upper and lower states,

$$\rho_{22} = \rho_{11} \exp\left(-\frac{E_2 - E_1}{kT}\right). \tag{4.6}$$

The system of equations (4.3) may be written concisely

$$\frac{\partial \rho}{\partial t} = \frac{1}{j\hbar}[\widehat{H}, \widehat{\rho}] - \widehat{\widehat{R}}\widehat{\rho}, \tag{4.7}$$

where the double operator sign over the relaxation matrix R signifies that its dimensionality is twice as high as that of matrix ρ. In the case of a two-level system matrix R has not 4 but 16 elements (even though many of them are zero, as in the above example).

It is worth mentioning here that, when analyzing the two-level systems, instead of four independent matrix elements ρ_{ij}, four other quantities, defined by

$$\widehat{\rho} = N\widehat{I} + W\widehat{\sigma}_z + P\widehat{\sigma}_x + Q\widehat{\sigma}_y, \tag{4.8}$$

are often used, where $\widehat{I}$ is the unit matrix and $\widehat{\sigma}_x, \widehat{\sigma}_y, \widehat{\sigma}_z$ are the Pauli matrices

$$\widehat{I} = \begin{pmatrix} 1 & 0 \\ 0 & 1 \end{pmatrix}, \quad \widehat{\sigma}_z = \begin{pmatrix} 1 & 0 \\ 0 & -1 \end{pmatrix}, \quad \widehat{\sigma}_x = \begin{pmatrix} 0 & 1 \\ 1 & 0 \end{pmatrix}, \quad \widehat{\sigma}_y = \begin{pmatrix} 0 & j \\ -j & 0 \end{pmatrix}.$$

Variables N and W in (4.8) represent the sum of the numbers of particles on the upper and lower levels and to their difference, respectively. Quantities P and Q are proportional to the real and imaginary parts of the particle polarization, respectively.

Let us now substitute (4.8) in (4.7) bearing in mind (4.4) and the commutation rules of Pauli's matrices, and equate the coefficients, multiplying the matrices. Then we obtain the following system of equations:

$$\dot{N} = 0,$$

$$\dot{W} = -(R_{1111} + R_{2222})W + (R_{2222} - R_{1111})N - \frac{\mu_{21}}{\hbar}\mathcal{E}Q,$$

$$\dot{P} = -R_{1212}P + \frac{E_2 - E_1}{\hbar}Q, \tag{4.9}$$

$$\dot{Q} = -R_{1212}Q - \frac{E_2 - E_1}{\hbar}P + \frac{\mu_{21}}{\hbar}EW.$$

Equations (4.9) are referred to as Bloch's equations. The total number of particles is seen to be constant, the difference of populations relaxing to its equilibrium value at the rate of $R_{1111} + R_{2222}$, and the polarization at the rate of R_{1212}. These rates are called longitudinal and transverse relaxation rates, respectively. Let us denote them by symbols $R_{\parallel}$ and $R_{\perp}$. Their reciprocal values, given by

$$T_1 = \frac{1}{R_{\parallel}} = \frac{1}{R_{1111} + R_{2222}}, \tag{4.10}$$

$$T_2 = \frac{1}{R_{\perp}} = \frac{1}{R_{1212}}, \tag{4.11}$$

are called longitudinal and transverse relaxation times.

The point is that if matrix ρ is the density matrix of a particle with a spin $\frac{1}{2}$, quantities W, P and Q will be proportional to the quantum-mechanical mean values of the z-, x- and y-components of the magnetization of the particle placed in a uniform magnetic field. The interaction of the particle with the radiation field should, naturally, be of a magneto-dipole type. Then the z-component of magnetization along the magnetic field direction would relax to its equilibrium value at the rate of $R_{\parallel}$, the transverse ones – x- and y- – at the rate of $R_{\perp}$.

This analogy between the behavior of a two-level system and that of a particle with spin $\frac{1}{2}$ gave rise to terms longitudinal and transverse relaxation rates (times). Of course, once we are dealing with optical range radiation, it would be more correct to speak of rate of relaxation of the above-thermal excitation ($R_{\parallel}$) and polarization relaxation rate ($R_{\perp}$). However, the terms *longitudinal* and *transverse* relaxation rates have taken roots in scientific terminology and we shall use them where they apply.

Equations of the (4.9) type were first formulated in the 1940s by F. Bloch. It was he who introduced the concepts of longitudinal and transverse relaxation times for the phenomenon of nuclear magnetic resonance (NMR), which, in short, may be described as the selective absorption of electromagnetic energy caused by the nuclear paramagnetism of a substance. The NMR is observed when the sample containing paramagnetic nuclei is exposed to a relatively weak field H_1 oscillating with radio frequency while placed in a strong uniform magnetic field H_0 perpendicular to H_1. The uniform field creates two energy levels corresponding to the spin of $\frac{1}{2}$, while the oscillating field induces transitions between them. The NMR lines may usually be observed in the range of 10^6–10^7 Hz.

The same equations are widely used in electron paramagnetic resonance (EPR) studies. This phenomenon is caused by the electron paramagnetism of atoms, molecules, ions, color centers in crystals, etc. with unpaired electrons. The experiment on EPR resembles that on NMR, while the range of resonant frequencies is quite different (10^9–10^{11} Hz). This effect was observed for the first time by E. K. Zavoisky in 1944.

From NMR and EPR the terms *longitudinal and transverse relaxation* have been adopted by radio spectroscopy, a more general field of physics that investigates the transitions between quantum-system energy levels induced by radio-frequency electromagnetic radiation. Quantum electronics has borrowed these terms from radio spectroscopy. In this more general context they have lost their initial meanings, remaining, however, rather convenient[*].

Let us now come back to the relaxation matrix. In order to correctly use equations like (4.7) which contain the relaxation additions, one should clearly realize the preconditions they are derived from. So, it seems reasonable to consider here, even though only schematically, how the density matrix equations of motion are derived. Procedures analogous to those which follow will be used later for the derivation of corresponding equations in more complicated situations.

Below we shall discuss two derivation procedures for (4.7)-type equations which result in the same final form of the density matrix equation. However, the relaxation matrix R involved is expressed through the parameters of the problem in different ways, in accordance with different physical situations.

Let us first consider a gaseous medium in which the molecules interact with each other by means of mutual collisions via a succession of short shocks, accidental in time the duration of which is small against an intercollision interval. The method we are going to use is known as Bogoljubov's method, or Bogoljubov's chain of equations. In this case, from the equation of motion for the density matrix of the whole gas $\mathcal{P}$,

$$j\hbar\dot{\mathcal{P}} = [\mathcal{H}, \mathcal{P}],\tag{4.12}$$

where the complete Hamiltonian $\mathcal{H}$ of the system as a whole is the sum of the Hamiltonians of isolated particles $\sum \widehat{H}$ and all possible interactions $\sum \widehat{V}$,

$$\mathcal{H} = \sum \widehat{H} + \sum \widehat{V}.\tag{4.13}$$

If we carry out the trace operation over all the molecules except one, then, for the density matrix of this particle we obtain

$$j\hbar\dot{\widehat{\rho}} = \left[\widehat{H}, \widehat{\rho}\right] + \mathrm{Sp}\left[\widehat{\widehat{V}}, \widehat{\widehat{\rho}}\right].\tag{4.14}$$

---

[*] Note that the widespread term *inhomogeneous broadening of a spectral line* also originates from NMR-spectroscopy, where the corresponding effect is accounted for by the spatial non homogeneity of the magnetic field H_0. In the optical range this term has not completely lost its initial, fairly transparent, meaning (see Lecture 6).

Here $\widehat{\widehat{\rho}}$ is a two-particle density matrix, i.e., the density matrix of the system of two particles – the chosen particle and the other particle colliding with it, $\widehat{\widehat{V}}$ is the Hamiltonian of their interaction. The trace operation is carried out over the second particle variables. Note that in case the density matrix $\widehat{\rho}$ is represented in matrix form ρ_{ij}, matrix $\widehat{\widehat{V}}$ and the density matrix $\widehat{\widehat{\rho}}$ in this representation have four indices corresponding to the states of two particles. The first term in the right-hand side of (4.14) describes the dynamics of the particle considered under the action of internal forces, the second term being responsible for the influence of collisions. The two-particle density matrix $\widehat{\widehat{\rho}}$, provided only pairwise interactions are essential, satisfies the equation which, in the interaction representation, has the form

$$ j\hbar \dot{\widehat{\widehat{\rho}}} = \left[\widehat{\widehat{V}}, \widehat{\widehat{\rho}} \right] . \tag{4.15} $$

For the variation of the density matrix $\widehat{\widehat{\rho}}$ in the case of a weak interparticle interaction, i.e., when the effect of the collision on the two-particle density matrix $\widehat{\widehat{\rho}}$ may be written retaining the first three terms (viz. the zero, the first and the second ones) of the expansion into a perturbation theory series over the time-dependent interaction $\widehat{\widehat{V}}(t)$, we have

$$ \widehat{\widehat{\rho}} - \widehat{\widehat{\rho}}_0 = \frac{1}{j\hbar} \int_{-\infty}^{+\infty} \left[\widehat{\widehat{V}}, \widehat{\widehat{\rho}}_0 \right] dt + \frac{1}{j\hbar} \int_{-\infty}^{+\infty} \left[\widehat{\widehat{V}}(t), \frac{1}{j\hbar} \int_{-\infty}^{t} \left[\widehat{\widehat{V}}(\tau), \widehat{\widehat{\rho}}_0 \right] d\tau \right] dt + \dots , \tag{4.16} $$

where $\widehat{\widehat{\rho}}_0$ is the two-particle density matrix before the collision that equals the direct product of the density matrices of colliding particles $\widehat{\widehat{\rho}}_0 = \widehat{\rho} \cdot \widehat{\rho}$. The variation of matrix $\widehat{\widehat{\rho}}$ in formula (4.16) is written for a particular collision with some specific parameters (the relative velocity, the impact parameter, etc.) which determine the dependence $\widehat{\widehat{V}}(t)$. The multiplication of (4.16) by the collision probability w followed by the integration over all the possible values of collision parameters $d\Gamma$ and the trace summation over the second-colliding-particle variables gives the second term on the right-hand side of (4.14) up to a constant factor $j\hbar$ (see (4.15)). Finally, (4.14) becomes

$$ j\hbar \dot{\widehat{\rho}} = [\widehat{H}, \widehat{\rho}] + \mathrm{Sp} \int w d\Gamma \int_{-\infty}^{+\infty} \left[\widehat{\widehat{V}}(t), \widehat{\rho} \cdot \widehat{\rho}' \right] dt $$
$$ + \frac{1}{j\hbar} \mathrm{Sp} \int w d\Gamma \int_{-\infty}^{+\infty} \left[\widehat{\widehat{V}}(t), \int_{-\infty}^{t} \left[\widehat{\widehat{V}}(\tau), \widehat{\rho} \cdot \widehat{\rho}' \right] d\tau \right] dt . \tag{4.17} $$

After the integration, the second term on the right-hand side of (4.17) containing operator $\widehat{\widehat{V}}(t)$ which, as a rule, oscillates rapidly in time in the interaction representation, usually vanishes. Thus, (4.17) appears to be equivalent to (4.7), where the four-index matrix with the following elements stands for $\widehat{\widehat{R}}$:

$$R_{abcd} = \frac{1}{\hbar^2} \int w d\Gamma \int_{-\infty}^{+\infty} dt \int_{-\infty}^{t} d\tau \sum_{k,l,s}$$

$$\times \left(\sum_j V_{ajkl}(t) V_{jclk}(\tau) \rho'_{kk} \delta_{bd} - V_{ackl}(t) V_{dblk}(\tau) \rho'_{ll} \right.$$

$$\left. - V_{dblk}(t) V_{ackl}(\tau) \rho'_{ll} + \sum_m V_{dmkl}(t) V_{mblk}(\tau) \rho'_{kk} \delta_{ac} \right). \quad (4.18)$$

Here we assume that the incident particle is characterized by the diagonal density matrix $\widehat{\rho}'$. In fact, this assumption implies the individual, unphased and statistical character of the motion of each particle in a gas. Finally, it leads to the irreversible relaxation equations of motion for the density matrix. The matrix elements V_{ijkl} involved in (4.18) are being written in the interaction representation and thus oscillate with a typical frequency of $(E_i - E_j + E_k - E_l)/\hbar$. This frequency should be small, since otherwise (4.18) vanishes. It is possible, provided that $i = j$ and $k = l$ (i.e., for the diagonal matrix elements V_{ijkl}), or the transition frequencies of the colliding particles are close $E_i - E_j \approx E_k - E_l$. This, in turn, corresponds to the common-practice case of the excitation of particles in the presence of an intermediate resonant or near-resonant buffer gas.

In the case of a two-level particle, what has been said above means that only 6 matrix elements of the interaction Hamiltonian are of interest – $V_{1111}, V_{1122}, V_{1221}, V_{2112}, V_{2211}, V_{2222}$. Substituting them in (4.18) we see that only the relaxation matrix elements presented in (4.4) are nonzero. The last equation of (4.4) appears to hold if the relaxant (resonant buffer gas) is in equilibrium at temperature T, in other words, it represents the thermostat at temperature T.

As a result, the elements of the relaxation matrix $\widehat{\widehat{R}}$ given by (4.4) are represented in terms of the matrix elements of the interaction Hamiltonian as follows:

$$R_{1111} = -R_{2211} = \frac{1}{\hbar^2} 2\mathrm{Re} \int w d\Gamma \int_{-\infty}^{+\infty} dt \int_{-\infty}^{t} d\tau V_{1221}(t) V_{2112}(\tau) \rho'_{22},$$

$$R_{2222} = -R_{1122} = \frac{1}{\hbar^2} 2\mathrm{Re} \int w d\Gamma \int_{-\infty}^{+\infty} dt \int_{-\infty}^{t} d\tau V_{2112}(t) V_{1221}(\tau) \rho'_{11},$$

$$R_{1212} = R_{2121} = R_{1111} + R_{2222} \qquad (4.19)$$

$$+ \frac{1}{\hbar^2} 2\mathrm{Re} \int w d\Gamma \int_{-\infty}^{+\infty} dt \int_{-\infty}^{t} d\tau \{ V_{1111}(t) V_{1111}(\tau) \rho'_{11}$$

$$+ V_{1122}(t) V_{1122}(\tau) \rho'_{22} + V_{2211}(t) V_{2211}(\tau) \rho'_{11} + V_{2222}(t) V_{2222}(\tau) \rho'_{22} \}.$$

The transverse relaxation rate $R_\perp = R_{1212} = R_{2221}$ is always seen to exceed the longitudinal relaxation rate $R_\parallel = R_{1111} + R_{2222}$:

$$R_\perp \geq R_\parallel. \qquad (4.20)$$

In other words, the longitudinal relaxation time T_1 always exceeds the transverse relaxation time T_2:

$$T_1 \geq T_2. \tag{4.21}$$

The fact that the transverse relaxation rate exceeds the longitudinal one is accounted for by the existence of relaxation processes which do not affect the population distribution. This means that the interaction associated with these processes does not result in an energy exchange between the molecules under study and the relaxant. It only shifts the energy level positions for the duration of collision, thus changing the off-diagonal elements of the density matrix on account of the corresponding change in the phase difference of the upper- and lower-level Ψ-functions. The act of a single collision brings about an additional phase shift of the off-diagonal density matrix element depending on the collision parameters: the velocity and the impact parameter. It is the difference in phase shifts which, after averaging over all the possible types of collisions, leads to the relaxation of the off-diagonal matrix elements ρ, i.e. to the relaxation of the polarization of the molecule being treated. The collisions affecting the population distribution (the longitudinal relaxation) also change the molecular polarization, thus leading, obviously, to a transverse relaxation as well. That is why the transverse relaxation rate always exceeds the longitudinal one.

Now let us discuss a slightly different physical situation. Consider a large system with internal motions so intricate that they may be treated as thermal. Let this system interact weakly with a relatively small dynamic system, the motion of which is of interest. Hereafter, the former system will be referred to as a thermostat, while the latter will simply be called a system. We are interested in the relaxation of the thermostat-contacting system.

Here we have to stress again that further consideration is based on the following three assumptions:

(i) The system is small against the thermostat (we mean, that its phase volume is much less than that of the thermostat) and only slightly affects the thermostat motion;

(ii) The interaction of the system with the thermostat is so small that the interaction practically Hamiltonian does not disturb the typical frequencies of the motion of the system;

(iii) The density matrix of the thermostat slightly differs from the equilibrium one, so that its diagonal elements are distributed in accord with Boltzmann's law at temperature T.

Therefore, we can exemplify this situation by the relaxation of the impurity ion energy levels in crystals and glasses. In this case the dynamic system is the system of ion energy levels, while the thermostat is a phonon reservoir surrounding the ions. Our consideration also applies to the single vibrational mode of a large polyatomic molecule. This mode is coupled with other modes serving as a thermostat by means of weak anharmonic interactions. We could give many more examples of this kind but let's now turn to the problem under consideration.

In accordance with the above assumptions we represent the complete Hamiltonian of the system and the thermostat $\widehat{\mathcal{H}}$ in the form

$$\widehat{\mathcal{H}} = \widehat{H}_s + \widehat{H}_t + \lambda \widehat{V}, \tag{4.22}$$

where $\widehat{H}_s$ is the system Hamiltonian, $\widehat{H}_t$, the thermostat Hamiltonian; $\lambda \widehat{V}$, their interaction Hamiltonian, $\lambda \ll 1$. The eigenvalues of the operator $\widehat{V}$ are of the same order as those of $\widehat{H}_s$.

For simplicity and convenience we shall hereafter use the Liouville operator determined by

$$\widehat{\widehat{\mathcal{L}}}\widehat{\mathcal{P}} = [\widehat{\mathcal{H}}, \widehat{\mathcal{P}}]. \tag{4.23}$$

Then (4.22) implies that

$$\widehat{\widehat{\mathcal{L}}} = \widehat{\widehat{\mathcal{L}}}_s + \widehat{\widehat{\mathcal{L}}}_t + \lambda \widehat{\widehat{\mathcal{V}}}, \tag{4.24}$$

where $\widehat{\widehat{\mathcal{V}}}$ is the Liouville operator of the Hamiltonian V.

In the case being considered the density matrix $\widehat{\mathcal{P}}$ has two pairs of indices – m, n and α, β. Let the Greek indices refer to the thermostat and the Latin, to the system. Let matrix $\widehat{\mathcal{P}}$ represent the summation of part $\widehat{\mathcal{X}}$ diagonal over Greek indices and the off-diagonal part $\widehat{\mathcal{Y}}$:

$$\mathcal{P}_{nm\alpha\beta} = \mathcal{X}_{nm\alpha\alpha}, \quad \alpha = \beta; \quad \mathcal{P}_{nm\alpha\beta} = \mathcal{Y}_{nm\alpha\beta}, \quad \alpha \neq \beta. \tag{4.25}$$

In the above assumptions the equation of motion of the density matrix

$$j\hbar \dot{\widehat{\mathcal{P}}} = \widehat{\widehat{\mathcal{L}}}\widehat{\mathcal{P}} \tag{4.26}$$

may be split in two

$$j\hbar \dot{\widehat{\mathcal{X}}} = \widehat{\widehat{\mathcal{L}}}_s \widehat{\mathcal{X}} + \lambda \widehat{\widehat{\mathcal{V}}}\widehat{\mathcal{Y}}, \tag{4.27}$$

$$j\hbar \dot{\widehat{\mathcal{Y}}} = \widehat{\widehat{\mathcal{L}}}_t \widehat{\mathcal{Y}} + \widehat{\widehat{\mathcal{L}}}_s \widehat{\mathcal{Y}} + \lambda \widehat{\widehat{\mathcal{V}}}\widehat{\mathcal{X}}. \tag{4.28}$$

The solution of (4.28) gives

$$\mathcal{Y}_{nm\alpha\beta} = \frac{\lambda}{j\hbar} \int_{-\infty}^{t} d\tau \exp\left[-j\frac{E_\alpha - E_\beta + E_n - E_m}{\hbar}(t - \tau)\right] \\ \times \sum_{\gamma, k, l} V_{nm\alpha\beta}^{kl\gamma\gamma} \mathcal{X}_{kl\gamma\gamma}(\tau). \tag{4.29}$$

Substituting (4.29) in (4.27) we can obtain

$$j\hbar \dot{\widehat{\mathcal{X}}} = \widehat{\widehat{\mathcal{L}}}_s \widehat{\mathcal{X}} + \int_{-\infty}^{t} d\tau \widehat{\widehat{R}}(t, \tau)\widehat{\mathcal{X}}(\tau), \tag{4.30}$$

where the matrix $\widehat{\widehat{R}}$ elements are given by

$$\mathcal{R}^{kl\gamma\gamma}_{nm\alpha\alpha} = \frac{1}{j\hbar}\lambda^2 \sum_{r,s,\delta,\kappa} V^{rs\delta\kappa}_{nm\alpha\alpha} V^{kl\gamma\gamma}_{rs\delta\kappa} \exp\left[-j\frac{E_\delta - E_\kappa + E_n - E_s}{\hbar}(t-\tau)\right].$$

$$(4.31)$$

The equation of motion of the system density matrix $\hat{\rho}$ may be obtained from (4.30). Since, by definition

$$\rho_{nm} \equiv \mathrm{Sp}_\alpha \mathcal{P}_{nm\alpha\beta} \equiv \sum_\alpha \mathcal{X}_{nm\alpha\alpha}, \qquad (4.32)$$

(4.30) leads to

$$j\hbar\dot{\hat{\rho}} = \left[\hat{H}_s, \hat{\rho}\right] + \mathrm{Sp}_\alpha \int_{-\infty}^{t} d\tau \widehat{\widehat{\mathcal{R}}}(t,\tau)\widehat{\mathcal{X}}(\tau). \qquad (4.33)$$

This equation describes the system dynamics up to the terms of the order of λ^3. Hence, it is valid owing to the weakness of the interaction. Now we can take advantage of the remaining two assumptions accepted to be basic in the given approach and rewrite the operator $\widehat{\mathcal{X}}$ matrix elements as

$$\mathcal{X}_{nm\alpha\alpha}(\tau) \approx \rho_{nm}(\tau)Z^{-1}\exp(-E_\alpha/kT), \qquad (4.34)$$

where

$$Z = \sum_\alpha \exp(-E_\alpha/kT)$$

is the statistical sum of the thermostat.

Expression (4.34) contains the undisturbed thermal matrix elements of the thermostat. This is worth explaining. The influence of the weak interaction λV on the system is much greater, since the system is relatively small with respect to the thermostat. Therefore, the thermostat density matrix differs from the thermal one only in the next order of the perturbation theory.

Substituting (4.34) in (4.33) we obtain the equation for matrix ρ:

$$j\hbar\dot{\hat{\rho}} = \left[\hat{H}_s, \hat{\rho}\right] + \int_{-\infty}^{t} d\tau \widehat{\widehat{R}}(t,\tau)\hat{\rho}(\tau), \qquad (4.35)$$

where the operator matrix elements are

$$R^{kl}_{nm}(t,\tau) = \frac{\lambda^2}{j\hbar} \sum_{\alpha,\gamma,\delta,\kappa,r,s} \frac{1}{Z}\exp\left(-\frac{E_\gamma}{kT}\right)$$

$$\times V^{rs\delta\kappa}_{nm\alpha\alpha} V^{kl\gamma\gamma}_{rs\delta\kappa} \exp\left[-j\frac{E_\delta - E_\kappa + E_r - E_s}{\hbar}(t-\tau)\right].$$

Some additional assumptions about the thermostat properties allow us to reduce the equation obtained to the form of (4.7). First, let the width of the frequency spectrum of the thermostat action on the system Δ exceed the interaction strength, as is usually the case:

$$\Delta \gg \lambda V/\hbar. \qquad (4.36)$$

This means that $\widehat{\widehat{R}}(t, \tau)$ in (4.35) varies in time much faster than $\widehat{\rho}(\tau)$ and, hence, the latter may be removed from the integrand.

Now let us assume that the energy spectrum of the thermostat is rather dense

$$\lambda V \gg E_\alpha - E_{\alpha-1}, \tag{4.37}$$

then (4.35) may be reduced to the form of (4.7) where the relaxation matrix $\widehat{\widehat{R}}$ elements are

$$R_{nm}^{kl} = \frac{\pi \lambda^2}{4} \int \left(\frac{d\Gamma}{dE} dE \right)^4 \frac{1}{Z} \exp\left(-\frac{E_\gamma}{kT} \right)$$
$$\times V_{nm\alpha\alpha}^{rs\delta\kappa} V_{rs\delta\kappa}^{kl\gamma\gamma} \delta(E_\delta - E_\kappa + E_r - E_s). \tag{4.38}$$

To shorten the notation we have introduced,

$$\left(\frac{d\Gamma}{dE} dE \right)^4 = \frac{d}{dE} \Gamma(E_\alpha) \frac{d}{dE} \Gamma(E_\gamma) \frac{d}{dE} \Gamma(E_\kappa) dE_\alpha dE_\gamma dE_\delta dE_\kappa, \tag{4.39}$$

and $d\Gamma(E_s)/dE$, in turn, stands for the density of states of the thermostat near E_s.

According to the definition of the Liouville operator,

$$V_{nm\alpha\beta}^{kl\delta\kappa} = V_{n\alpha}^{k\delta} \delta_m^l \delta_\beta^\kappa - V_{m\beta}^{l\kappa} \delta_n^k \delta_\alpha^\delta. \tag{4.40}$$

Substituting the so written Liouville operator matrix elements in (4.38), we derive the expression for the relaxation matrix elements R_{ik}^{jl}. Sometimes it is more convenient to write these elements as we did earlier with all subscripts R_{ijkl}. Therefore we have

$$R_{ijkl} = 2J_{iklj} - \sum_q J_{qkiq} \delta_{jl} - \sum_q J_{qjlq} \delta_{ki}, \tag{4.41}$$

where

$$J_{iklj} = \pi \lambda^2 \int \left(\frac{d\Gamma}{dE} dE \right)^2 V_{i\alpha}^{k\beta} V_{l\beta}^{i\alpha} \frac{1}{Z} \exp\left(-\frac{E_\beta}{kT} \right)$$
$$\times \delta(E_l - E_j + E_\alpha - E_\delta)(\delta_{ik}\delta_{lj} + \delta_{ij}\delta_{kl}),$$

$$\left(\frac{d\Gamma}{dE} dE \right)^2 = \frac{d}{dE} \Gamma(E_\alpha) \frac{d}{dE} \Gamma(E_\beta) dE_\alpha dE_\beta.$$

Now let us turn directly to a two-level system without further discussion of this expression. As in the above case of relaxation in the gas-kinetic collisions only 6 matrix elements represented by (4.4) appear to be nonzero. However, in contrast to (4.19) these elements are determined not by the parameters of the collision interaction, but by the parameters of the interaction with the thermostat.

So, the longitudinal relaxation rate is

$$R_{\parallel} = 2\pi\lambda^2 Z^{-1}\left[1 + \exp\left(-\frac{\hbar\omega_{21}}{kT}\right)\right]$$

$$\times \int V_{1\beta2\alpha}^2 \delta(E_\beta - E_\alpha - \hbar\omega_{21})\frac{d}{dE}\Gamma(E_\alpha + \hbar\omega_{21})$$

$$\times \exp\left(-\frac{E_\alpha}{kT}\right)\frac{d}{dE}\Gamma(E_\alpha)dE_\alpha dE_\beta. \tag{4.42}$$

This expression may be elucidated as follows: the value

$$\lambda^2\langle V^2\rangle = \lambda^2 \int V_{1\beta2\alpha}^2 \delta(E_\beta - E_\alpha - \hbar\omega_{21})dE_\beta \tag{4.43}$$

is the square of the matrix element of the operator of the relaxation transition from state 2 to state 1 averaged over all the final thermostat states β. This transition is accompanied by the thermostat transition from state α to state β, energy E_β exceeding energy E_α by $\hbar\omega_{21}$. The multiplication of the mean-square transition matrix element $\lambda^2\langle V^2\rangle$ by the level density $d\Gamma(E_\alpha + \hbar\omega_{21})/dE$ in the vicinity of the final state gives the probability of this transition. The integration of the product obtained over dE_α, with allowance for the thermal function of the thermostat distribution among its initial states $1/Z \exp(-E_\alpha/kT)d\Gamma(E_\alpha)/dE$, gives the net probability R_{2222} of the relaxation decay of the upper state. Taking into account the last equation of (4.4) and the definition (4.10) we obtain (4.42).

The transverse relaxation rate of a two-level system is

$$R_{\perp} = R_{\parallel} + \frac{\pi\lambda^2}{Z}\int (V_{1\beta1\alpha}^2 + V_{2\alpha2\beta}^2)\delta(E_\alpha - E_\beta)$$

$$\times \frac{d}{dE}\Gamma(E_\beta)\frac{d}{dE}\Gamma(E_\alpha)\exp\left(-\frac{E_\alpha}{kT}\right)dE_\alpha dE_\beta. \tag{4.44}$$

The addition to $R_{\parallel}$ in the expression for $R_{\perp}$ allows for the thermostat transitions not resulting in a system energy change but bringing about the relaxation of the off-diagonal elements of the system density matrix. This relaxation is accounted for by the chaotic modulation of the system transition frequency due to the thermal motion of the thermostat.

Thus we have shown how the system equations of motion of type (4.3) or (4.7) with the relaxation matrix (4.4) may be derived. We have carried out this derivation for a quantum system interacting with the surroundings in the mode of gas-kinetic collisions with the particles of a resonant or near-resonant buffer gas and for the case of a weak interaction with a large thermostat. We have shown that for the rates of longitudinal (4.10)- and transverse (4.11) relaxation the inequality (4.20) is valid for both cases considered. We emphasize again that the above consideration does not represent a derivation of all these relations, but only serves to illustrate the derivation procedure. The methods schematically presented have become widespread in the analysis of complex systems. In particular, of special interest is the case of a photon reservoir serving as a thermostat.

Let us consider this case in greater detail. Imagine a two-level quantum system interacting with a large number of identical photons, comprising the radiation field of an ideal laser. The number of identical photons corresponding to a single field oscillator (a single oscillation type, a single mode) is known to be unlimited. The resulting state of the radiation field is determined by the numbers of photons in each mode. The radiation of an ideal laser occupies only one of the variety of possible modes. But the same two-level quantum system may both absorb and emit the photons of other modes (field oscillators). There are many of these modes and in free space there are still more. The aggregate of these modes is the photon thermostat. In this situation the system relaxation appears to correspond to the transition of the system from an upper level to a lower one accompanied by the radiation of a photon of one of the thermostat modes different from the laser mode. Then $R_\parallel = R_\perp$ is the probability of the spontaneous emission of the photon. The inversion of either $R_\parallel$ or $R_\perp$ gives the so-called natural lifetime of the particle in the excited state. The process of spontaneous decay assumes specific features when the typical interparticle distance becomes less than the radiation wavelength (see Sect. 1 of "Appendix").

Lecture Five

Susceptibility of a Two-Level System

Solution of density matrix equations of motion in resonant approximation, stationary solution. Difference of populations. Dipole moment. Absorption. Susceptibility. Line width and homogeneous broadening. Saturation effect. Field broadening. Dispersion relations.

Let us consider the use of the density matrix equations of motion of type (4.7) and calculate the susceptibility of a two-level system with the relaxation times T_1 and T_2. We represent an external field as previously in the form

$$\mathcal{E}(t) = (\mathcal{E}_0/2)(e^{j\omega t} + e^{-j\omega t}) \qquad (5.1)$$

and consider the case of the resonant interaction

$$\Delta = \omega_{21} - \omega \ll \omega, \omega_{21}. \qquad (5.2)$$

In the frame of the resonant approximation similar to that developed in lecture 2, when analyzing the coherent interaction of radiation with matter, we try the solution of system (4.3) in the form (see (4.1))

$$\rho_{11} = \sigma_{11}, \quad \rho_{22} = \sigma_{22}, \quad \rho_{12} = \sigma_{12}e^{j\omega t}, \quad \rho_{21} = \sigma_{21}e^{-j\omega t}, \qquad (5.3)$$

where $\sigma_{12} = \sigma_{21}^*$.

In other words, as is often done in the theory of vibrations, we try the solution oscillating with the frequency of the driving external field. Then, substituting (5.3) in (4.3) and omitting the terms that oscillate with double frequency in the resonant approximation, we obtain a set of equations of motion for a two-level system with the relaxation

$$
\begin{aligned}
\frac{\partial \sigma_{11}}{\partial t} &= j\frac{\mu_{21}\mathcal{E}_0}{2\hbar}(\sigma_{21} - \sigma_{12}) - (R_{1111}\sigma_{11} + R_{1122}\sigma_{22}), \\
\frac{\partial \sigma_{12}}{\partial t} &= j\frac{\mu_{21}\mathcal{E}_0}{2\hbar}(\sigma_{22} - \sigma_{11}) - (R_{1212} - j(\omega_{21} - \omega))\sigma_{12}, \\
\frac{\partial \sigma_{21}}{\partial t} &= j\frac{\mu_{21}\mathcal{E}_0}{2\hbar}(\sigma_{11} - \sigma_{22}) - (R_{2121} + j(\omega_{21} - \omega))\sigma_{12}, \\
\frac{\partial \sigma_{22}}{\partial t} &= j\frac{\mu_{21}\mathcal{E}_0}{2\hbar}(\sigma_{12} - \sigma_{21}) - (R_{2222}\sigma_{22} + R_{2211}\sigma_{11}).
\end{aligned}
\qquad (5.4)
$$

These equations resemble the resonant approximation for the amplitudes of wavefunctions ψ_1 and ψ_2 (2.19) we used when considering the coherent interaction effects (lectures 2 and 3). In contrast to (2.19) Eqs. (5.4) make allowance for the relaxation.

Let us start with the case of $R_\perp = R_{2121} = R_{1212} \gg \mu_{21}\mathcal{E}_0/\hbar$. Then σ_{12} and σ_{21} rapidly relax to their stationary values determined by the second and the third equations of (5.4)

$$
\begin{aligned}
\sigma_{12} &= j\frac{\mu_{21}\mathcal{E}_0}{2\hbar}\frac{\sigma_{22}-\sigma_{11}}{R_\perp - j(\omega_{21}-\omega)}\,, \\
\sigma_{21} &= \sigma_{12}^* = j\frac{\mu_{21}\mathcal{E}_0}{2\hbar}\frac{\sigma_{11}-\sigma_{22}}{R_\perp + j(\omega_{21}-\omega)}\,.
\end{aligned}
\tag{5.5}
$$

Substituting (5.5) into the first and the fourth equations of (5.4), we find

$$
\begin{aligned}
\frac{\partial\sigma_{11}}{\partial t} &= -W(\omega)(\sigma_{11}-\sigma_{22}) - (R_{1111}\sigma_{11} - R_{2222}\sigma_{22})\,, \\
\frac{\partial\sigma_{22}}{\partial t} &= -W(\omega)(\sigma_{22}-\sigma_{11}) - (R_{2222}\sigma_{22} - R_{1111}\sigma_{11})\,,
\end{aligned}
\tag{5.6}
$$

where

$$
W(\omega) = \left(\frac{\mu_{21}\mathcal{E}_0}{2\hbar}\right)^2 \frac{2R_\perp}{R_\perp^2 + (\omega_{21}-\omega)^2}
\tag{5.7}
$$

is the probability of the radiation-induced absorption or emission of the photon normalized to a unit time interval (the probability of induced transition).

Now let us consider the stationary regime, assuming the left-hand sides of Eqs. (5.4) equal to zero. We rewrite $\sigma_{21} = \sigma_{12}^*$ in the form $\sigma_{21} = \mathrm{Re}\,\sigma_{21} + j\mathrm{Im}\,\sigma_{21}$ and substitute it into (5.4). The requirement for Eqs. (5.4) to be valid for the real and imaginary parts separately, along with the normalization condition $\sigma_{22} + \sigma_{11} = 1$, gives

$$
\mathrm{Re}\,\sigma_{21} = T_2(\omega_{21}-\omega)\mathrm{Im}\,\sigma_{21}\,,
$$

$$
T_2(\omega_{21}-\omega)\mathrm{Re}\,\sigma_{21} + \mathrm{Im}\,\sigma_{21} = -\frac{\mu_{21}\mathcal{E}_0}{2\hbar}(\sigma_{22}-\sigma_{11})\,,
\tag{5.8}
$$

$$
\sigma_{11} + \sigma_{22} = 1\,.
$$

Deriving Eqs. (5.8), we take into account that according to (4.4) and (4.11), $R_{1122} = -R_{2222}$, $R_{2211} = -R_{1111}$ and $R_{1212} = R_{2121} = 1/T_2$.

The equations obtained are easy to solve. One should just remember that according to (4.10), $R_{1111} + R_{2222} = 1/T_1$. Then, for the relative difference of populations we obtain

$$
\sigma_{11} - \sigma_{22} = \frac{1 + (\omega_{21}-\omega)^2 T_2^2}{1 + (\omega_{21}-\omega)^2 T_2^2 + \left(\frac{\mu_{21}\mathcal{E}_0}{\hbar}\right)^2 T_2 T_1} \frac{R_{2222} - R_{1111}}{R_{2222} + R_{1111}}\,.
\tag{5.9}
$$

In addition

$$\operatorname{Im}\sigma_{21} = \frac{\mu_{21}\mathcal{E}_0}{2\hbar}T_2\frac{1}{1+(\omega_{21}-\omega)^2T_2^2+\left(\frac{\mu_{21}\mathcal{E}_0}{\hbar}\right)^2T_2T_1}\frac{R_{2222}-R_{1111}}{R_{2222}+R_{1111}},\tag{5.10}$$

$$\operatorname{Re}\sigma_{21} = \frac{\mu_{21}\mathcal{E}_0}{2\hbar}\frac{(\omega_{21}-\omega)T_2}{1+(\omega_{21}-\omega)^2T_2^2+\left(\frac{\mu_{21}\mathcal{E}_0}{\hbar}\right)^2T_2T_1}\frac{R_{2222}-R_{1111}}{R_{2222}+R_{1111}}.$$

The relative difference of the diagonal elements of the relaxation matrix in (5.9) and (5.10) in accordance with (4.4) and (4.6) determines the relative difference of populations in equilibrium

$$(\sigma_{11}-\sigma_{22})_{\text{equil}} = \frac{R_{2222}-R_{1111}}{R_{2222}+R_{1111}} = \frac{1-e^{-\hbar\omega_{21}/kT}}{1+e^{-\hbar\omega_{21}/kT}}.\tag{5.11}$$

This difference at $\hbar\omega_{21}\gg kT$ turns to unity. Bearing in mind the optical range we rewrite (5.9), (5.10) in the form

$$\sigma_{11}-\sigma_{22} = \frac{1+(\omega_{21}-\omega)^2T_2^2}{1+(\omega_{21}-\omega)^2T_2^2+\left(\frac{\mu_{21}\mathcal{E}_0}{\hbar}\right)^2T_2T_1},\tag{5.12}$$

$$\operatorname{Im}\sigma_{21} = \frac{\mu_{21}\mathcal{E}_0}{2\hbar}T_2\frac{1}{1+(\omega_{21}-\omega)^2T_2^2+\left(\frac{\mu_{21}\mathcal{E}_0}{\hbar}\right)^2T_2T_1},\tag{5.13}$$

$$\operatorname{Re}\sigma_{21} = \frac{\mu_{21}\mathcal{E}_0}{2\hbar}T_2\frac{(\omega_{21}-\omega)T_2}{1+(\omega_{21}-\omega)^2T_2^2+\left(\frac{\mu_{21}\mathcal{E}_0}{\hbar}\right)^2T_2T_1}.\tag{5.14}$$

The relative difference of the populations of levels 1 and 2 allows us to find the population difference per unit volume in an evident way

$$\delta N = N_1 - N_2 = N(\sigma_{11}-\sigma_{22}),\tag{5.15}$$

where N is the density of the resonant particles.

If we know both the real and the imaginary parts of the off-diagonal elements of the density matrix corresponding to the system considered, we can evaluate the polarization. According to (4.2) the quantum-mean dipole moment of a two-level particle is (see also (5.3))

$$\langle\mu\rangle = \mu_{21}(\sigma_{21}^*e^{j\omega t}+\sigma_{21}e^{-j\omega t}).\tag{5.16}$$

Then, using the Euler equations we can readily obtain

$$\langle\mu\rangle = 2\mu_{21}(\operatorname{Re}\sigma_{21}\cos\omega t+\operatorname{Im}\sigma_{21}\sin\omega t).\tag{5.17}$$

Hence, the macroscopic polarization, i.e., the dipole moment per unit volume $P = N\langle\mu\rangle$, is

$$P = N\mu_{21}\frac{\mu_{21}\mathcal{E}_0}{\hbar}T_2\frac{T-2(\omega_{21}-\omega)\cos\omega t+\sin\omega t}{1+(\omega_{21}-\omega)^2T_2^2+\left(\frac{\mu_{21}\mathcal{E}_0}{\hbar}\right)^2T_2T_1}.\tag{5.18}$$

Now let us find the power spent by the field per unit volume of medium for the periodic change in its polarization. According to Poynting's theorem this power equals $E\partial P/\partial t$. At $E = \mathcal{E}_0 \cos \omega t$ and $P(t)$ given for this particular dependence $E(t)$ by formula (5.18), the value of $\dot{Q}$, averaged over a high-frequency oscillation period, where Q is the spatial density of energy, equals

$$\langle \dot{Q} \rangle = \left\langle E \frac{\partial P}{\partial t} \right\rangle = \frac{N}{2} \left(\frac{\mu_{21} \mathcal{E}_0}{\hbar} \right)^2 \frac{T_2 \hbar \omega}{1 + (\omega_{21} - \omega)^2 T_2^2 + \left(\frac{\mu_{21} \mathcal{E}_0}{\hbar} \right)^2 T_2 T_1}. \quad (5.19)$$

As one should expect, the presence of energy losses at a cyclic change of polarization is associated with the existence of the relaxation processes. The spatial density of the energy losses may be easily related to the running-wave absorption coefficient $\alpha = dI/Idz$, where I is the wave intensity,

$$\alpha = \frac{1}{c} \frac{1}{Q} \langle \dot{Q} \rangle. \quad (5.20)$$

Then, bearing in mind that the specific energy is $Q = \mathcal{E}_0^2/8\pi$, we obtain

$$\alpha = \frac{8\pi}{c} \frac{N}{2} \left(\frac{\mu_{21}}{\hbar} \right)^2 \frac{T_2 \hbar \omega}{1 + (\omega_{21} - \omega)^2 T_2^2 + \left(\frac{\mu_{21} \mathcal{E}_0}{\hbar} \right)^2 T_2 T_1}. \quad (5.21)$$

Now, without discussing how the absorption coefficient α depends upon the parameters of the problem involved, we simply note that the energy lost in re-polarization is determined by the sinusoidal component of polarization, as follows from a comparison of (5.21) or (5.19) with (5.18). This component is $\pi/2$ out of phase with respect to the exciting field $\mathcal{E}_0 \cos \omega t$. The cosinusoidal component or the component phased with the field is responsible for the dispersive properties of a medium. Indeed, induction $D = \varepsilon E = E + 4\pi P$. Hence, $\varepsilon = 1 + 4\pi P/E$. After averaging over the oscillation period, (5.18) gives

$$\varepsilon = 1 + 4\pi N \left(\frac{\mu_{21}}{\hbar} \right)^2 T_2 \frac{T_2 (\hbar \omega_{21} - \hbar \omega)}{1 + (\omega_{21} - \omega)^2 T_2^2 + \left(\frac{\mu_{21} \mathcal{E}_0}{\hbar} \right)^2 T_2 T_1}. \quad (5.22)$$

The addition to the dielectric constant, $\varepsilon_0 = 1$, is seen to depend on the frequency and vanish if the system is exactly in resonance with the field.

For further analysis it is expedient to introduce the susceptibility χ of the medium exposed to irradiation, which is usually written as

$$\chi = \chi' - j\chi''. \quad (5.23)$$

An accumulated phase shift of the electromagnetic wave propagating in a medium is determined by the value of χ' and the absorption, by χ''. The frequency dependence of χ'' determines the dispersive properties of the substance.

Indeed, the index of refraction is known to be $n = \sqrt{\varepsilon} = \sqrt{1 + 4\pi \chi}$. At small χ, i.e., when the energy losses and the accumulated phase shift induced by the substance on the wavelength distance are small, expression (5.23) yields that a plane wave of the form

$$E(t, z) = \mathcal{E}_0 \exp[j(\omega t - 2\pi n z/\lambda)],$$

propagating along the z-axis and having wavelength λ may be represented as

$$E(t, z) = \mathcal{E}_0 \exp\left(-4\pi^2 \chi'' \frac{z}{\lambda}\right) \times$$
$$\times \exp\left[j\left(\omega t - 2\pi(1 + 2\pi\chi')\frac{z}{\lambda}\right)\right]. \tag{5.24}$$

With the aid of (5.22) and (5.21) this expression allows linkage of χ' and χ'' with the coefficients that multiply the cosinusoidal and sinusoidal components of polarization (5.18), respectively. Assuming by definition

$$\alpha = 8\pi^2 \chi''/\lambda, \tag{5.25}$$
$$\sqrt{\varepsilon} = 1 + 2\pi\chi', \quad 2\pi\chi' \ll 1, \tag{5.26}$$

where ε and α are given by (5.22) and (5.21), we obtain

$$\chi'' = N\frac{\mu_{21}^2}{\hbar}T_2\frac{1}{1 + (\omega_{21} - \omega)^2 T_2^2 + \left(\frac{\mu_{21}\mathcal{E}_0}{\hbar}\right)^2 T_2 T_1}, \tag{5.27}$$

$$\chi' = N\frac{\mu_{21}^2}{\hbar}T_2\frac{(\omega_{21} - \omega)T_2}{1 + (\omega_{21} - \omega)^2 T_2^2 + \left(\frac{\mu_{21}\mathcal{E}_0}{\hbar}\right)^2 T_2 T_1}. \tag{5.28}$$

Remember that polarization (5.18) is written down in the form

$$P(t) = \mathcal{E}_0(\chi' \cos \omega t + \chi'' \sin \omega t). \tag{5.29}$$

Then the comparison of (5.29) with (5.17) shows that the real part of the susceptibility is proportional to $\operatorname{Re}\sigma_{21}$, while the imaginary part is proportional to $\operatorname{Im}\sigma_{21}$.

From the relations obtained the shape of the absorption line is first of all seen to be of the Lorentz form. The line half-width at $\mathcal{E}_0 \to 0$ is determined by the relaxation time T_2 of the density matrix off-diagonal elements and equals

$$\Delta\omega = 2/T_2. \tag{5.30}$$

So, the transverse relaxation determines the broadening of the resonant line of transition. Note that the method based on the phenomenological balance equations for the populations, i.e., for the diagonal elements of the density matrix, ignores the role of the off-diagonal elements. Hence, it cannot be used for the determination of the transition line width. The resonant denominator in (5.21), (5.27) and (5.28) corresponds to the so-called homogeneous broadening. The latter notion has the following simple meaning. The Lorentz form of the frequency dependence of the absorption line (5.21) or (5.27) follows from formula (5.18) for the polarization. It is derived from the solution of equations for the density matrix (5.4) which describe an isolated atom (molecule) with relaxation times T_1 and T_2. The medium polarization is obtained by mere multiplication of a single-particle dipole moment (5.17) by the density of those particles supposed to be identical. This is equivalent to the assumption of a completely homogeneous medium. The frequency characteristic of a medium coincides, as a result, with that of a single-particle polarization. Hence, the broadening is only homogeneous if it is caused by the relaxation processes acting identically on all the particles of the medium that have the same transition frequency $\omega_{21} = (E_2 - E_1)/\hbar$.

The spontaneous radiative transitions responsible for the so-called intrinsic line width, the nonradiative transitions caused by inelastic collisions of the particle under consideration with like particles, with particles of some other sort and with quasi particles, as well as phase relaxation due to elastic events all refer to the class of relaxation processes that yield homogeneous broadening.

Equations (5.27) and (5.28) as well as (5.18) and (5.21) clearly show the presence of the saturation effect leading to the decrease of both χ'' and χ' with field intensity. Introducing the saturation condition by means of equating the field term in the denominators of (5.27) or (5.28) with unity, we can easily obtain the saturating specific energy $E_{\text{sat}}^2/8\pi$, and, hence, for the system in resonance, the saturating radiation intensity is

$$I_{\text{sat}}^{\text{res}} = \frac{\mathcal{E}_{\text{sat}}^2}{8\pi} c = \frac{c}{8\pi} \frac{\hbar^2}{\mu_{21}^2} \frac{1}{T_2 T_1}, \qquad (5.31)$$

where c is the velocity of light. The saturating flux is inversely proportional to the squared matrix element and to the product of the longitudinal and the transverse relaxation times. The last formula (5.31) may be simplified, if we introduce the absorption cross section σ by the following relation

$$\alpha = N\sigma \qquad (5.32)$$

and make use of (5.25) for the absorption coefficient α. Then, at $\omega = \omega_{21}$ by using (5.27) for χ'' and performing simple transformations, (5.31) may be reduced to the familiar form:

$$I_{\text{sat}}^{\text{res}} = \hbar\omega/2\sigma T_1, \qquad (5.33)$$

which coincides with (1.3) from lecture 1, where we derived it using plain considerations based on the balance equations for the populations. In this form the longitudinal relaxation rate, i.e., the total relaxation rate of the density matrix diagonal elements, determines the saturation effect in the resonant system. If the frequency detuning is nonzero, the saturating intensity increases by the factor $1 + (\omega_{21} - \omega)^2 T_2^2$. In the case of a deep saturation, when

$$(\mu_{21}\mathcal{E}_0/\hbar)^2 T_2 T_1 \gg 1, \qquad (5.34)$$

the relative difference of populations $\sigma_{11} - \sigma_{22}$ becomes small

$$\sigma_{11} - \sigma_{22} \approx \frac{1}{(\mu_{21}\mathcal{E}_0/\hbar)^2 T_2 T_1}.$$ (5.35)

From (5.19) the power spent per unit volume of a substance is seen to be

$$\langle \dot{Q} \rangle = \frac{N}{2} \frac{\hbar\omega}{T_1}.$$ (5.36)

The meaning of this formula is evident. In order to support the system in the state with two levels equally populated ($N/2$ particles in either state), within the longitudinal relaxation time interval T_1 (the time of the population relaxation) one should pump into the system $N/2$ quanta of energy $\hbar\omega$. The maintenance of such a half-transparent state requires an expenditure of energy that eventually passes over to the thermostat in the process of the longitudinal relaxation.

The saturation manifests itself in many ways. It cannot be reduced to just the attenuation of absorption. Equations (5.27) and (5.28) reveal another important field effect. In contrast to the saturation in the literal sense of the word its presence is not evident from an analysis of the balance equations. We mean the field broadening. A rather intense radiation field may be easily seen to broaden the Lorentz profile of the already homogeneously broadened line enlarging the half-width from $\Delta\omega = 2/T_2$ in the zero field to

$$\Delta\omega_E = \Delta\omega[1 + (\mu_{21}\mathcal{E}_0/\hbar)^2 T_2 T_1]^{1/2}.$$ (5.37)

This expression may be written more simply:

$$\Delta\omega_I = \Delta\omega(1 + I/I_{\text{sat}}^{\text{res}})^{1/2}.$$ (5.38)

It is worth noting that in (5.27) and (5.28) as well as in all the equivalent relations and their consequences there is a detached parameter of the frequency dimension $\mu_{21}\mathcal{E}_0/\hbar$ depending linearly on the field strength. It is this parameter that the Rabi frequency repeatedly discussed above is equal to (see (1.4) and (2.36)).

Let us treat this field-broadening effect in more detail. Formulae (5.27) and (5.28),which have the same resonant denominator, are known as the Carplus-Schwinger expressions. Three extremes will be considered here. We are well acquainted with the first two of them. In a weak field the line width is determined by the transverse relaxation time T_2 (see (5.30)), which is completely consistent with the traditional views of the linear vibration theory. In a strong field at a Rabi oscillation period shorter than the relaxation times for the case of $T_2 = T_1$ and the typical times exceeding the time of relaxation (see (5.30)), the resonant width coincides with the field broadening for the pure coherent interaction. The third extreme is less evident. If $T_2 \ll T_1$, then in a strong field the field broadening is $(T_1/T_2)^{1/2}$ times wider than in the purely coherent case,

$$\Delta\omega_E \approx \frac{\mu_{21}\mathcal{E}_0}{\hbar}\left(\frac{T_1}{T_2}\right)^{1/2}.$$ (5.39)

The presence of an additional factor $(T_1/T_2)^{1/2}$ may be treated as follows. The field induced particle polarization $\mu_{21}\mathcal{E}_0$ persists in magnitude during the longitudinal relaxation time interval (T_1), while its phase remains unchanged only during the transverse relaxation time interval (T_2). In other words, the polarization vector changes its direction T_1/T_2 times during its lifetime. As in the case of any erratic (Brownian) motion the particle polarization averaged over its lifetime is proportional to the square root of the number of random events (the number of steps, the number of phase-smashing knocks). It is this quantity that plays the part of the energy of interaction with the field in the particle's lifetime, i.e., determines the spectral width of the transition line.

Before we proceed to considering the spectral dependence of χ' and χ'' from a slightly different viewpoint, we should stress that Eqs. (5.27) and (5.28) are companion expressions to one obtained in the theory of magnetic resonance when solving the Bloch equations. The relaxation times T_1 and T_2 encountered in these equations have already been said to obviously refer to the longitudinal and transverse relaxation times, respectively.

Let us now come back to χ' and χ''. Their spectral dependencies differ dramatically. Factor $\omega_{21} - \omega$ in the expression for χ' turns it to zero when the system is precisely in resonance. The spectral dependencies of χ' and χ'' normalized to the maximum value of χ'' are shown in Fig. 5.1. Their correlation reflects the well-known fact in optics that resonant absorption is accompanied by abnormal dispersion. Changing the sign of the population difference σN, i.e., inverting the populations, makes χ'' change its sign, too. Therefore (see (5.25)), the absorption becomes negative, i.e., it is transformed into reinforcement. This also inverts the dispersion curve. The retardation in phase space caused by the presence of a medium is now positive not at $\omega < \omega_{21}$ but at $\omega > \omega_{21}$ and is still vanishing in resonance.

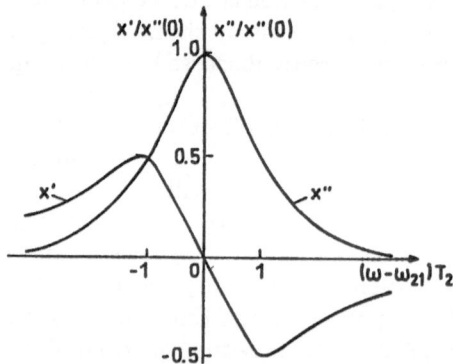

Fig. 5.1. Spectral dependencies of real ξ' and imaginary ξ'' parts of linear susceptibility

The correspondence between the real and the imaginary parts of the susceptibility of a two-level system with homogeneous broadening is of a rather general type. In complex-value analysis there is a theorem which states that should function $f(z)$ be analytic, then there is a connection between the real and the imaginary parts of this function. If this function is analytic in the upper or in the lower half-plane, $\operatorname{Re} f(z)$ and $\operatorname{Im} f(z)$ are connected by the so-called dispersion relations (sometimes called the Gilbert transformations). These relations are consequences of the Cauchy theorem, of the assumptions concerning the absence of singularities in the chosen half-plane as well as of the hypothesis about the fast decrease of $f(z)$ at $z \to \infty$ (faster than $1/|z|$).

As far as these relations are concerned, the relation between real and imaginary parts of the complex susceptibility $\chi(\omega)$ (5.23) taken at $\chi(\infty) = 0$ has acquired the name of the Kramers-Kronig relations. It is usually written in the form

$$\chi' = \frac{1}{\pi} \!\!\!\!\!\!\int_{-\infty}^{\infty} \frac{\chi''(\omega')d\omega'}{\omega' - \omega}\,, \quad \chi'' = \frac{1}{\pi} \!\!\!\!\!\!\int_{-\infty}^{\infty} \frac{\chi'(\omega')d\omega'}{\omega' - \omega}\,. \tag{5.40}$$

The crossed sign means the principal value of the integral.

The general derivation of the Kramers-Krönig relations and an explanation of their connection to such fundamental concepts as the causality principle are given in §62 of "The Electrodynamics of Continuous Media" by Landau and Lifshitz. For our purposes it is sufficient to declare that the substitution of (5.27) and (5.28) into (5.40) followed by direct calculation shows that the Kramers-Krönig dispersion relations are valid for the susceptibility of a two-level quantum system with two relaxation times.

Lecture Six

Saturation of Inhomogeneously Broadened Lines

Homogeneous bunches. Distribution in frequencies. Lorentz and Gaussian line shapes. Doppler broadening. Saturation of homogeneously and inhomogeneously broadened lines. Burning out the gap. Assortment in velocities. Intra-Doppler spectroscopy. A solid.

Now let us consider inhomogeneous broadening. In practice the atoms (molecules) interacting with laser light are either not quite identical or exist under not quite identical conditions. As a result different particles have different eigenfrequencies of transition. One speaks of inhomogeneous broadening if the line width of the aggregate of such particles exceeds a homogeneously broadened line. We can exemplify inhomogeneous broadening by impurity ions in dielectric crystals and glasses. In these systems the inhomogeneous broadening is caused by the ever present crystal imperfections (the absence of long-range order in glasses). Another example is gases at low pressures where the lines are broadened due to the Doppler effect caused by the thermal motion of molecules.

Inhomogeneous broadening is noticeable if the homogeneous broadening of isolated particles is much narrower than the total line width of the ensemble of these particles. A group of particles is called a homogeneous bunch if their eigenfrequencies lie within the interval less than the homogeneous line width $\Delta\omega$ (5.30). Then, the medium may be treated as an aggregate of homogeneous bunches and its polarization is determined by the sum of the polarizations of all the bunches. Hence, the resulting shape of an inhomogeneously broadened line is the superposition of lines of homogeneous bunches which comprise the system. The distribution $f(\omega_{21})$ of eigenfrequencies ω_{21} of the bunches is, therefore, of primary importance.

Let us pay more attention to this problem.

From (5.25) it follows that a homogeneously broadened absorption line may be represented as

$$\alpha(\omega) = 8\pi^2 \chi''/\lambda. \tag{6.1}$$

Substituting (5.27) for χ'' in this expression we obtain

$$\alpha(\omega) = \frac{8\pi^2}{\lambda} N \frac{\mu_{21}^2}{\hbar} T_2 \frac{1}{1 + (\omega_{21} - \omega)^2 T_2^2 \left(\frac{\mu_{21}\mathcal{E}_0}{\hbar}\right)^2 T_2 T_1}. \tag{6.2}$$

Initialy we shall confine ourselves to a linear regime (small $\mathcal{E}_0$). Then

$$\alpha(\omega) = 8\pi^3 \lambda N \frac{\mu_{21}^2}{\hbar} g_0(\omega, \omega_{21}),\qquad (6.3)$$

where $g_0(\omega, \omega_{21})$ is the normalized shape of a homogeneously broadened line

$$g_0(\omega, \omega_{21}) = \frac{1}{\pi} \frac{\Delta\omega}{(\omega - \omega_{21})^2 + (\Delta\omega/2)^2},\qquad (6.4)$$

and $\Delta\omega = 2/T_2$ (see (5.7) and (5.30)). The normalization condition is of the usual form:

$$\int_{-\infty}^{\infty} g_0(\omega)d\omega = 1.\qquad (6.5)$$

If the eigenfrequencies of the homogeneous bunches ω_{21} are spread, the absorption line $\alpha(\omega)$ may be written as

$$\alpha(\omega) = \frac{8\pi^3}{\lambda} N \frac{\mu_{21}^2}{\hbar} \int_{-\infty}^{\infty} g_0(\omega, \omega_{21}) f(\omega_{21}) d\omega_{21},\qquad (6.6)$$

where the distribution function $f(\omega_{21})$ is also normalized per unity:

$$\int_{-\infty}^{\infty} f(\omega_{21}) d\omega_{21} = 1.\qquad (6.7)$$

Hence, the form-factor of the superposition shape of the homogeneous lines $g(\omega)$ is given by the integral

$$g(\omega) = \int_{-infty}^{\infty} g_0(\omega, \omega_{21}) f(\omega_{21}) d\omega_{21}.\qquad (6.8)$$

Now let us pay attention to some particular cases.

Let the distribution $f(\omega_{21})$ be much narrower than the homogeneous line. Then $g_0(\omega, \omega_{21})$ may be removed from the integrand:

$$g(\omega) = g_0(\omega, \omega_0) \int_{-\infty}^{\infty} f(\omega_{21}) d\omega_{21} = g_0(\omega, \omega_0),\qquad (6.9)$$

where ω_0 is the central frequency of the distribution coinciding with the center of the homogeneously broadened transition line. This case is of no interest, since the shape of the absorption curve of the ensemble coincides, as one should expect, with the homogeneous line. The opposite extreme is when distribution $f(\omega_{21})$ is much wider than line $g_0(\omega, \omega_{21})$. Then, due to the same dependence of $g(\omega, \omega_{21})$ on ω_{21} and on ω, we obtain similarly

$$g(\omega) = f(\omega) \int_{-\infty}^{\infty} g_0(\omega_{21}, \omega) d\omega_{21} = f(\omega).\qquad (6.10)$$

The distribution of homogeneous bunches in frequencies is transferred into the resulting line shape of the ensemble. This is inhomogeneous broadening.

In an important particular case of the normal distribution of bunch frequencies, the normalized distribution has the form

$$f(\omega) = \frac{1}{\Delta\omega_\Gamma} \sqrt{\frac{4\ln 2}{\pi}} \exp\left[-4\ln 2 \left(\frac{\omega - \omega_0}{\Delta\omega_\Gamma}\right)^2\right], \qquad (6.11)$$

where ω_0 is the center of distribution and $\Delta\omega_\Gamma$ is its half-width. Since this distribution is assumed to be much wider than a single homogeneous bunch and, hence, contains many homogeneous lines, we omit subscript "21" in the notation of the frequency variable.

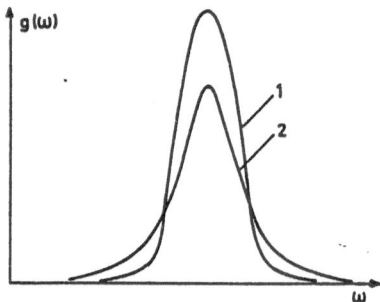

Fig. 6.1. The Gaussian and Lorentz lines of the same half-width and of the same area norm

In the case of the Doppler broadening in a gas of molecules of mass M distributed in velocities according to Maxwell's formula, expression (6.11) is valid while

$$\Delta\omega_\Gamma = \Delta\omega_D = 2\omega_0 \left(\frac{2kT}{Mc^2}\ln 2\right)^{1/2}. \qquad (6.12)$$

The shape of the Gaussian line differs considerably from that of the Lorentz line. Function (6.11) is higher in maximum than (6.4) and has steeper edges while their half-widths and areas are the same (Fig. 6.1).

The relations obtained determine the frequency dependence of the imaginary part of susceptibility χ'' in the case of inhomogeneous broadening. These relations allow us to determine the spectral dependence of the real part of susceptibility χ' in the case of inhomogeneous broadening using the Kramers-Krönig dispersion relations (Lecture five).

The difference between the homogeneously and inhomogeneously broadened absorption lines is most striking in intense resonant interactions where the saturation effect should be taken into account.

We have already discussed to a large extent in the previous lecture the saturation of a homogeneously broadened line. Here we should just mention that the expression for the absorption coefficient of the (6.3) type allowing for the field effect leads to the line shape form-factor dependent on the radiation intensity:

$$\Gamma_0(\omega, \omega_{21}), I_\omega) = \frac{1}{\pi} \frac{\Delta\omega/2}{(\omega - \omega_{21})^2 + (\Delta\Omega/2)^2(1 + I_\omega/I_{\mathrm{sat}}^{\mathrm{res}})}, \qquad (6.13)$$

where the radiation intensity at frequency ω equals $I_\omega = c\mathcal{E}_0^2/8\pi$, and $I_{\text{sat}}^{\text{res}} = c\hbar^2/8\pi\mu_{21}^2 T_2 T_1 = \hbar\omega_{21}/2\sigma T_1$ is the saturating intensity at $\omega = \omega_{21}$. First of all we should emphasize one evident fact. The saturation affects the line shape. Preserving the Lorentz form the line changes its width which according to (5.38) is

$$\Delta\omega_I = \Delta\omega(1 + I_\omega/I_{\text{sat}}^{\text{res}})^{1/2}. \tag{6.14}$$

The normalization also changes:

$$\int_{-\infty}^{\infty} g_0(\omega, \omega_{21}, I_\omega)d\omega = (1 + I_\omega/I_{\text{sat}}^{\text{res}})^{-1/2}. \tag{6.15}$$

The saturating effect of the radiation depends on the field frequency. Equating

$$g_0(\omega, \omega 21, I_\omega) = \frac{1}{2}g_0(\omega, \omega_{21}, I_\omega = 0),$$

from (6.13) and (6.5) one can easily obtain that

$$I_{\omega,\text{sat}} = I_{\text{sat}}^{\text{res}}\left(1 + \frac{(\omega - \omega_{21})^2}{(\Delta\omega/2)^2}\right). \tag{6.16}$$

The decrease in the absorption (amplification) at the resonant frequency is proportional to $(1 + I^{\text{res}}/I_{\text{sat}}^{\text{res}})^{-1}$. This may also be seen from formulae (5.15) and (5.8) for the difference of populations:

$$\delta N = (1 + I^{\text{res}}/I_{\text{sat}}^{\text{res}})^{-1} \quad \text{at} \quad \omega = \omega_{21}. \tag{6.17}$$

To visualize the "flattening" of a homogeneously broadened transition line under the action of monochromatic radiation of any frequency I_ω, we substitute the saturating intensity value $I_{\omega,\text{sat}}$ from (6.16) for $I_{\text{sat}}^{\text{res}}$ in (6.13). Simple transformations yield

$$g_0(\omega, \omega_{21}, I_\omega) = \frac{g_0(\omega, \omega_{21}, I_\omega = 0)}{1 + I_\omega/I_{\omega,\text{sat}}}. \tag{6.18}$$

So, we can state in a general form that the decrease in absorption (amplification) of a homogeneously broadened line caused by saturation is proportional to $(1 + I_\omega/I_{\omega,\text{sat}})^{-1}$.

The situation is quite different in the case of inhomogeneous broadening. Radiation I_ω interacts directly with a single homogeneous bunch. The line of this bunch is saturated like the homogeneous one, i.e., it becomes wider and lower. This leads to a decrease of absorption of the whole inhomogeneously broadened line due to the leak of particles through the ever-widening homogeneous bunch. It is convenient to consider the integral effect of monochromatic light saturating one of the homogeneous bunches on the whole ensemble comprising an inhomogeneously broadened line by introducing, as before, distribution function $f(\omega_{21})$ of homogeneous bunches in eigenfrequencies normalized according to (6.7). Then, for an inhomogeneously broadened line we have

$$g(\omega, I_\omega) = \int_{-\infty}^{\infty} g_0(\omega, \omega_{21}, I_\omega) f(\omega_{21}) d\omega_{21} \,. \tag{6.19}$$

In the case of an explicit inhomogeneous broadening function, $f(\omega)$ is much wider than $g_0(\omega)$ and may be removed from the integrand. Besides, we can omit subscript "21" of its argument. In accordance with (6.15) we have

$$g(\omega, I_\omega) = f(\omega) \int_{-\infty}^{\infty} g_0(\omega, \omega_{21}, I_\omega) d\omega_{21} = \frac{f(\omega)}{1 + I_\omega / I_{\text{sat}}^{\text{res}})^{1/2}} \,. \tag{6.20}$$

Since in the present case of substantially inhomogeneous broadening $g(\omega) = f(\omega)$ (see (6.10)), (6.20) is equivalent to

$$g(\omega, I_\omega) = \frac{g(\omega, I_\omega = 0)}{(1 + I_\omega / I_{\text{sat}}^{\text{res}})^{1/2}} \,. \tag{6.21}$$

The comparison of (6.21) with (6.18) shows first of all that the inhomogeneously broadened line is saturated with an intensity much slower than the homogeneously broadened one. Besides, in contrast to the homogeneous case the typical value of saturation is independent of the light frequency if the latter has just entered the region covered by distribution $f(\omega)$. All this may be easily explained by the character of inhomogeneous broadening where many homogeneously broadened lines distributed in some spectral interval are superimposed on each other to form a single spectral line.

The difference in the saturation character of homogeneously and inhomogeneously broadened lines becomes extremely apparent in the absorption (amplification) spectrum of a weak signal in the presence of intense, i.e., saturating, radiation.

Let us first consider homogeneous broadening. In this case, due to the character of saturation, spectral peculiarities in the absorption line of a weak signal are not expected to be observed in the presence of strong radiation. Indeed, the absorption coefficient of a weak signal at frequency ω' whose intensity is small enough not to induce any noticeable changes in the populations of levels 1 and 2 may be represented in the form

$$\alpha(\omega') = A\delta N g_0(\omega') \,, \tag{6.22}$$

where A is a constant factor; $g_0(\omega')$, the shape of the Lorentz absorption line (6.4); and δN, the difference in populations. However, in the presence of a strong field of frequency ω one should not forget that $\delta N = \delta N(I_\omega)$. Using (5.15) for δN, introducing, as usual, intensities I_ω and $I_{\text{sat}}^{\text{res}}$ and taking into account the relation between $I_{\text{sat}}^{\text{res}}$ and $I_{\omega,\text{sat}}$ (6.16), we can easily obtain (see (6.16) and (6.17)):

$$\delta N = \frac{N}{1 + I_\omega / I_{\omega,\text{sat}}} \,. \tag{6.23}$$

Then

$$\alpha(\omega', I_\omega) = AN \frac{g_0(\omega', \omega_{21}, I_\omega = 0)}{1 + I_\omega / I_{\omega,\text{sat}}} = \frac{\alpha(\omega', I_\omega = 0)}{1 + I_\omega / I_{\omega,\text{sat}}} \tag{6.24}$$

and differs from the case of the absence of the saturating radiation at frequency ω just by factor $(1 + I_\omega/I_{\omega,\text{sat}})^{-1}$.

The case of inhomogeneous broadening is quite another thing. Taking into account, as was done in (6.8), that the homogeneous bunches compose an inhomogeneously broadened line, and using (6.18), we can represent the absorpton coefficient of the weak signal in the presence of intense radiation at frequency ω in a form analogous to (6.22):

$$\alpha(\omega', I_\omega) = AN \int_{-\infty}^{\infty} \frac{f(\omega_{21})g_0(\omega', \omega_{21}, I_\omega = 0)}{1 + I_\omega/I_{\omega,\text{sat}}} d\omega_{21}, \qquad (6.25)$$

where $I_{\omega,\text{sat}}$ is given by (6.16). However, the saturating radiation affects just one of the homogeneous bunches. Its effect on the weak signal is only pronounced if the weak-signal frequency ω' approaches that of the strong field ω. So, we ought to replace ω_{21} by ω' in (6.16) and substitute it into (6.25). As a result, (6.25) becomes

$$\alpha(\omega', I_\omega) = AN \int_{-\infty}^{\infty} \frac{[(\omega' - \omega)^2 + (\Delta\omega/2)^2]g_0(\omega', \omega_{21}, I_\omega = 0)f(\omega_{21})}{(\omega' - \omega)^2 + (\Delta\omega/2)^2(1 + I_\omega/I_{\text{sat}}^{\text{res}})} d\omega_{21}. \qquad (6.26)$$

Taking into account (6.8) and using the definition of $\alpha(\omega)$ similar to (6.22), we eventually obtain that for an inhomogeneously broadened line the weak signal absorption (amplification) coefficient in the presence of a strong field may be represented as

$$\alpha(\omega', I_\omega) = \alpha(\omega', I_\omega = 0) \frac{(\omega' - \omega)^2 + (\Delta\omega/2)^2}{(\omega' - \omega)^2 + (\Delta\omega/2)^2(1 + I_\omega/I_{\text{sat}}^{\text{res}})}. \qquad (6.27)$$

This coefficient differs considerably from the case of homogeneous broadening (6.24) by its sharp spectral dependence. The latter becomes apparent as the gap of relative depth $(1 + I_\omega/I_{\text{sat}}^{\text{res}})^{-1}$ and width,

$$\Delta\omega_{\text{gap}} = \Delta\omega(1 + I_\omega/I_{\text{sat}}^{\text{res}})^{1/2}, \qquad (6.28)$$

near the saturation field frequency ω. This gap is superposed, generally speaking, on the smooth background of a typical spectral dependence $\alpha(\omega')$ that remains practically unchanged and is characteristic of the considered inhomogeneously broadened line in the absence of saturation.

The above-discussed distinction in between saturation of homogeneously and inhomogeneously broadened absorption (amplification) lines is illustrated in Fig. 6.2. The spectral dependence (6.27) has become known as the phenomenon of "burning out the gap" in an inhomogeneously broadened line by intense monochromatic irradiation.

The phenomenon of burning out the gap has many spectroscopic applications.

The broadening of spectral lines evidently limits the resolving ability of spectroscopy. The Doppler broadening of spectral lines in gases at close-to-room temperatures reaches approximately $(1-2)\cdot 10^9$ Hz in the visible range

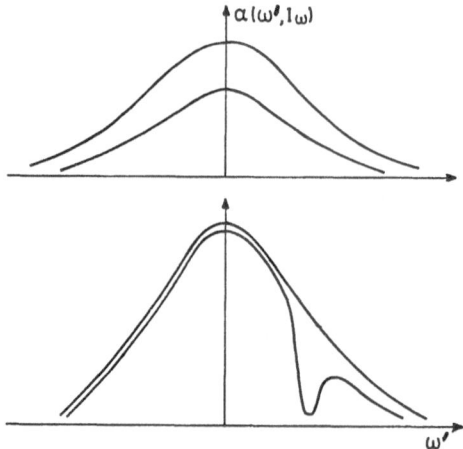

Fig. 6.2. The saturation of homogeneously (top) and inhomogeneously (bottom) broadened absorption (amplification) lines by monochromatic signal of frequency ω

of the spectrum and 10^8–10^7 Hz in the IR range (see (6.12)). At the same time the intrinsic line width of allowed transitions in the visible range of the spectrum does not, as a rule, exceed 10^8 Hz, while that of the forbidden but easily observable transitions may comprise 10–10^5 Hz. The typical values of intrinsic broadening of molecules in the IR range are 10–10^3 Hz. Hence, the Doppler broadening seriously hinders the investigation of the transition structure by methods of linear spectroscopy and prevents approaching the limit of the resolving ability determined by spontaneous emission.

Intrinsic broadening is homogeneous. However, one should not forget that considerable contribution to homogeneous broadening results from the collisions of the gas particles. The collisional width of line is determined by dephasing collisions. The cross section of these collision events is usually much larger than that of gas-kinetic collisions. At the gas pressure of 1 torr the collisions of molecules result in homogeneous broadening of $(1$–$30)\cdot10^6$ Hz. A pressure decrease to $10^{-3} \div 10^{-4}$ torr reduces the collisional width to its intrinsic limit. However, at low pressures the free path of molecules increases, and this leads accordingly to an enhancement of the time-of-flight mechanism of homogeneous broadening. In the gaseous phase at room temperature the gas-cuvette diameter or the light-beam width of several centimeters correspond to a time-of-flight broadening of 10^4 Hz. In the IR range this exceeds natural broadening but remains, however, quite small in comparison with the Doppler broadening. Hence, in a low pressure gas the spectral lines are broadened in an essentially inhomogeneous way. Under the envelope of such a broadening these lines may contain the spectral details which in principle are resolvable.

Burning out the gap in an inhomogeneously broadened line by intense monochromatic radiation is the method of spectral investigation in the region covered by the Doppler envelope.

Let the travelling light wave

$$E(t) = \mathcal{E}_0 \cos(\omega t - (kv)) \qquad (6.29)$$

interact with molecules moving with different velocities. The totality of these interactions forms a Doppler broadened line (6.11) of width (6.12). The inhomogeneous width $\Delta\omega_0$ considerably exceeds the line width of the homogeneous bunch $\Delta\omega$, $\Delta\omega_0 \gg \Delta\omega$. The molecule motion with velocity v may be easily taken into consideration by passing into the coordinate system associated with this molecule. This results in replacing ω by $\omega - (kv)$. Then the analysis of the laser-light interaction with matter given in previous lectures and the consequences derived there remain valid. We only have to modify the resonant condition for the particle. The question of which modification is to be made is rather clear. The field of a travelling plane wave interacts with molecules occupying the homogeneous spectral interval $\Delta\omega$ if the field frequency falls into this interval shifted by the Doppler effect:

$$|\omega_{21} - \omega + (kv)| \leq \Delta\omega/2 , \qquad (6.30)$$

where ω_{21} is the transition eigenfrequency of the resting molecule. Hence, the plane light wave (or, more generally, the light wave of small divergence) interacts with a small fraction of the molecules inside the Doppler envelope. Thus, it is capable of assorting the molecules by velocity according to the correspondence of molecule velocity to the resonant condition (6.30). This assortment leads to a shortage of molecules on the lower level of their distribution by velocity while on the upper level there is an excess of molecules satisfying the resonant condition (6.30) (Fig. 6.3). The gap and peak in these distributions are only noticeable when the intensity approaches the saturating value. Their width equals the homogeneous width corrected for the field broadening (6.28). In accordance with (6.30) the gap and peak correspond to the molecule velocities determined by the field frequency:

$$v_{\text{res}} = \frac{\omega - \omega_{21}}{\omega} c . \qquad (6.31)$$

Thus, the monochromatic light wave of sufficient intensity and high collimation modifies the distribution of the resonant molecules among the levels. The resulting nonisotropic distribution of molecules by velocity distorts the envelope of the absorption line broadened by the Doppler effect. It reveals the gap at the expense of the molecules forced to the excited state. The gap width (see (6.28)) is determined by the width of the homogeneous bunch, as was previously discussed, and may be many times narrower than the Doppler width. It is this fact that shows the advantages of intra-Doppler spectroscopy.

Note that the modification of the molecular distribution by velocity under the action of intense laser light may be applied to the selective control of molecular kinetic processes. But let us postpone this question until later and turn to spectroscopy.

We shall not discuss here the spectral properties of an important class of lasers in which the resonator is filled with a resonantly absorbing nonlinear

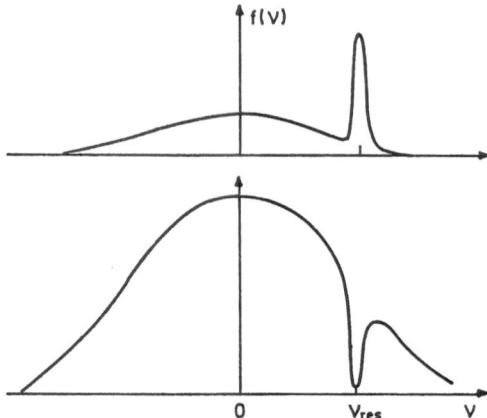

Fig. 6.3. The distribution of molecules by velocity on the upper (top) and lower (bottom) energy levels under the saturating action of the radiation with frequency ω. The value of v_{res} is given by formula (6.31)

substance, even though they make it possible to burn out the gap in an inhomogeneously broadened absorption line. We only stress here that narrow resonances are most often observed outside the resonator. The trial counter-wave method is very important for such experiments. It has become the most widely used method in non-linear laser spectroscopy of superhigh resolution.

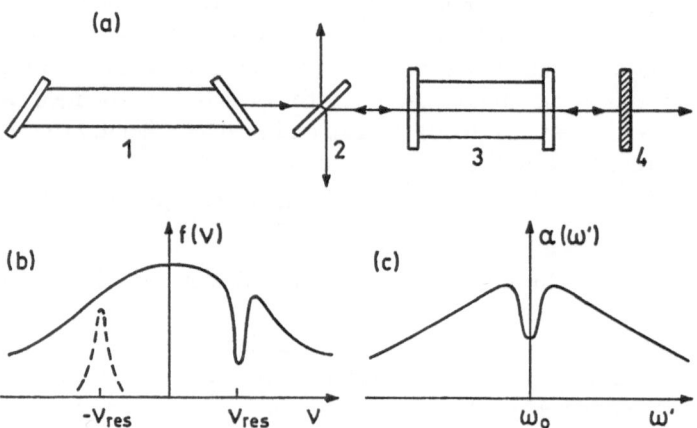

Fig. 6.4. The counter-waves method: (a) experimental setup, 1 – laser, 2 – light-dividing plate, 3 – cuvette with a gas, 4 – semitransparent mirror; (b) the distribution by velocity caused by the strong field of a direct wave. The dotted line shows the homogeneous bunch of molecules, interacting with the weak field of the trial (reflected) wave; (c) the absorption of the trial wave (falls in the case of precise resonance)

Its essence is as follows. A strong travelling wave (Fig. 6.4) makes the molecules with the velocity component in the direction of the wave propagation that satisfies condition (6.31) transit to the upper molecular state. A trial wave of the same frequency propagates in the opposite direction and, hence, interacts with other molecules. Should the frequencies of these waves not coincide with the center of the Doppler envelope, the weak trial wave does not feel the presence of the strong main wave. Otherwise, the trial wave interacts with the molecules of the homogeneous bunch with the absorption which has already been decreased by the strong main wave field. The resulting trial-wave absorption spectrum has a narrow resonant minimum precisely in the center of the absorption line broadened by the Doppler effect.

The method of strong and weak counter waves of the same frequency should apply when extending expression (6.29) for the travelling light wave to include this case:

$$E(t) = \mathcal{E}_0 \cos(\omega t - (kv)) + \mathcal{E} \cos(\omega t + (kv)), \qquad (6.32)$$

where the wave of amplitude $\mathcal{E}_0$ is the strong one and saturates the homogeneous bunch corresponding to its frequency while the counter wave of amplitude E is weak and does not lead to the saturation. The spectral dependence of the weak-field absorption coefficient may be easily found using the methods developed in the previous lecture, the result obtained completely corresponding to the qualitative considerations given above.

This method has been implemented with the use of dye-, Ar-, He-Ne- and CO_2-lasers to investigate the fine structure of the spectrum for many molecules and some atoms. One of the important experimental details one should pay attention to is the necessity of decoupling of gas cuvette from the laser resonator to prevent the inverse wave from getting trapped in the laser. Should a slight angular disalignment of the inverse wave with respect to the direct one occur (a method the decoupling), allowance should be made for the additional broadening of the resonance for geometric considerations by the value of

$$\Delta\omega_{\text{geom}} \approx (\langle v \rangle / \lambda)\theta, \qquad (6.33)$$

where $\langle v \rangle$ is the mean molecule velocity, and θ, the disalignment angle. At $\theta \approx \lambda/a$, where a is the diameter of the light beam, i.e., if the disalignment is of the order of the diffraction angle, $\Delta\omega \approx \langle v \rangle / a$, i.e., it coincides with the time-of-flight broadening.

The weak trial wave may propagate colinearly to the strong one as well. However, in this case, for the spectrum to be recorded, its frequency should be scanned independently. The gap occurs now at the frequency of the strong field and does not necessarily coincide with the center of the Doppler envelope. The advantage of this method (Fig. 6.5) is its ability to reveal the whole superfine structure hidden under the Doppler envelope, should it be present. The complication involved is the necessity of a smooth and independent frequency adjustment of both lasers.

Thus we ought to say that narrow nonlinear resonances resulting from the saturation of homogeneous bunches constituting the Doppler broadened resonant lines of gases, i.e., arising from the intense resonant interaction of

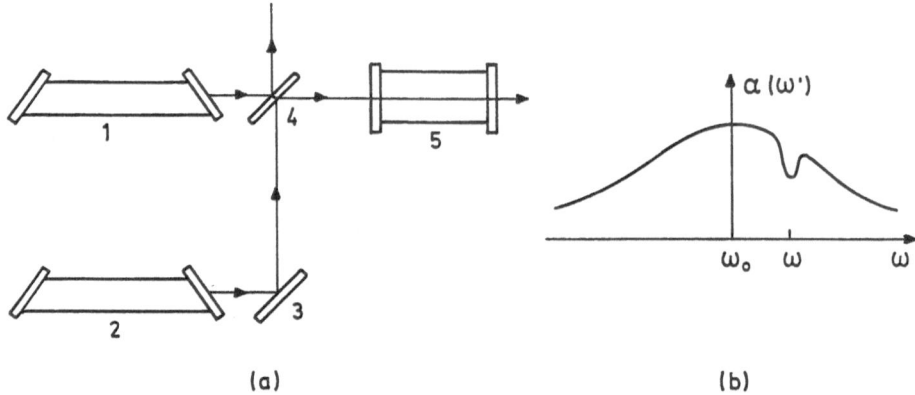

Fig. 6.5. The method of colinear strong and weak (trial) waves: (**a**) arrangement of the experiment, 1 and 2 – lasers, 3 – turning mirror, 4 – light-dividing plate, 5 – cuvette with a gas, (**b**) absorption spectrum. The absorption of the weak wave decreases at the frequency of the strong one, ω

radiation with gases, have provided the basis for a new field of superhigh-resolution spectroscopy ($\omega/\Delta\omega \approx 10^9 \div 10^9$).

However, the inhomogeneous broadening of spectral lines can be found not only in low-pressure gases but in solids as well. Tensions, dislocations, stacking faults, and undesirable impurities are always present in doped luminescent crystals. Due to this fact not all optically active ions exist under identical conditions. They are distributed among locations subjected to some disturbances. The local crystalline field determines the ion transition frequency. This results in an inhomogeneous broadening of the resonant absorption and luminescence lines of the impurity ions in a solid. Of course, inhomogeneous broadening is noticeable only for the transitions whose homogeneous broadening may be neglected. Without considering the particular mechanism of homogeneous broadening (spontaneous decay, electron-phonon interaction) we only mention that even for crystals of the highest quality, the inhomogeneous width (fractions of wavenumber) as a rule considerably exceeds the homogeneous one. In glasses, inhomogeneous broadening may be as large as hundreds of inverse centimeters. Note, that inhomogeneous broadening in solids does not directly depend on the transition wavelength, as distinct from Doppler broadening in glasses, and usually exceeds the latter by several orders of magnitude. Only for the best crystals does the inhomogeneous width value approach that of gases at room temperature in the visible range of the spectrum ($0.1\,\mathrm{cm}^{-1}$ is equivalent to 3 GHz).

For solids, in contrast to gases, the non-Doppler nature of inhomogeneous broadening does not allow an assortment of particles by the saturation of homogeneous bunches. However, burning out the gap in an inhomogeneous line is still possible and may be used for high-resolution spectroscopic studies. Thanks to the specific features of doped luminescent crystals (in glasses) the so-called "method of narrowing the luminescent line" is commonly used for these substances. We shall describe its essence below.

The broad-band source of light, an arc lamp, for instance, excites all the impurity ions, irrespective of their local surroundings, with the same probability. The resulting luminescence shows the inhomogeneously broadened line typical of the crystal transition being excited. Narrow-band laser radiation excites only the ions comprising a single homogeneous bunch, i.e., the ions whose surroundings are almost the same. The luminescent radiation of this homogeneous bunch of ions is concentrated in the laser light band. This effect has acquired the name "contraction of the luminescent line." It is widely used to study the phenomena which are usually masked in solids by a large inhomogeneous broadening. The excitation of ions by narrow-band pulsed radiation of short duration with the possibility of a frequency readjustment allows study of the time evolution of spectral peculiarities of the radiative and nonradiative processes of energy transfer in solids.

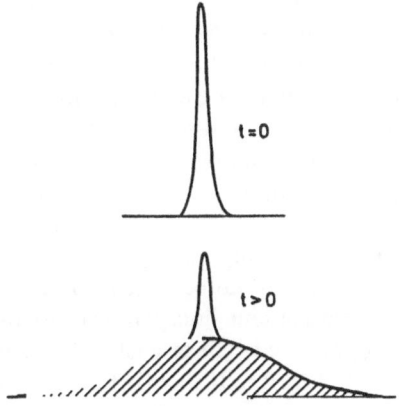

Fig. 6.6. The scheme of the time evolution of the luminescence spectrum. The diffusion of the homogeneous bunch may be observed

Figure 6.6 schematically illustrates the time evolution of the luminescent spectrum. The presence and development of wide-band luminescence emitted by ions which are not directly excited makes it possible to trace the process of the excitation energy migration between different active centers of a luminescent crystal. The advantages of this method of contracting the luminescent line compared to the absorption spectroscopy of burning out the gap are explained by a higher sensitivity of the radiative methods. The main drawback is the necessity of using a high-resolution spectrometer for the luminescent spectrum analysis.

In general, this method has turned out to be rather suitable for the study of the fine structure of the spectrum of doped crystals and glasses as well as in

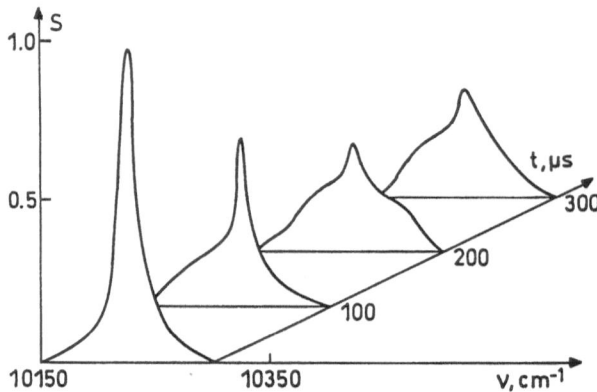

Fig. 6.7. The kinetics of the luminescence spectrum of a Yb^{3+} ion in Ba-Al-phosphate glass at 80 K (T. T. Basiev – Preprint of the Lebedev Institute, USSR, 1983, no. 160)

spectral kinetics. Dye lasers and color-center lasers (mainly F_2^+-, F_2^-- centers in LiF) are widely used in such experiments.

Figure 6.7 shows an example of the spectral migration of selective excitation energy. It manifests itself as the diffusion of an initially narrow luminescent line to the envelope of an inhomogeneously broadened line of the transition considered. The given example corresponds to an Yb^{3+} ion in barrium-alumophosphate glass at a temperature of 80 K. The decrease in temperature from nitrogen to helium considerably changes the migration of energy in the spectrum. This fact may be easily detected by the above-mentioned method of the contraction of the luminescent line. Another example of the same sample of the barrium-alumophosphate glass with Yb, irradiated with the same color-center laser, is given in Fig. 6.8. It illustrates the kinetics of a luminescent spectrum but at a temperature of 4.2 K. Without discussing the importance of the results thus obtained for solid state physics we only mention that at $T = 4.2$ K, the migration of the excitation energy occurs along the inhomogeneous contour of the line considered, i.e., between Yb^{3+} ions with different surroundings but directed in the Stokes side.

In cases in which the interaction of radiation with a homogeneous bunch brings about an irreversible change of the particles constituting this bunch, the radiation can burn out the gap that persists for a considerable time interval. As an example we can consider the photochemical burning out of the gap in an inhomogeneously broadened line of frozen solutions of such complex molecules as phtalocianin, tetracen, chlorophille and their derivatives. Deep cooling (1.5–2 K) is necessary both to suppress the phonon broadening of the homogeneous bunch line and to conserve the products of photo-decomposition.

We should mention that the photochemical burning out of a homogeneous gap in an inhomogeneously broadened line differs from that due to the saturation of transition. In the former case the magnitude of this gap is

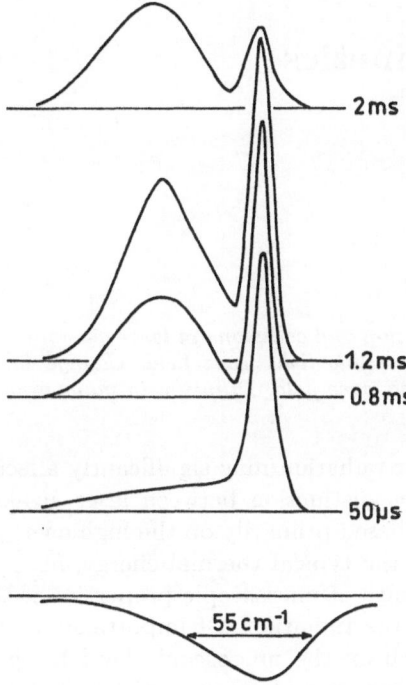

Fig. 6.8. The kinetics of the luminescence spectrum of Yb^{3+} ion in Ba-Al-phosphate glass at 4.2 K (T. T. Basiev – Preprint of the Lebedev Institute, USSR, no. 160)

determined by the density of the radiation energy, not by its intensity. (Photodecomposition results from photon absorption, and the amount of decomposition products is determined by the irradiation dose obtained.) So, the lifetime of a gap corresponds to the lifetime of the photo-dissociation fragments with respect to the channel of their association.

Note that selective photochemical burning out the gap in a narrow spectral interval (10^{-3} cm^{-1}) on the background of a wide inhomogeneous line (10^2–$5 \cdot 10^2$ cm^{-1}) may provide a method of densely recording of a spectral information and lead to holographic applications.

Lecture Seven

Laser Gas-Kinetics

Basic equation. Relaxation and collisions in laser gas-kinetics. Acceleration of molecules (atoms) by a travelling light wave field. Change in the motion of resonant particles in the standing wave field. Diffusion in momentum space.

Intense resonant laser radiation may significantly affect the character of gas-kinetic processes. The distinction between laser gas-kinetics and the traditional gas-kinetics is based primarily on the high energy of a laser light quantum as compared to the typical thermal energy, $\hbar\omega \gg kT$. This may result in a considerable change of microscopic properties of the substance that has resonantly absorbed the radiation. Of importance may be the selectivity of the laser action which on the microscopic level brings about a pronounced difference between resonant and other particles, despite their initial identity. The only reason for such a distinction is that nonresonant particles do not absorb radiation quanta. This difference manifests itself ever more completely as laser light is absorbed by the resonant particles. Such an absorption results in the formation of an essentially nonthermal distribution of gas particles among the states. The nonthermal action of laser light on a gas is also possible in the process of momentum exchange between the laser-radiation photons and the gas molecules.

Nonthermal changes in the character of gas-kinetic processes are, of course, accompanied by thermal ones with corresponding thermal distribution functions. As a rule, the thermal action of laser light is not to be found among the topics constituting the subject of intense resonant interactions of laser light with matter. The only exception is resonant laser thermal action in which the flow of a process in a multicomponent mixture depends on the specific component chosen to be an inlet for the thermal energy provided by the laser. For example, this may be the case for gas-phase chemical reactions. However, we are now interested in nonthermal laser gas-kinetics.

Here we use an approximation which assumes that translational degrees of freedom of atoms and molecules may be described by classical mechanics, whereas electron, rotational and vibrational (i.e., internal) ones should be treated quantum mechanically. First, let us consider a basic equation for describing laser gas-kinetics.

Traditional classical kinetics uses the Boltzmann equation for the distribution function $f(x, t, p)$:

$$-\partial f/\partial t = v \partial f/\partial x + F \partial f/\partial p, \tag{7.1}$$

where p stands for momentu; v, for velocity; $F = \partial H/\partial x$, for force; and H, is a Hamiltonian.

Quantum mechanics is based on the equation for the density matrix $\hat{\rho}$ which is known to be a quantum-mechanical analogue of the classical distribution function. This equation has the form

$$\frac{\partial \hat{\rho}}{\partial t} = \frac{1}{j\hbar} \left[\hat{H}, \hat{\rho}\right]. \tag{7.2}$$

An expedient way of describing the processes of laser gas-kinetics is to use an equation combining the typical features of Eqs. (7.1) and (7.2). This should be an equation for the density matrix $\hat{\rho}(x, t, p)$, all its elements being functions of classical coordinates and momenta. We may write this equation as

$$\frac{\partial \hat{\rho}}{\partial t} = -v\frac{\partial \hat{\rho}}{\partial x} + \frac{1}{j\hbar}\left[\hat{H}, \hat{\rho}\right] + \frac{1}{2}\left\{\frac{\partial \hat{H}}{\partial x}, \frac{\partial \hat{\rho}}{\partial p}\right\}, \tag{7.3}$$

where $\hat{H}$ is the quantum Hamiltonian of the system depending, however, on classical coordinates; the braces denote the anticommutator $\{\hat{A}, \hat{B}\} = \hat{A}\hat{B} + \hat{B}\hat{A}$.

For a proper understanding of the processes of laser gas-kinetics, one should clearly realize how the basic Eq. (7.3) is derived. We are confined to a schematic derivation since the complete procedure is rather lengthy.

The derivation is based on the quantum-mechanical equation of motion for a density matrix of the type of (7.2) (or (4.12)) in the Schrödinger representation. Each matrix element is the function of two variables x and x' corresponding to the coordinates of a molecule in its translational motion. This equation should also allow for a Hamiltonian describing the kinetic energy of the translational motion $\hat{H}_{tr}$. The commutator of the Hamiltonian and the operator $\hat{\rho}$ is of the form

$$[\hat{H}_{tr}, \hat{\rho}] = -\frac{\hbar^2}{2m}\left(\frac{\partial^2}{\partial x^2} - \frac{\partial^2}{\partial (x')^2}\right)\hat{\rho}, \tag{7.4}$$

where m is the mass of a particle.

Now we shall employ the so-called Wigner method. Replacing x and x' by new variables,

$$X = (x + x')/2, \quad Y = (x - x')/2. \tag{7.5}$$

Now, by means of a Fourier transform over Y, we perform the transition to momentum representation. As a result, (7.4) gives the first term on the right-hand side of (7.3).

The quasi-classical treatment of translational motion assumes the molecular wavefunction to be well localized and to occupy a space region considerably smaller than the typical range of interaction potentials. Hence, the expansion of Hamiltonian $\hat{H}$ from commutator $[\hat{H}, \hat{\rho}]$ into a series may be restricted to the terms linear in Y. After the transition to momentum representation, the terms of this expansion independent of Y give the second term on the right-hand side of (7.3) and those dependent on it give the third.

Equations (7.3) is valid for a closed system. If we ignore the collisions between atoms and molecules as well as the spontaneous emission, it also holds for the density matrix of a single isolated particle. Allowance for the collisions and for the spontaneous emission brings about both the additional relaxation terms of the $\hat{\hat{R}}\hat{\rho}$ type to the one-particle density matrix equation (see (4.7)) and the $\hat{\text{St}}(\hat{\rho}r)$ term equivalent to the Boltzmann collision integral in classical kinetics.

Taking these terms into account, the basic equation may be written as

$$\frac{\partial \hat{\rho}}{\partial t} = -v \frac{\partial \hat{\rho}}{\partial x} + \frac{1}{j\hbar} [\hat{H}, \hat{\rho}] + \frac{1}{2} \left\{ \frac{\partial \hat{H}}{\partial x}, \frac{\partial \hat{\rho}}{\partial p} \right\} + \hat{\text{St}}(\hat{\rho}) - \hat{\hat{R}}\hat{\rho}, \qquad (7.6)$$

where $\hat{\text{St}}(\hat{\rho})$ is the collision operator (alias *Stoss* operator).

Let us discuss the physics of (7.6). The time evolution of the density matrix at a fixed point in space depends upon several factors, each of them assigned a corresponding term on the right-hand side of (7.6). The first term corresponds to the translational motion at which particles that have been previously occupying the considered point eventually leave it, others arriving in their place. So, this is the origin of the first right-hand term in (7.6). The second one describes the internal changes in particles caused by quantum motions. It includes the terms representing the interaction of the internal degrees of freedom with the external fields. The third term determines the acceleration of particles by the external fields, including that acceleration affecting their internal states. The fourth and the fifth terms describe the variation of the density matrix caused by binary collisions of the particles. In this semiclassical approach the effect of collisions happens to have two independent aspects. The first is the collisional relaxation that slightly affects the translational motion. The second is the gas-kinetic collisions that change the translational motion, practically without affecting the energy of the internal degrees of freedom. It is worth noting that this relative independence of the changes in the translational and internal motions during collision is the consequence of a quasi-classical approach. Such an approach has a serious drawback. At $\hbar\omega \gg kT$ it does not provide for the description of ET-processes. These processes are responsible for the relaxation of the internal energy (of the order of $\hbar\omega$) into the energy of translational motion (of the order of kT). The ET-process is analogous to quantum-mechanical tunneling and so cannot be described in the framework of classical mechanics.

The relaxation and collision terms may be obtained with Bogoljubov's method from (7.3) written for the complete system, just as Eq. (4.7), containing the relaxation term, has been derived from (4.12). Such a derivation shows that the term $[\mathcal{H}, \mathcal{P}]$ yields addition $\hat{\hat{R}}\hat{\rho}$. The matrix elements of $\hat{\hat{R}}$ turn out to coincide with (4.18). The cross terms originating from the commutator and anticommutator are absent from (7.6). The anticommutator brings about the $\widehat{St}(\widehat{\rho})$ expression whose structure is analogous to the last term in (4.17), which leads to the expression (4.18) for the relaxation matrix:

$$\widehat{St}(\widehat{\rho}) = \int w d\Gamma \int_{-\infty}^{+\infty} dt \left\{ \frac{\partial \widehat{\widehat{H}}(t)}{\partial x_i}, \int_{-\infty}^{t} d\tau \frac{1}{4} \left\{ \frac{\partial \widehat{\widehat{H}}(\tau)}{\partial x_j}, \frac{\partial^2 \widehat{\rho}\widehat{\rho}'}{\partial p_i \partial p_j} \right\} \right\}, \quad (7.7)$$

where $\widehat{\rho}$ and $\widehat{\rho}'$ are the density matrices of the colliding particles, and p is the momentum of relative motion.

Here we emphasize once more that the given expressions for the collision integral (7.7) and the relaxation matrix (4.18) are valid under the same assumptions concerning the weakness and binary character of interactions in the gas considered. We have already adopted these assumptions in Lecture 4, when analyzing the collisional relaxations. In case these assumptions are invalid, (7.6) holds, still remaining the basic equation of laser gas-kinetics as long as the semiclassical approach applies. The only difference is significantly more involved expressions for matrices $\widehat{\widehat{R}}$ and $\widehat{St}(\widehat{\rho})$. From the discussion in Lecture 4 we know that the relaxation matrix is determined only by the spontaneous transitions unless the collisions are essential and the operator $\widehat{St}(\widehat{\rho})$ in (7.6) may be neglected.

An example of a gas-kinetic process whose progress is determined by the spontaneous decay rather than the collisions is the process of the acceleration of gas molecules (atoms) by a travelling light wave (the light pressure).

Let us consider Eq. (7.6) for the case in which the Hamiltonian of the system corresponds to an atom (molecule) with two energy levels $E_2 > E_1$ placed in the resonant external field of the travelling wave $E(t, x) = \mathcal{E}_0 \cos(\omega t - kx)$ with the frequency exactly coinciding with the transition frequency, $\hbar\omega = E_2 - E_1$.

Let us represent the system Hamiltonian in the form

$$H = \begin{pmatrix} E_2 & 0 \\ 0 & E_1 \end{pmatrix} + \begin{pmatrix} 0 & -\mu_{21}\mathcal{E}_0 \cos(\omega t - kx) \\ -\mu_{21}\mathcal{E}_0 \cos(\omega t - kx) & 0 \end{pmatrix}. \quad (7.8)$$

Then using the Pauli matrices (see (4.8)) and performing the transition to the rotating-wave representation by means of the matrix

$$\widehat{u} = \exp\left[-j\frac{\omega}{2}\widehat{\sigma}_z(t - kx) \right], \quad (7.9)$$

from (7.6) in the resonant approximation we obtain

$$\frac{\partial \widehat{\rho}}{\partial t} + v\frac{\partial \widehat{\rho}}{\partial x} - jkv[\widehat{\sigma}_z, \widehat{\rho}] + j\frac{\mu_{21}\mathcal{E}_0}{2\hbar}[\widehat{\sigma}_x, \widehat{\rho}] + \frac{k\mu_{21}\mathcal{E}_0}{4}\left\{ \widehat{\sigma}_y, \frac{\partial \widehat{\rho}}{\partial p} \right\} = \widehat{\widehat{R}}\widehat{\rho}. \quad (7.10)$$

Now let us agree upon a number of simplifying assumptions. Assuming the space to be homogeneous, we neglect the dependence of $\widehat{\rho}$ on the coordinates, i.e., $\frac{\partial \widehat{\rho}}{\partial x} = 0$. Then, considering the external field to be rather intense, we neglect the Doppler shift kv against the Stark broadening $\mu_{21}\mathcal{E}_0/\hbar$; that is, we presume that $\mu_{21}\mathcal{E}_0 \gg kv$.[*]) In addition, we presume, as usual, that $\hbar\omega \gg kT$. Then (7.10) yields the following system:

$$\dot{\rho}_{11} + j\frac{\mu_{21}\mathcal{E}_0}{2\hbar}(\rho_{21} - \rho_{12}) + j\frac{\mu_{21}\mathcal{E}_0}{4}k(\rho_{21}' - \rho_{12}') = \frac{\rho_{22}}{T_1},$$

$$\dot{\rho}_{22} + j\frac{\mu_{21}\mathcal{E}_0}{2\hbar}(\rho_{12} - \rho_{21}) + j\frac{\mu_{21}\mathcal{E}_0}{4}k(\rho_{21}' - \rho_{12}') = -\frac{\rho_{22}}{T_1},$$

$$\dot{\rho}_{12} + j\frac{\mu_{21}\mathcal{E}_0}{2\hbar}(\rho_{22} - \rho_{11}) + j\frac{\mu_{21}\mathcal{E}_0}{4}k(\rho_{22}' + \rho_{11}') = -\frac{\rho_{12}}{T_2},$$

$$\dot{\rho}_{21} + j\frac{\mu_{21}\mathcal{E}_0}{2\hbar}(\rho_{11} - \rho_{22}) - j\frac{\mu_{21}\mathcal{E}_0}{4}k(\rho_{11}' + \rho_{22}') = -\frac{\rho_{21}}{T_2}.$$

(7.11)

In these equations the dots above the letters signify the time derivative, while the primes denote the derivative with respect to momentum. It is useful to draw the reader's attention to the difference between the equations obtained and those for matrix elements (5.4) used in Lecture 5, when we calculated the susceptibility of a two-level system. The former include the terms containing the derivatives with the respect to momenta. In Bloch's notation, $N = \rho_{11} + \rho_{22}$, $W = \rho_{11} - \rho_{22}$, $jQ = \rho_{12} - \rho_{21}$ (see (4.8)), (7.11) may be written more compactly (see (4.9)-(4.11)):

$$\dot{N} + \mu_{21}\mathcal{E}_0\frac{kQ'}{2} = 0,$$

$$\dot{Q} - \frac{\mu_{21}\mathcal{E}_0}{\hbar}W + \frac{\mu_2 1\mathcal{E}_0}{2}kN' + \frac{Q}{T_2} = 0,$$

(7.12)

$$\dot{W} + \frac{\mu_{21}\mathcal{E}_0}{\hbar}Q + \frac{W - N}{T_1} = 0.$$

The terms which are new compared to (4.9) reflect the fact that the polarization and the population of the particles with a fixed velocity in the radiation field may be changed not only due to their direct excitation but also due to the acceleration-deceleration processes changing the velocities of the particles.

A preliminary analysis of the solutions of (7.12) may be carried out as follows. Note that the rates of changes in populations and polarization considerably exceed the translational motion velocity (divided by the wavelength). Hence, we can use the time-independent solutions of the last two equations of system (7.12). Moreover, the term with the derivative N_p' entering these equations may be omitted, as it is small in the parameter $k/\hbar p$. This parameter characterizes the ratio of the radiation wavelength to the de Broglie wavelength of the translational motion of the particle. Under these conditions the last two equations of (7.12) lead to

$$Q = \frac{(\mu_{21}\mathcal{E}_0/\hbar)T_2}{1 + (\mu_{21}\mathcal{E}_0/\hbar)^2T_2T_1}N,$$

(7.12 a)

---

[*]) The allowance for the Doppler shift brings about the forces of friction. These forces originate from the different energies of quanta emitted in the direction of motion and in the opposite direction, the former being harder.

(compare with the term that is $\frac{\pi}{2}$ out of phase with respect to the field $\mathcal{E}_0 \cos \omega t$ in (5.18)). The substitution of this expression into the first equation of (7.12) leads to the conclusion that the following force is exerted upon the particle (compare with (7.1))

$$F = \frac{\mu_{21}\mathcal{E}_0}{2} \frac{(\mu_{21}\mathcal{E}_0/\hbar)T_2}{1 + (\mu_{21}\mathcal{E}_0/\hbar)T_2T_1} . \qquad (7.12\,\mathrm{b})$$

A suitable method to solve (7.12) exactly combined with the initial conditions, is the Laplace transform over time and momenta. Let us denote the Laplace variables corresponding to time and momenta by the symbols r and q, respectively. Suppose that initially, all the particles are at rest, i.e., the initial condition for system (7.12) has the form $N(t = 0) = \delta(p)$. Then, expressed in terms of the Laplace variables, the system reduces to

$$rN + \frac{\mu_{21}\mathcal{E}_0}{2} kqQ = 1,$$

$$rQ - \frac{\mu_{21}\mathcal{E}_0}{\hbar} W + \frac{\mu_{21}\mathcal{E}_0}{2} kqN = 0, \qquad (7.13)$$

$$rW + \frac{\mu_{21}\mathcal{E}_0}{\hbar} Q + (W - N)T_1 = 0,$$

and, therefore,

$$N = \left[r + \frac{\frac{1}{2}\left(\frac{\mu_{21}\mathcal{E}_0}{\hbar}\right)^2 \hbar k q \frac{1}{T_1} - \frac{1}{4}\mu_{21}^2 \mathcal{E}_0^2 k^2 q^2 \left(r + \frac{1}{T_1}\right)}{\left(r + \frac{1}{T_2}\right)\left(r + \frac{1}{T_1}\right)\left(\frac{\mu_{21}\mathcal{E}_0}{\hbar}\right)^2} \right]^{-1} . \qquad (7.14)$$

At times exceeding the longitudinal relaxation time T_1 we may assume that $r \ll 1/T_2,\ 1/T_1$. Then

$$N = \left[r + \frac{1}{2}\left(\frac{\mu_{21}\mathcal{E}_0}{\hbar}\right)^2 \frac{T_2\hbar kq(1 - \hbar kq/2)}{1 + (\mu_{21}\mathcal{E}_0/\hbar)^2 T_2 T_1} \right]^{-1} . \qquad (7.15)$$

Expression (7.15) contains the product $\hbar kq$ which can be evaluated by the ratio of the photon momentum $\hbar k$ to the momentum of a particle (atom, molecule) $p \approx 1/q$. For visible light photons and the particles of noticeable rest mass this ratio is always much less than unity ($\hbar kq \ll 1$). Then

$$N = \rho_{11} + \rho_{22} = \left[r + \frac{1}{2}\left(\frac{\mu_{21}\mathcal{E}_0}{\hbar}\right)^2 \frac{T_2\hbar kq}{1 + (\mu_{21}\mathcal{E}_0/\hbar)^2 T_2 T_1} \right]^{-1} . \qquad (7.16)$$

By the inverse Laplace transform, (7.16) reduces to

$$N = \rho_{11} + \rho_{22} = \delta\left(2p - \left(\frac{\mu_{21}\mathcal{E}_0}{\hbar}\right)^2 \frac{T_2\hbar kt}{1 + (\mu_{21}\mathcal{E}_0/\hbar)^2 T_2 T_1} \right) . \qquad (7.17)$$

This, in turn, means that the particle momentum grows linearly with time:

$$p = \frac{1}{2} \left(\frac{\mu_{21}\mathcal{E}_0}{\hbar} \right)^2 \frac{T_2 \hbar k q}{1 + (\mu_{21}\mathcal{E}_0/\hbar)^2 T_2 T_1} \,. \tag{7.18}$$

The rate of the momentum transfer from the radiation field to a particle, i.e., the force exerted on a particle by the field, is seen to be determined by the photon momentum $\hbar k$ and by the rate of radiation absorption (Lecture 5). If the absorbing transition is saturated, the force acting on the particle comprises

$$\dot{p} = \hbar k / 2T_1 \,. \tag{7.19}$$

Therefore, the light pressure of a travelling wave on the resonant particle is accounted for by the photon momentum and by the presence of relaxation processes. If time T_1 equals T_2, both of them being determined by the same spontaneous lifetime τ $(T_1 = T_2 = \tau)$, the light pressure phenomenon may be easily explained as follows.

An unexcited particle absorbs a laser radiation photon and changes its momentum, acquiring the momentum of the photon. Then the particle returns to its unexcited state, emitting spontaneously and isotropically a photon distinct from the laser photon. Such a process is equivalent to the emission of a spherically symmetric wave by a particle. Note, that such a wave carries away no momentum; thus the particle momentum remains unchanged. So, the particle accumulates momentum at the rate of (7.19) due to the multiple events of absorption followed by re-emission. Factor $\frac{1}{2}$ is natural, since in a saturated state only half the particles occupy the upper level. It is worth noting here that the light pressure may be directed opposite to the incident light beam. If the particles are artificially maintained in the inverted state by some external pumping source that transfers no momentum to them, the incident coherent light causes stimulated emission. The corresponding recoil momentum $\hbar k$ directed against the beam is transferred to the particle which has previously emitted the photon. Now let us turn to the motion of an atom (molecule) in the field of a standing light wave.

We shall assume the atom to be a two-level quantum system with frequency detuned from resonance by Δ, $\hbar\Delta = E_2 - E_1 - \hbar\omega$. In the standing wave field the interaction Hamiltonian is

$$\widehat{V} = -\mu_{21}\mathcal{E}_0\widehat{\sigma}_x \cos kx \cos \omega t \,. \tag{7.20}$$

The light field of a standing wave may exert pressure on the particle in the presence as well as in the absence of the relaxation processes. We can now analyze the process, which is similar to the one we have just considered. In order to do so, let us substitute the following Hamiltonian into (7.10):

$$H = \begin{pmatrix} E_2 & 0 \\ o & E_1 \end{pmatrix} + \begin{pmatrix} 0 & -\mu_{21}\mathcal{E}_0 \cos \omega t \cos kx \\ -\mu_{21}\mathcal{E}_0 \cos \omega t \cos kx & 0 \end{pmatrix} \,.$$

In the rotating wave representation obtained here by the transformation matrix $\widehat{u} = \exp\left[-j\frac{\omega}{2}\widehat{\sigma}_z t\right]$, using the resonant approximation and the Bloch variables, we get

$$\dot{N} + vN'_x + \frac{\mu_{21}\mathcal{E}_0}{2}k\sin(kx)P'_p = 0\,,$$

$$\dot{W} + vW'_x + \frac{\mu_{21}\mathcal{E}_0}{\hbar}\cos(kx)Q + \frac{W-N}{T_1} = 0\,,$$

$$\dot{P} + vP'_x + \Delta Q + \frac{\mu_{21}\mathcal{E}_0}{2}k\sin(kx)N'_p + \frac{p}{T_2} = 0\,,$$

$$\dot{Q} + vQ'_x + \Delta P + \frac{\mu_{21}\mathcal{E}_0}{\hbar}\cos(kx)W + \frac{Q}{T_2} = 0\,.$$

(7.21)

Note that these equations have a number of essential distinguishing features compared to Eq. (7.12). First of all, the standing wave field is spatially non-homogeneous, the latter being taken into account by the derivatives of N, W, P and Q with respect to the coordinate. Then, the nonzero detuning $\Delta = E_2 - E_1 - \hbar\omega$ is basic to the process of standing-wave light pressure. Still another distinction is that the first of Eqs. (7.21) contains P'_p, not Q'_p. If, by analogy with the analysis of (7.12), we take advantage of the slowness and classical character of the translational motion, the last three equations may be treated as stationary. In complete accordance with (5.18) they give

$$P = \frac{\Delta T_2^2(\mu_{21}\mathcal{E}_0/\hbar)\cos kx}{1 + \Delta^2 T_2^2 + (\mu_{21}\mathcal{E}_0/\hbar)^2 T_2 T_1 \cos^2 kx}N\,.$$

The substitution of this expression into the first equation of (7.21) yields

$$\dot{N} + vN'_x + \frac{\mu_{21}\mathcal{E}_0}{2}\sin(kx)\frac{T_2\Delta(\mu_{21}\mathcal{E}_0/\hbar)\cos kx}{1 + \Delta^2 T_2^2 + (\mu_{21}\mathcal{E}_0/\hbar)^2 T_2 T_1 \cos^2 kx}N'_p = 0\,.$$

This equation is, in fact, the well-known kinetic equation for the distribution function of noncolliding particles (7.1), with force $F = PE$, and (5.18) used for polarization. The factor multiplying the derivative N'_p is the gradient of the following potential:

$$\Delta(T_2/T_1)\hbar\ln\left(1 + \Delta^2 T_2^2 + (\mu_{21}\mathcal{E}_0/\hbar)^2 T_2 T_1 \cos^2 kx\right)\,.$$

At small frequency detuning $\Delta T_2 \ll 1$ and field strength $\mu_{21}\mathcal{E}_0 \ll \hbar T_1^{-1}$, the logarithmand may be expanded into a power series. Then the potential of a two-level particle in the field happens to be equal to $T^2\hbar\Delta(\mathcal{E}_0\mu_{21}/\hbar)^2\cos^2 kx$. This expression gives the potential of light pressure forces exerted upon the particle in the standing wave field. Note that the presence of an intense light field accompanied by the processes we mentioned above while discussing (7.10) and (7.11) may lead to the capture of the particle in the standing wave antinode.

We now proceed to consider the light pressure phenomenon in the absence of relaxation processes, i.e., we assume the relaxation times to be longer than all the characteristic times of the problem, such as $\hbar/\mu_{21}\mathcal{E}_0$, $1/\Delta$, $1/vk$, $\hbar/vp$. In other words, solving (7.6) for the case being considered, we neglect not only $\widehat{\text{St}}(\hat{\rho})$ but $\widehat{\widehat{R}}\hat{\rho}$ as well.

Suppose further, that the beam of atoms moving with the same velocity v along the x-axis with the coordinate-uniform distribution function comes into the region occupied by the standing wave field.

Let the external forces only slightly disturb the spatial uniformity of the distribution function, i.e., $vp \ll \mu_{21}\mathcal{E}_0$, $\hbar\Delta$. In addition, let the Doppler shift be assumed to be much less than both the interaction energy (as in the previous case) and the frequency detuning.

Then, in terms of the Bloch variables (see (4.8)), (7.6) gives the system of Eqs. (7.21) without the relaxation terms:

$$\frac{dN}{dt} + v\frac{dN}{dx} - \frac{k\mu_{21}\mathcal{E}_0}{2}\sin(kx)\frac{dP}{dp} = 0\,,$$

$$\frac{dP}{dt} + v\frac{dP}{dx} - \delta Q - \frac{k\mu_{21}\mathcal{E}_0}{2}\sin(kx)\frac{dN}{dp} = 0\,,$$

$$\frac{dQ}{dt} + v\frac{dQ}{dx} + \Delta P + \frac{\mu_{21}\mathcal{E}_0}{\hbar}\cos(kx)W = 0\,,$$

$$\frac{dW}{dt} + v\frac{dW}{dx} - \frac{\mu_{21}\mathcal{E}_0}{\hbar}\cos(kx)Q = 0\,.$$

(7.21 a)

we insert the variables $y = k(vt - x)/2$ and $z = k(vt + x)/2$. Then,

$$\frac{dN}{dz} - \frac{\mu_{21}\mathcal{E}_0}{2v}\sin(y - z)\frac{dP}{dp} = 0\,,$$

$$\frac{dP}{dz} - \frac{\Delta Q}{kv} - \frac{\mu_{21}\mathcal{E}_0}{2v}\sin(y - z)\frac{dN}{dp} = 0\,,$$

$$\frac{dQ}{dz} - \frac{\Delta P}{kv} + \frac{\mu_{21}\mathcal{E}_0}{\hbar kv}\cos(y - z)W = 0\,,$$

$$\frac{dW}{dz} - \frac{\mu_{21}\mathcal{E}_0}{\hbar kv}\cos(y - z)Q = 0\,.$$

(7.22)

This substitution is seen to lead to a set of equations which depend only on y as a parameter. Thus the set of linear differential Eqs. (7.22) under consideration actually contains only two independent variables, z and p. Parameter y is important only when imposing initial and/or boundary conditions.

The physical meaning of the equations obtained is as follows. The first one describes the particle motion in the flow under the action of the force caused by jointed effect of the standing wave field and the polarization. The set of the last three equations in (7.22) is, in fact, Bloch's set and describes the internal (quantum) motion of a two-level particle. The displacement of the particle in the standing wave field adds a term to the first of these three equations. When solving them, this term may be considered as the right-hand side that causes a change in polarization under the action of an external source.

We assume that $\mu_{21}\mathcal{E}_0/\hbar, \Delta \gg kv$. It follows that while a particle travels from one antinode of a standing wave to another, it undergoes a great number of Rabi oscillations. The difference in populations and the polarization of the particle oscillate rapidly. Thus solving the last three equations of (7.22) we can make use of the so-called adiabatic approximation.

The essence of the approximation is as follows. Suppose that we have a system of linear differential equations,

$$\frac{d}{dz}\boldsymbol{X}(z) = \widehat{A}(z)\boldsymbol{X}(z), \qquad (7.23)$$

where $\boldsymbol{X}$ is the n-dimensional vector, while $\widehat{A}(z)$ is the matrix of coefficients which are both functions of the variable z. Eigenvectors $\boldsymbol{a}_n(z)$ and eigenfrequencies $\omega_n(z)$ of this matrix vary slowly with z:

$$d\omega_n/dz \ll \omega_n^2, \quad d|\boldsymbol{a}_n|/dz \ll \omega_n|\boldsymbol{a}_n|, \qquad (7.24)$$

Then the adiabatic approximation of the solution of the set of linear equations suggests that n linearly independent solutions of (7.23) may be represented in the form

$$\boldsymbol{X}_n = \boldsymbol{a}_n \exp\left(\int_{-\infty}^{+\infty} \omega_n(z)dz\right). \qquad (7.25)$$

Indeed, let us substitute (7.25) in (7.23) and get

$$\dot{\boldsymbol{a}}_n(z)\exp\left(\int_{-\infty}^{+\infty}\omega_n(z)dz\right) + \omega_n \boldsymbol{a}_n(z)\exp\left(\int_{-\infty}^{+\infty}\omega_n(z)dz\right)$$
$$= \widehat{A}(z)\boldsymbol{a}_n(z)\exp\left(\int_{-\infty}^{+\infty}\omega_n(z)dz\right). \qquad (7.26)$$

Given the slowness condition (7.24), the first term on the left-hand side of (7.26) is small compared to the second one and may be omitted. Then (7.26) only keeps the terms yielding the identity that determines the eigenfrequency ω_n.

If we know the solutions of the homogeneous set (7.23), we can build up its response function, which is required for the solution of an inhomogeneous set; the last three equations of (7.22) comprise such a set.

The consistent implementation of the procedure of the adiabatic solution of Eqs. (7.22) allows us to obtain the following expression for polarization P:

$$P = \int_{-\infty}^{z} \frac{(A+B)\frac{\mu_{21}\mathcal{E}_0}{2v}\sin(y-z')\frac{\partial N}{\partial p}dz'}{\left\{\left[\Delta^2 + \left(\frac{\mu_{21}\mathcal{E}_0}{\hbar}\right)^2\cos(y-z)\right]\left[\Delta^2 + \left(\frac{\mu_{21}\mathcal{E}_0}{\hbar}\right)^2\cos(y-z')\right]\right\}^{1/2}},$$

$$A = \Delta^2 \cos\left(\int_{z'}^{z}\left[\left(\frac{\mu_{21}\mathcal{E}_0}{\hbar kv}\right)^2\cos(y-t) + (\Delta/kv)^2\right]^{1/2}\right)dt, \qquad (7.27)$$

$$B = \left(\frac{\mu_{21}\mathcal{E}_0}{\hbar}\right)^2\cos(y-z)\cos(y-z').$$

This expression resulting from the development of coherent processes differs considerably from (5.18). Here the first fraction in the integrand represents the matrix elements with index PP of the response function of the Bloch equation system in the standing-wave field.

Note that the adiabatic approximation concerns the behavior of a system with several discrete energy levels subjected to slowly varying external field, a problem one often comes across in quantum mechanics. This field makes the values of eigenenergies and the eigenfunctions vary slowly with no level transitions.

Let us now turn back to Eqs. (7.22) and substitute (7.27) into the first of these. Generally speaking, this substitution leads to the integro-differential equation containing $\partial N/\partial z$ and $\partial^2 N/\partial p^2$. However, $\partial N/\partial p$ as a function of z varies slowly (the adiabatic approximation) and may be removed from the integrand in (7.27). Then, the first of Eqs. (7.22) gives the diffusion equation

$$\frac{\partial N}{\partial z} = \left(\frac{\mu_{21}\mathcal{E}_0}{2v}\right)^2 F\frac{\partial^2 N}{\partial p^2}, \tag{7.28}$$

where the value of F is determined from (7.27) and (7.22), taking into account the slowness of the dependence of $\partial N/\partial p$ upon z. The value of F obtained by this method has both constant component and components oscillating with z. When determining P these oscillations were considered to be slow, but from the viewpoint of the diffusion processes they should treated as fast, and may be omitted. It is the constant component of F that determines the population diffusion in momenta. Since $\partial N/\partial z = (1/k)\partial N/\partial x + (1/kv)\partial N/\partial t$, we have $\partial N/\partial z = (1/kv)\partial N/\partial t$ in the case of spatial homogeneity. So we get

$$\frac{\partial N}{\partial t} = kv\left(\frac{\mu_{21}\mathcal{E}_0}{2v}\right)^2 F\frac{\partial^2 N}{\partial p^2}. \tag{7.29}$$

Let us dwell on the physics of F.

The two-level quantum particle placed in a near-resonant field is characterized by two values: the total energy of the particle and that of the field. These are called quasi-energy levels[*]. The difference in quasi-energy levels equals $2\hbar\Omega$, where the frequency of population oscillations Ω is given by formula (2.35) and is determined by the Rabi frequency $\Omega_0 = \mu_{21}\mathcal{E}_0/\hbar$ and by the frequency detuning Δ. In the standing wave field the quasi-energy levels are periodically modulated in space. If the motion of the particle in such a field is slow, the probability of the transition of levels is small. Such transitions, should they occur, are referred to as nonadiabatic. Their probability cannot be found in the framework of an adiabatic approximation when solving the Schrödinger equations for the two-level system considered. However, this probability is introduced into the consideration by the right-hand side of the equation for the polarization (the second of Eqs. (7.22)) when solving the Bloch equations in the bounds of an adiabatic approximation. It is these adiabatic transitions that the small quantity F is responsible for. As a measure of the nonadiabaticity, quantity F is exponentially small, $\ln(F) = O(\Delta/kv, \mu_{21}\mathcal{E}_0/\hbar kv)$, the

---

[*] In Lecture 9 this subject will be considered in detail.

notation $O(a)$ standing for the value of the order of a. This implies that the force entering (7.1) for the distribution function itself depends on the type of the particle distribution in momenta. The reason is that the probability of the transition between the quasi-energy levels leading to the nonzero polarization, phased with the field, is itself dependent on the velocity of the motion of the particles.

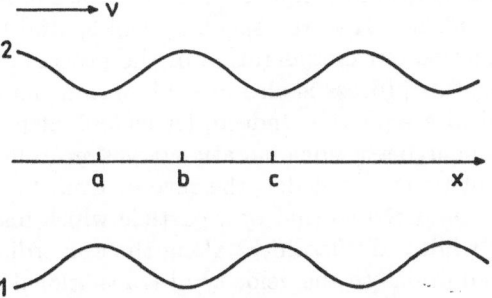

Fig. 7.1. Illustration of diffusion over momenta at nonadiabatic transitions. The dependence of the quasi-energy levels of the particle in the standing-wave field on the spatial coordinate is shown. The energy of level 1 is $-\hbar[\Delta^2 + (\mu_{21}\mathcal{E}_0/\hbar)^2 \cos^2 kx]^{1/2}$, and the energy of level 2 is $+\hbar[\Delta^2 + (\mu_{21}\mathcal{E}_0/\hbar)^2 \cos^2 kx]^{1/2}$. The particle is moving with velocity v in the positive direction of the x-axis. In the $a - b$ interval the particle in state 2 decelerates changing its momentum by $v\delta p = \hbar(-\Delta + [\Delta^2 + (\mu_{21}\mathcal{E}_0/\hbar)^2]^{1/2})$, and in the $b - c$ interval it accelerates with the momentum change modulo the same. For a particle in state 1 the stages of deceleration and acceleration are in opposite sequence. If a particle is promoted from state 2 to state 1 at point a and comes back at point b, its momentum differs from the momentum of the particle that has not undergone any transitions by $2(\hbar/v)[(\Delta^2 + (\mu_{21}\mathcal{E}_0/\hbar)^2)^{1/2} - \Delta]$

Qualitatively, the process of the particle diffusion in momenta (7.29) can be explained in the following way. While the particle travels through period $(kv)^{-1}$ of the standing-wave field it makes the transition from one quasi-energy level to another and back with probability F. Such transitions break the regular alternation of acceleration and deceleration by the field. Such a regular alternation is typical for the particle which, at any moment in time, occupies the same quasi-energy level. In the standing wave field two quasi-energy levels differ in the sign of the force exerted on the particle by the field (Fig. 7.1). Then the temporary transition of the particle from one quasi-energy level to another may cause the prolongation (or the shortening) of the acceleration (or the deceleration) interval of the particle by the field. This brings about a change in the momentum of the particle moving with velocity v by a typical value of $\pm\mu_{21}\mathcal{E}_0/v$.

Therefore, when a particle moves in a standing wave field, the initial distribution of particles in momenta diffuses. Note that this irreversible-in-time process is attained if we ignore the presence of relaxations. Something similar takes place in the process of diffusion of a wave packet propagating in a dispersive medium. Coherent damping (without relaxations, see Lecture 2) is of the same type. In the latter case such effects are accounted for by averaging over the ensemble of the systems with different eigenfrequencies.

Indeed, a moving particle is a packet of the de Broglie waves. Non-adiabatic transitions cause velocity-dependent changes in the energy of the system. This, in turn, leads to the quadratic dispersion law of these waves. The dispersion of de Broglie waves brings about wave-packet diffusion. From the viewpoint of the ensemble of particles this is equivalent to diffusion in momentum space.

It should be said here that the suggestion of spatial homogeneity made previously has canceled our consideration of the process of particle diffusion in space. However, this process is also possible if nonadiabatic transitions in a standing-wave field are present. Indeed, let us look at Fig. 7.1 and consider a particle which undergoes a nonadiabatic transition from state 2 to state 1 at point a, while at point c it makes the reverse transition. The momentum of such a particle compared to that of a particle which makes no transitions remains the same while a displacement along the x-coordinate caused by the different signs of additions to the velocity of translational motion in states 2 and 1 takes place. The transitions of this type lead to spatial diffusion of the particles.

In conclusion, it would be good to discuss again the question of whether or not the processes of the interaction of radiation with matter are reversible. The processes in which relaxation is important (Lecture 4) are irreversible, since in this case the influence of a thermostat of any type, i.e., the influence of a large uncontrolled system, is essential. It is the involved thermostat motion that explains irreversible relaxation. When we deal with a closed system, the relaxation of this type is not necessarily present. However, if this closed system is complex, it can demonstrate the processes which we are not able to control, due to the involved type of appropriate motions. In this case the system dynamics may turn out to be irreversible only owing to the fact that we have no suitable method to invert the phases of quantum-mechanical states. A fortunate exception is the photon echo (Lecture 3).

Lecture Eight

Laser Gas-Kinetics (continued)

Light-induced drift. Asymmetry of distribution in momenta. Drift flows. Atoms and molecules. Difference in transport cross sections of excited and unexcited states. Detuning from line center. Saturation.

Due to the Doppler effect the probability of the excitation of an atom or a molecule by laser light is also determined by the velocity of the particle being excited. As we have mentioned at the beginning of the previous lecture the cross sections of the gas-kinetic collisions of excited particles with those of a buffer gas should be expected to differ from the cross sections of unexcited particles. Then the selective excitation of the particles with a certain velocity may lead to directed macroscopic motions in the gas stimulated by laser light. In this case the laser light beam acts as the Maxwell demon. It stabilizes the thermodynamically non-equilibrium state of the gas in which the resonant and buffer components move toward each other. This effect is referred to as the light-induced drift (LID).

Therefore, the effect of the light-induced drift is based on the velocity-sensitive excitation of atoms or molecules and on the difference in the cross sections of the gas-kinetic collisions of these particles with the particles of a buffer gas in excited and unexcited states.

In Lecture 6 it has been shown that, if the resonant transition is broadened by the Doppler effect, intense laser light creates a gap in the velocity distribution of the particles on the lower level of transition and an excessive peak on the upper level (Fig. 6.3). However, one should not forget that the narrow peak and gap in the corresponding distributions are produced by the velocity-sensitive absorption of resonant photons and are observed in the absence of the collisional relaxation of the particle momentum. But for the effect of the light-induced drift the collisions changing the particle momentum are significant. Such collisions prevent the appearance of a narrow peak and gap in the distributions of the particles by velocity despite the sharp resonant dependence of the absorption coefficient. The peak and the gap do not appear. The relaxation of the molecule velocities smooths out the sharp non uniformity of the distribution but forms a small and smooth difference of the distribution function from the thermal one.

The simplest description of the light-induced drift may be obtained as follows. The particle occupying the lower level and moving with velocity v in the direction of the light beam propagation resonates with the radiation field (in case the Doppler shift of the transition frequency kv compensates for the laser frequency ω_l detuning, $kv = \omega_l - \omega$) and is excited by this radiation. The gas-kinetic cross section of the collisions of excited particles becomes higher. If the collision along with the change in the momentum brings about the relaxation of internal energy of the particle, it returns to its initial quantum state but with the value of its momentum changed. In other words, the laser light selectivity shortens the mean free path of the particles moving in a certain direction. The change in the mean free path means the difference between the diffusion coefficient $D = (1/3)\lambda\langle v \rangle$ of the particles whose motion is aligned with the light beam and that of the particles oppositely aligned with the light beam. In other words, $(1/3)n\langle v \rangle$ particles cross the unit area surface from right to left and travel through distance λ, while the same number of particles flying in the opposite direction travel through distance λ'. Hence, this process results in a displacement of the particles (the drift flow) of $J = (1/3)(\lambda - \lambda')n\langle v \rangle$. Taking into account that $\delta/\lambda = -\delta\sigma/\sigma$, where σ is the cross section, we get $J = -(\delta\sigma/\sigma)Dn$.

Of course, if the relaxation of the internal energy were a slow process, it would require many collisions. It is this fact that leads to an involved distribution of the excited particles by velocity. The asymmetry of this distribution caused by laser radiation accounts for the effect in this case. No doubt, this effect is significantly smaller than the one we have just considered, but it is quite observable.

And now let us consider the light-induced drift effect in more detail. We shall begin with an analysis of the collision integral (7.7). In the rotating wave representation (see (7.9)) the off-diagonal matrix elements of $\partial\widehat{H}/\partial x$ are rapidly oscillating functions, so we shall ignore their effect on the light-induced drift phenomenon. Then the collision integral (7.7) affects only the diagonal matrix elements of the density matrix $\widehat{\rho}$. This, in fact, implies the difference in the gas-kinetic cross sections of excited and unexcited particles. We represent the two-particle density matrix describing the interaction of the two-level particle under consideration with a buffer gas particle by the direct product $\rho_{ij}(\boldsymbol{p_1})f(\boldsymbol{p_2})$. Here we take into account that the buffer gas particles are considered to have no internal degrees of freedom (a non structured buffer gas). So their density matrix is, in fact, a scalar and corresponds to the classical gas-kinetic distribution function of the particles in momentum space. Then (7.7) yields[*]

$$\widehat{\mathrm{St}}(\widehat{\rho}) = \begin{pmatrix} A & 0 \\ 0 & B \end{pmatrix},$$

---

[*] One should not forget that every time we speak of the two-particle density matrix, coordinate x denotes the distance between the particles.

$$A = \frac{1}{4}\sum_{ij}\int \omega \int_{-\infty}^{\infty} dt \int_{-\infty}^{t} d\tau \frac{\partial H_{22}(t)}{\partial x_i}\frac{\partial H_{11}(\tau)}{\partial x_j}\frac{\partial^2}{\partial p_i \partial p_j}\rho_{22}(\boldsymbol{p}_1)f(\boldsymbol{p}_2)\,,$$

$$B = \frac{1}{4}\sum_{ij}\int \omega \int_{-\infty}^{\infty} dt \int_{-\infty}^{t} d\tau \frac{\partial H_{11}(t)}{\partial x_i}\frac{\partial H_{22}(\tau)}{\partial x_j}\frac{\partial^2}{\partial p_i \partial p_j}\rho_{11}(\boldsymbol{p}_1)f(\boldsymbol{p}_2)\,.$$

(8.1)

This implies that the collision integral is split into two different collision integrals: one for excited, and the other for unexcited, particles.

Expression (8.1) can be significantly simplified if we take into account that the changes in the momenta of colliding particles always occur along the vector drawn between the particles at the moment of their closest approach. Let us denote the mean square momentum transferred per unit time in the collisions of the buffer gas particles with excited resonant particles by q_2, and that with unexcited ones, by q_1. Then, assuming f to be the Boltzmann distribution, and after integration over phase volume Γ and over times t and τ, we get

$$\widehat{\mathrm{St}}(\widehat{\rho}) = \begin{pmatrix} A & 0 \\ 0 & B \end{pmatrix}\,,$$

$$A = q_2\left[\frac{1}{4}\nabla_p^2 + \frac{1}{mkT}[(\boldsymbol{p}\nabla)+1]\right]\rho_{22}(\boldsymbol{p})\,,$$

$$B = q_1\left[\frac{1}{4}\nabla_p^2 + \frac{1}{mkT}[(\boldsymbol{p}\nabla)+1]\right]\rho_{11}(\boldsymbol{p})\,,$$

(8.2)

where m is the mass of the particle and T is the gas temperature. Matrix elements (8.2) contain the terms corresponding to the diffusion in momentum space (∇_p^2) and to the viscous friction $(\boldsymbol{p}\nabla_p)$.

In other words, the collisions of both the excited and unexcited particles with the buffer gas particles have two consequences: diffusive broadening of the distribution function and viscous friction. The distinction between the excited and unexcited particles within the framework of this model becomes apparent only in the different rates of these processes. Note that the above relation between diffusion and viscous friction in the absence of external forces leads to the establishment of the equilibrium Maxwell distribution of particles in velocities in accord with the Boltzmann equation. The collision event operator $\mathrm{St}(\widehat{\rho})$ turns out to be the differential operator of second order in momentum space. This fact may be explained by the weakness of the collisions and is an instance of a more general law of diffusion in momentum space which is known as Fokker-Planck diffusion.

Substituting (8.2) in the main equation of laser gas-kinetics (7.6) and neglecting terms $(1/2)\{\partial \widehat{H}/\partial x, \partial \widehat{\rho}/\partial p\}$, which are responsible for the light pressure, we obtain in the Bloch notations

$$\frac{\partial P}{\partial t} - (\Delta - (\boldsymbol{kv}))Q + R_\perp P = 0\,,$$

$$\frac{\partial Q}{\partial t} + (\Delta - (\boldsymbol{kv}))P + \frac{\mu_{21}\mathcal{E}_0}{\hbar}W + R_\perp Q = 0\,,$$

$$\frac{\partial W}{\partial t} - \frac{\mu_{21}\mathcal{E}_0}{\hbar}Q + R_\| W - R_\| \frac{1 - \exp(-\hbar\omega_{21}/kT)}{1 + \exp(-\hbar\omega_{21}/kt)}$$

$$- (q_1 + q_2)\widehat{D}W - (q_2 - q_1)\widehat{D}N = 0, \qquad (8.3)$$

$$\frac{\partial N}{\partial t} - (q_1 + q_2)\widehat{D}N - (q_2 - q_1)\widehat{D}W = 0.$$

Here, operator $\widehat{D} = (1/4)\nabla_p^2 + [(\boldsymbol{p}\nabla_p) + 1]/mkT$.

The first three of Bloch's equations describe the polarization and the population of the particles with velocity v. Besides the changes that the quantum dynamics accounts for, the derivatives of these quantities over time are also determined by the number of particles found in the interval of velocities $v + dv$ due to the collisions, and by the values of W, P, Q corresponding to these particles. The fourth equation is pertains to the one-particle distribution function (7.1), which contains the diffusion flows and the viscous friction owing to the collisions.

We now consider the stationary regime. Then the first two equations of (8.3) give

$$Q = -\frac{R_\perp \mu_{21}\mathcal{E}_0/\hbar}{R_\perp^2 + (\Delta - (\boldsymbol{kv}))^2}W. \qquad (8.4)$$

Substituting (8.4) into the third of Eqs. (8.3) we get

$$\left(R_\| + \frac{R_\perp (\mu_{21}\mathcal{E}_0/\hbar)^2}{R_\perp^2 + (\Delta - (\boldsymbol{kv}))^2}\right)W - (q_1 + q_2)\widehat{D}W$$

$$- R_\| \frac{1 - \exp(-\hbar\omega_{21}/kT)}{1 + \exp(-\hbar\omega_{21}/kT)}N - (q_2 - q_1)\widehat{D}N = 0. \qquad (8.5)$$

But in the steady-state conditions the fourth of Eqs. (8.3) states that

$$(q_2 - q_1)\widehat{D}W + (q_2 + q_1)\widehat{D}N = 0. \qquad (8.6)$$

Then (8.5) takes on the form

$$\left(R_\| + \frac{R_\perp (\mu_{21}\mathcal{E}_0/\hbar)^2}{R_\perp^2 + (\Delta - (\boldsymbol{kv}))^2}\right)W$$

$$- \left(R_\| \frac{1 - \exp(-\hbar\omega_{21}/kT)}{1 + \exp(-\hbar\omega_{21}/kT)} - \frac{4q_1 q_2}{q_2 - q_1}\widehat{D}\right)N = 0. \qquad (8.7)$$

Equation (8.6) testifies to the fact that the quantities W and N are proportional to each other up to the function $W_0(p)$, satisfying the equation

$$\widehat{D}W_0(p) = 0. \qquad (8.8)$$

Hence, we have for W

$$W = -\frac{q_1 - q_2}{q_1 + q_2}N + a\exp(-p^2/2mkt). \qquad (8.9)$$

Then the substitution of (8.9) into (8.7) allows us to obtain the equation for N which contains the quantity A determined by the normalization. It is convenient to solve this equation by representing the distribution function in momentum space $N(p)$ as the sum of the Boltzmann distribution function $N_0 \exp(-p^2/2mkt)$ and function $N^1(p)$ orthogonal to it:

$$N = N_0 \exp\left(-\frac{p^2}{2mkT}\right) + N^1(p) \,,$$

$$\int N_0 N^1(p) \exp\left(-\frac{p^2}{2mkT}\right) dp = 0 \,. \qquad (8.10)$$

Then constant A may be excluded from the consideration and the non equilibrium part of distribution $N^1(p)$ satisfies the equation

$$
\begin{aligned}
G(p)N^1(p) & \\
+ \frac{G(p) + R_\parallel}{G_0 + R_\parallel} & \frac{\exp(-p^2/2mkT)}{(\pi mkT)^{1/2}} \int G(p)N^1(p) \exp(-p^2/2mkT)dp \\
+ R_\parallel N^1(p) & + (q_1 + q_2)\widehat{D}N^1(p) \\
= R_\parallel & \frac{G(p) - G_0}{G_0 + R_\parallel} \frac{q_2 - q_1}{q_1 + q_2} N_0 \frac{\exp(-p^2/2mkT)}{(2\pi mkT)^{1/2}} \,, \qquad (8.11)
\end{aligned}
$$

where we make allowance for $q_2 - q_1 \ll q_1 + q_2$ and suppose that $\hbar\omega_{21} \gg kT$. We denote the probability of the excitation of the particle with momentum p by

$$G(p) = \frac{R_\perp(\mu_{21}\mathcal{E}_0/\hbar)^2}{R_\perp^2 + (\Delta - (\boldsymbol{kv}))^2} \,, \qquad (8.12)$$

and the total probability of the excitation of the Boltzmann gas particles by

$$G_0 = \int G(p) \exp(-p^2/mkT)(\pi mkt)^{-1/2}dp \,. \qquad (8.13)$$

The equation obtained shows that N^1 is proportional to N and comprises a small fraction of it. The latter is caused by the relative smallness of the difference of momenta $q_2 - q_1$ transferred per unit time from excited and unexcited resonant particles to the particles of a buffer gas by the collisions compared to the sum of these momenta $q_1 + q_2$:

$$|q_2 - q_1| \ll q_1 + q_2 \,.$$

The distribution of the particles in momentum space thus derived is always distinct from the equilibrium distribution, given that the rate of particle excitation depends on their momenta.

Let us consider (8.11) in greater detail. The first two terms on the left-hand side are responsible for the changes of N^1 caused by the radiative transitions. The origin of the first term is obvious. The second term is the consequence of the orthogonality condition (8.10). The third term describes the change in N^1

due to longitudinal relaxation, while the fourth describes the change in N^1 resulting from a diffusion of populations over momenta, caused by the collisions. While the first two terms change the above-thermal distribution function N^1 at a rate of the order of the typical rate of the upper-level radiative occupation $(\mu_{21}\mathcal{E}_0/\hbar)/R_\perp$, and the third, at the rate of the longitudinal relaxation $R_\parallel$, the fourth term that corresponds to the population diffusion over momenta leads to a change of N^1 at the typical rate of momentum relaxation $R_p = (q_1 + q_2)/mkT$. In the case of moderate fields,

$$(\mu_{21}\mathcal{E}_0/\hbar)^2 \ll R_\parallel R_\perp \quad \text{or} \quad R_p R_\perp, \tag{8.14}$$

i.e., either far from the saturation of the homogeneously broadened transition or when the diffusion over momenta determines the population of an upper level, we can omit the first two terms ($G(p)$ is small) on the left of (8.11). In other words, if (8.14) holds, the inhomogeneously broadened transition under consideration is not saturated. The correlation of the two remaining terms depends on the physical situation. In particular, term $R_\parallel N^1$ dominates for the atoms with relatively short radiative lifetimes of the transitions in the visible and UV ranges. The opposite situation is characteristic of molecules on the vibrational (vibrational-rotational) transitions in the IR range, i.e., term $(q_1 + q_2)\widehat{D}N^1$ dominates.

Consider now the first of these extremes. In accordance with what was said above, from (8.11) we obtain

$$N^1 = \frac{q_2 - q_1}{q_1 + q_2}(G(p) - G_0)\frac{N_0}{R_\parallel}\exp\left(-\frac{p^2}{2mkT}\right)(2\pi mkT)^{-1/2}. \tag{8.15}$$

Thus an addition to the equilibrium distribution is seen to be asymmetric in momenta (see(8.12)). This addition reaches its maximum for the atoms with momenta corresponding to the maximum of $G(p)$:

$$p_k = \frac{m\Delta}{|\boldsymbol{k}|}. \tag{8.16}$$

The value of N^1 (8.15) allows us to find the flux of the particles of the resonantly excited gas corresponding to this above-thermal distribution of atoms in momenta:

$$J = \int_{-\infty}^{\infty} vN^1(p)dp = \frac{N_0}{R_\parallel}\frac{q_2 - q_1}{q_1 + q_2}\int_{-\infty}^{\infty} G(p)\frac{p}{m}$$
$$\times \exp\left(-\frac{p^2}{2mkT}\right)(2\pi mkT)^{-1/2}dp. \tag{8.17}$$

Without a detailed general analysis of the integral (8.17) we shall only specify the extremes. When the Doppler width is much narrower than homogeneous broadening, i.e., when $R_\perp \gg |\boldsymbol{k}|(2kT/m)^{1/2}$, function $G(p)$ in first-order perturbation theory over p may be replaced by its value at zero, $G(0)$, and removed from the integrand (8.17). Thus this integral turns to zero. At this extreme the effect of the light-induced drift proves to be small. Its magnitude is determined from the linear term of the series expansion of $G(p)$ over p and comprises

$$J = 4\frac{q_2 - q_1}{q_1 + q_2}\left(\frac{\mu_{21}\mathcal{E}_0}{\hbar}\right)^2 \frac{N_0\Delta}{R_\parallel R_\perp^3}|k|\frac{kT}{m}. \tag{8.18}$$

In a more interesting case of essentially inhomogeneous broadening, when $R_\perp \ll |k|(2kT/m)^{1/2}$, we can take advantage of the sharp resonant behavior of $G(p)$ removing function $p\exp(-p^2/2mkT)$ taken at the point $p = m\Delta/|bfk|$ from the integrand (8.17). As a result,

$$J = \frac{q_2 - q_1}{q_1 + q_2}\left(\frac{\mu_{21}\mathcal{E}_0}{\hbar}\right)^2 \frac{N_0\Delta}{R_\parallel}\frac{1}{|k|^2}\exp\left(-\frac{m\Delta^2}{2kT|k|^2}\right)\left(\frac{2\pi m}{kT}\right)^{1/2}. \tag{8.19}$$

Now let us turn back to (8.11). If (8.14) holds, in the second of the extremes mentioned for this equation when the diffusion of molecules in momenta is dominating, we have

$$\hat{D}N^1 = R_\parallel\frac{q_2 - q_1}{(q_1 + q_2)^2}\frac{G(p) - G_0}{G_0 + R_\parallel}N_0\exp\left(-\frac{p^2}{2mkT}\right)(2\pi mkT)^{-1/2}, \tag{8.20}$$

where $\hat{D} = (1/4)\nabla_p^2 + [(p\nabla_0) + 1]/mkT$ (see (8.3)). The eigenfunction of operator $\hat{D}$ can be expressed through the Hermit polynomials

$$u_n(p) = \exp\left(-\frac{p^2}{2mkT}\right)(2\pi mkT)^{-1/2}H(n)\left[\frac{p}{(2mkT)^{1/2}}\right], \tag{8.21}$$

their corresponding eigenvalues being

$$\lambda_n = n/mkT. \tag{8.22}$$

Representing functions $N'(p)$ as an expansion over functions u_n,

$$N^1(p) = \sum_{n=1}^{\infty} N_n u_n(p), \tag{8.23}$$

we obtain the following expressions for the flux J:

$$J = \sum_{n=1}^{\infty} N_n \int_{-\infty}^{\infty} \frac{p}{m}\dot{H}_n\left[\frac{p}{(2mkT)^{1/2}}\right]\exp\left(-\frac{p^2}{2mkT}\right)$$
$$\times (2\pi mkT)^{-1/2}dp \approx N_1\int_{-\infty}^{\infty}\frac{p}{m}2p(2mkT)^{-1/2}$$
$$\times \exp\left(-\frac{p^2}{2mkT}\right)(2\pi mkT)^{-1/2}dp = N_1\left(\frac{2kT}{m}\right)^{1/2}. \tag{8.24}$$

We restrict the expansion in (8.24) to the first term, since other nonzero terms, i.e., the odd ones, decrease with n as a factorial of n. Now we can find $N(p)$ as follows. Let us substitute (8.23) and (8.21) into (8.20) and make use of the orthogonality of the Hermit polynomials. Then, after the integration over momenta, we obtain

$$N_1 = \frac{q_2 - q_1}{(q_1 + q_2)^2} \frac{R_{\parallel}}{G_0 + R_{\parallel}} N_0 mkT \int_{-\infty}^{\infty} \exp\left(-\frac{p^2}{mkT}\right) (2\pi mkt)^{-1/2}$$

$$\times H_1 \left[\frac{p}{(2mkT)^{1/2}}\right] G(p) dp.$$
(8.25)

As a result,

$$J = \frac{q_2 - q_1}{(q_1 + q_2)^2} \frac{1}{R_p} \frac{R_{\parallel}}{G_0 + R_{\parallel}} N_0 \int_{-\infty}^{\infty} \frac{2p}{m} \exp\left(-\frac{p^2}{mkT}\right) (2\pi mkT)^{-1/2} G(p) dp,$$
(8.26)

where we have taken into account that $R_p = (q_1 + q_2)/mkT$.

The integral in (8.26) is similar to that in (8.17). So the analysis of (8.26) is analogous to the case of the dominating longitudinal relaxation (the first of the extremes of (8.14), atomic systems) considered above. Without discussing the case of small inhomogeneous broadening which is usually of no particular interest, we immediately pass over to the essentially inhomogeneously broadened line. Here we make allowance for the sharp resonant behavior of $G(p)$ not only in the integral of (8.26) but in the integral of (8.13), too. The latter integral determines the value G_0 that may in general be comparable to the value of $R_{\parallel}$ in the case under consideration:

$$G_0 = \pi \left(\frac{\mu_{21}\mathcal{E}_0}{\hbar}\right)^2 \frac{1}{|\mathbf{k}|} \left(\frac{m}{kT}\right)^{1/2} \exp\left(-\frac{\Delta^2 m}{|\mathbf{k}|^2 kT}\right).$$
(8.27)

Then, for the flux J we obtain

$$J = \sqrt{\frac{2}{\pi}} \frac{q_2 - q_1}{q_1 + q_2} \frac{N_0 \Delta R_{\parallel}}{|\mathbf{k}|^2 R_p}$$

$$\times \frac{\pi(\mu_{21}\mathcal{E}_0/\hbar)^2 (m/kT)^{1/2} \exp(-\Delta^2 m/|\mathbf{k}|^2 kT)}{R_{\parallel} + \pi(\mu_{21}\mathcal{E}_0/\hbar)^2 (m/kT)^{1/2} \exp(\Delta^2 m/|\mathbf{k}|^2 kT)/|\mathbf{k}|}.$$
(8.28)

Before we proceed to discussing the expressions obtained, we shall make a note of the fact that each expression includes the factor $(q_2 - q_1)/(q_1 + q_2)$. This simple ratio distinguishes the collisions of excited particles with particles of the buffer gas from the collisions of unexcited particles. It equals the ratio of the difference of the corresponding transport cross sections to their sum $\delta\sigma/2\sigma$.

Let us now discuss in more detail expressions (8.19) and (8.28) for the flux of atoms and molecules in the light-induced drift under the conditions of essentially homogeneous broadening of the resonant transition line. These expressions have much in common.

The proportionality of the effect to the ratio $\delta\sigma/\sigma$ is observed in both cases. The flux of atoms and molecules follows the sign of the detuning and vanishes at $\Delta = 0$, i.e., when the system is precisely in resonance. In both cases the effect is inversely proportional to the rate of the fastest relaxation process: either to the longitudinal relaxation rate or to the rate of the momentum relaxation. In each case the effect is proportional to the density of resonant particles. In a weak field the magnitude of the effect is directly proportional to the laser field intensity, to the form-factor of the Doppler line on the frequency of radiation $\exp[-\Delta^2 m/|\mathbf{k}|^2 kT]$ and inversely proportional to the width of this Doppler line $|\mathbf{k}|(kT/m)^{1/2}$.

The difference between the expressions describing the light-induced drift of atoms and molecules becomes apparent with the increase of the radiation intensity. The flux of the light-induced drift is saturated with the intensity in both the atomic and the molecular case. For the molecules we can trace the saturation process which is achieved at (see (8.14) and (8.28))

$$R_\perp R_p > \left(\frac{\mu_{21}\mathcal{E}_0}{\hbar}\right)^2 > R_\parallel |\mathbf{k}| \left(\frac{kT}{m}\right)^{1/2} \exp\left(\frac{\Delta^2 m}{|\mathbf{k}|^2 kT}\right) \pi^{-1}. \qquad (8.29)$$

Under the conditions (8.29) we have

$$J_{\text{sat}} = (2\pi)^{-1/2} \frac{\delta\sigma}{\sigma} \frac{\Delta}{|\mathbf{k}|} \frac{R_\parallel}{R_p} N_0. \qquad (8.30)$$

In the atomic case the point of saturation lies outside the bounds of the applicability of the present consideration limited by the first of the conditions (8.14).

Formula (8.30) gives the upper limit of the drift flux of the molecules. According to the definition of a gas-kinetic collision, the relaxation rate of the molecular momentum R_p is inversely proportional to the free path time of this molecule. Generally speaking, the longitudinal relaxation rate $R_\parallel$ can be both less and greater than R_p. However, due to the second of the conditions in (8.14), in the case under consideration the ratio of these rates is less than unity. This implies that our consideration is extended to include the case of energy relaxation through only several gas-kinetic collisions. Nevertheless, for the upper estimate we assume that $R_\parallel/R_p \sim 1$. The ratio $\Delta/|\mathbf{k}|$ cannot exceed a certain extreme value since the resonant saturation conditions (8.29) have to be satisfied. At fairly reasonable intensities the value of $\Delta/|\mathbf{k}|$ cannot considerably exceed the mean thermal velocity $(kT/m)^{1/2}$. Therefore the drift flow velocity J_{sat}/N_0 comprises no more than a $\delta\sigma/\sigma$ fraction of the mean thermal velocity.

One should remember that because of the momentum conservation law, the drift flux of a resonant gas is accompanied by the counter flux of the buffer gas. The light-induced drift of the particles may lead to the establishment of observable concentration gradients under stationary conditions. And so here we end our discussion of the process of light-induced drift.

Let us now turn our attention to the following circumstance of considerable importance. In the problems of laser gas-kinetics discussed above, an essential component is the longitudinal relaxation rate $R_\parallel$. We know that longitudinal relaxation may occur due to the spontaneous decay of excited states and due to collisions in a gas phase. The cases of spontaneous decay and of relaxation in collisions with resonant buffer gas particles have already been considered in Lecture 4. But in the consideration of the light-induced drift we have been taking into account the collisions of resonant particles with those of a structureless buffer gas. In this process all the internal energy of the resonant particle passes over to its translational degrees of freedom and to those of a buffer gas in the course of the collision. The probability of this happening is very small due to the smallness of the matrix element of the transition between two semiclassical translational states before and after the collision, which are characterized by significantly different kinetic energies.

Despite the fact that the transition in this relaxation process occurs between two semiclassical translational states, it is a purely quantum phenomenon analogous to the process of tunnelling a potential barrier. So, analyzing the longitudinal relaxation process, we cannot utilize the approach developed in the previous lecture for the present case since it is based on the assumption of the classical character of translational motion.

A consistent quantum-mechanical description of the process considered is analogous to a description of the scattering processes and is performed in the following way.

Let us write down the Schrödinger equation in the center-of-mass system for a two-level particle colliding with an unstructured one. Let us denote the reduced mass of the system by m; the wavefunction of the system corresponding to the excited state of a two-level particle, by ψ_2; the wavefunction corresponding to the ground state, by ψ_1; and the amplitude of the probability of the transition between levels 2 and 1 per unit time induced by the presence of the collision interaction, by $V(x)$. Then

$$
\begin{aligned}
j\hbar\dot{\Psi}_1 &= -\frac{\hbar^2}{2m}\nabla^2\Psi_1(x) + V(x)\Psi_2(x)\,, \\
j\hbar\dot{\Psi}_2 &= -\frac{\hbar^2}{2m}\nabla^2\Psi_2(x) + E\Psi_2(x) + V(x)\Psi_1(x)\,,
\end{aligned}
\tag{8.31}
$$

where $E = \hbar\omega_{21}$ is the energy difference between levels 2 and 1. At $V(x) = 0$ the transitions between states 2 and 1 are absent and the solution of system (8.13) is well-known:

$$
\begin{aligned}
\Psi_1 &= a\exp[-j\kappa x - j(\kappa^2\hbar^2/2m)t]\,, \\
\Psi_2 &= b\exp[-j\kappa x + j(\kappa^2\hbar^2/2m)t - jEt/\hbar]\,,
\end{aligned}
\tag{8.32}
$$

where κ is the wave number corresponding to the reduced momentum.

It is this solution at $a = 0$ that is used as the zero approximation when solving system (8.31) with the help of perturbation theory over the interaction $V(x)$. First order perturbation theory allows us to obtain the expression for the probability of transition $2 \rightarrow 1$ in the process of translational relaxation. For this purpose the methods considered in the chapters on the collision theory in "Quantum Mechanics" by L. D. Landau and E. M. Lifshitz may be used.

$$w = \frac{m^2}{4\pi^2 \hbar^4 \kappa \kappa'} \left| \int V(x) \exp[j(\kappa - \kappa')x] dx \right|^2, \qquad (8.33)$$

where $(\kappa')^2 = \kappa^2 + 2mE/\hbar^2$ is the squared wave number corresponding to the reduced momentum after the collision event.

The expression obtained describes, generally speaking, the improbable process of the excitation energy relaxation into the translational degrees of freedom (the X–T-relaxation). Indeed, should we deal with the visible and infrared ranges, the typical difference of energy levels would considerably exceed the typical energy of the thermal motion, $\hbar\omega_{21} \gg kT$. Therefore, a single relaxation event brings about a considerable change in the translational motion energy. This is why the change in the wave number κ is large, $\kappa \ll \kappa' - \kappa \simeq \kappa' \simeq (2mE/\hbar^2)^{1/2}$. Under these conditions the value of wavefunction κ' is large; hence, the corresponding value of de Broglie wavelength is small as compared to typical molecular dimensions. So the integral in (8.33) is small since the typical molecular dimensions and those of the interaction potential are of the same order. It follows that the probability of the X–T-relaxation per collision event is small and the cross section of this process is substantially less than that of the gas-kinetic one. This deduction may be illustrated by an example of translational relaxation caused by the collision of a two-level particle with a charged unstructured one. Suppose a charged particle flies at distance d from a two-level particle. Then $V(x)$ may be expressed through the matrix element of the transition operator μ_{21} and the charge of the buffer particle q as follows:

$$V(x) = q\mu_{21}/(d^2 + x^2). \qquad (8.34)$$

In this case (8.33) takes on the value

$$w = \frac{m^2}{\hbar^4 \kappa (2mE/\hbar^2)^{1/2}} \frac{\mu_{21} q^2}{d^2} \exp\left[-2d \left(\frac{2mE}{\hbar^2} \right)^{1/2} \right]. \qquad (8.35)$$

The value of d cannot be less than the molecular dimensions, so, due to the smallness of the de Broglie wavelength, the value w is exponentially small as compared to d. The averaging of (8.35) over velocities and impact parameters leads to the conclusion that the cross section of the X–T-relaxation is $exp[-2d(2mE/\hbar^2)^{1/2}]$ (where d is the molecular dimension) times less than the gas-kinetic cross section. This result remains qualitatively valid for other kinds of interaction potentials, too, i.e., for the X–T-relaxation to occur, a large number of gas-kinetic collisions is necessary. This is why in molecular systems in the infrared range of the spectrum, where the longitudinal relaxation is always some kind of XT-relaxation, e.g., vibrational-translational (VT) and rotational-translational (RT) relaxations, condition $R_p \gg R_\parallel$ as a rule remains valid.

Lecture Nine

Multiphoton Processes

Quasi-energy method. Two-photon excitation of a three-level system. Two-photon resonance. Composite matrix element. Field dependence of resonant frequency. Multilevel system. Total composite matrix element. Multiphoton resonance in a two-level system. Polychromatic irradiation. Frequency sweep. Relay-type occupation.

In the previous lecture we have been considering the interaction of a two-level quantum system with resonant radiation, the radiation frequency closely coinciding with the frequency of transition. However, when we speak about a rather intense interaction, the concept of resonance acquires a much wider meaning. It is extended to include not only the resonance between the main frequency of a system and the field frequency but also the coincidence of the field harmonics with the frequencies corresponding to the transitions between different pairs of energy levels of the quantum system. Thus, multilevel quantum systems are involved in the consideration.

However, the behavior of multilevel quantum systems in which multiphoton transitions are possible resembles, in general, the behavior of two-level systems. The former can also demonstrate the coherent Rabi oscillations, and their mean values of populations show a Lorentz-type dependence on the radiation frequency. However, there are significant differences between them. The point is that multiphoton transitions are in some aspects analogous to the process of escaping through a potential barrier, since the intermediate quantum states occupied in the transition process have an energy noticeably different from that of both the initial and the final states. This drastically modifies the matrix elements of the multiphoton transition operator and changes the very conditions of resonance.

The multiphoton transitions considered in this lecture are coherent processes. They should be distinguished from a sequence of incoherent single-photon transitions in a cascade process. This process accounted for by the finite relaxation linewidths can occur even in poor resonant conditions. This implies that the consideration of multiphoton processes should be carried out on the basis of the Schrödinger equation rather than the density matrix equation with additional relaxation terms.

A consistent quantum-mechanical analysis assumes the quantization of both the internal particle motions and the radiation field. Considering the intense resonant interaction, the straightforward approach to this problem turns out to be rather complicated, since the number of identical photons is too large. So, the approximate approaches based on the quasi-classical description of the radiation field are used for the purpose.

It is convenient to describe the quantum system excitation dynamics by the so-called quasi-energy method. Its essence may be easily explained for the simplest case of harmonic action.

Let us write down the Schrödinger equation of a system placed in an external harmonic field of frequency ω:

$$j\hbar\dot{\Psi}_n = \sum_{n'} H_{nn'}\Psi_{n'} + \cos\omega t \sum_{n'} V_{nn'}\Psi_{n'}, \qquad (9.1)$$

where $\widehat{H}$ is the Hamiltonian of the system in the absence of the field and $\widehat{V}$ is the operator of the interaction with the field. Let us represent the ψ-functions of the system in the form

$$\Psi_n(t) = \sum_{k=-\infty}^{\infty} \Psi_{n,k}(t)e^{jk\omega t}. \qquad (9.2)$$

Substituting (9.2) into (9.1) and equating the terms containing the same harmonics we get

$$j\hbar\dot{\Psi}_{n,k} = k\hbar\omega\Psi_{n,k} + \sum_{n'} H_{nn'}\Psi_{n',k}$$

$$+ \frac{1}{2}\sum_{n'} V_{nn'}\Psi_{n',k+1} + \frac{1}{2}\sum_{n'} V_{nn'}\Psi_{n',k-1}. \qquad (9.3)$$

Thus, with the aid of the representation (9.2) we have reduced the system of equations with variable coefficients to one with constant coefficients but of a higher order. The latter can be solved by standard methods. The main difficulty is the appropriate choice of initial conditions for $\Psi_{n,k}$, since their number exceeds that of initial conditions $\Psi_n(t=0)$ given for the original Eq. (9.1). There are no general recommendations as to how to choose the initial conditions for (9.3) except the evident requirement of the correspondence to the initial conditions for (9.1),

$$\Psi_n(t=0) = \sum_{k=-\infty}^{\infty} \Psi_{n,k}(t=0). \qquad (9.4)$$

The initial conditions should be individually chosen in every particular case for reasons of convenience, bearing in mind, naturally, that (9.4) must be fulfilled.

In the absence of an external disturbance, $V_{nn'} = 0$, the eigenvalues of the energy levels of Hamiltonian $\widehat{H}$ denoted by E_n correspond to system (9.1), while eigenvalues $E_{n,k}$

$$E_{nk} = E_n + k\hbar\omega, \qquad (9.5)$$

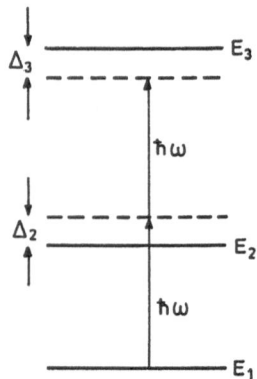

Fig. 9.1. Two-photon transition in a three-level system

called quasi-energy levels correspond to system (9.3). These levels are none other than the total energy levels of the system and of the field quanta. Nevertheless, the quasi-energy approach considers the external field to be classical with no allowance for the photon statistics. So, the wavefunctions $\Psi_{n,k}$ corresponding to the quasi-energy levels $E_{n,k}$ differ, generally speaking, from the amplitudes of the probability of detecting the system in the nth state at a fixed quantum state of the field (the initial number plus k photons). In other words, each quasi-energy state is attributed not to a single state with a fixed number of photons but to their superposition. The adjacent quasi-energy levels correspond to the superpositions of states differing in the mean number of photons by unity. For these states, contrary to the states with a fixed number of photons, it turns out to be possible to determine both the amplitude and the phase of the field. Sometimes a certain coherent state of the field oscillator (oscillators) is said to correspond to each quasi-energy level.

We now employ the quasi-energy method for the description of the dynamics of the excitation of the two-photon resonance in a three-level system. The situation is illustrated in Fig. 9.1, where the figures denote the level numbers $E_3 > E_2 > E_1$, $E_2 - E_1 \neq E_3 - E_2$, while the arrows of the same length schematically represent the energy of field quanta $\hbar\omega$ responsible for the two-photon transition $1 \to 3$. This scheme may be described by the following system of equations:

$$j\hbar\dot{\Psi}_1 = E_1\Psi_1 + \mu_{21}\mathcal{E}_0\Psi_2 \cos\omega t\,,$$
$$j\hbar\dot{\Psi}_2 = E_2\Psi_2 + \mu_{21}\mathcal{E}_0\Psi_1 \cos\omega t + \mu_{23}\mathcal{E}_0\Psi_3 \cos\omega t\,, \qquad (9.6)$$
$$j\hbar\dot{\Psi}_3 = E_3\Psi_3 + \mu_{32}\mathcal{E}_0\Psi_2 \cos\omega t$$

with the initial condition

$$\Psi_1(t = 0) = 1\,, \qquad (9.7)$$

where E_1, E_2, E_3 are the energies of levels 1, 2 and 3 in Fig. 9.1 and $\mathcal{E}_0$ is the amplitude of the external field $E(t) = \mathcal{E}_0 \cos\omega t$.

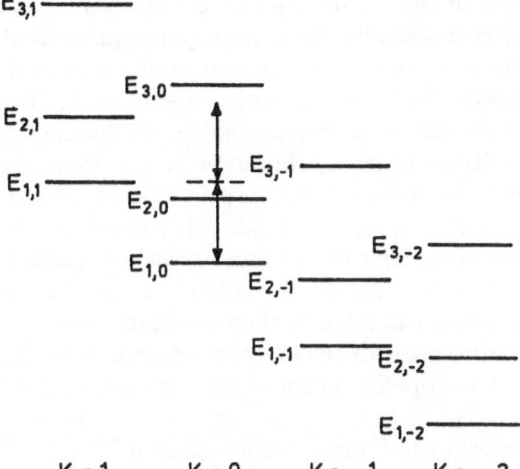

Fig. 9.2. A three-level system in the quasi-energy representation

Now let us turn to the quasi-energetic representation and build up the quasi-energy levels considered in the zero external field, $\mathcal{E}_0 = 0$ (Fig. 9.2). We shall consider the case when the frequencies of transitions $1 \to 2$ and $2 \to 3$ differ from the external field frequency but are close to it:

$$(E_2 - E_1 - \hbar\omega) \ll \hbar\omega \,,$$
$$(E_3 - E_2 - \hbar\omega) \ll \hbar\omega \,. \tag{9.8}$$

Then the initial condition for the quasi-energetic ψ-functions corresponding to the initial condition (9.7) of the original set can be chosen in the form

$$\Psi_{1,0}(t = 0) = 1 \,. \tag{9.9}$$

For the consideration that follows we keep only three equations from set (9.3) corresponding to the three close quasi-energy levels $E_{1,0}$, $E_{2,-1}$, $E_{3,-2}$, which contain ψ-functions corresponding to these levels:

$$j\hbar\dot{\Psi}_{1,0} = E_1\Psi_{1,0} + \frac{\mu_{12}\mathcal{E}_0}{2}\Psi_{2,-1} \,,$$
$$j\hbar\dot{\Psi}_{2,-1} = (E_2 - \hbar\omega)\Psi_{2,-1} + \frac{\mu_{21}\mathcal{E}_0}{2}\Psi_{1,0} + \frac{\mu_{23}\mathcal{E}_0}{2}\Psi_{3,-2} \,, \tag{9.10}$$
$$j\hbar\dot{\Psi}_{3,-2} = (E_3 - 2\hbar\omega)\Psi_{3,-2} + \frac{\mu_{32}\mathcal{E}_0}{2}\Psi_{2,-1} \,.$$

The rest of the quasi-energetic ψ-functions are small. They are merely non-resonant, so the consideration of equations containing them is meaningless. Giving up these equations when passing over from (9.3) to (9.10) is called the resonant approximation. This is analogous to the resonant approximation in a two-level case (Lecture 2).

For the solution of set (9.10) one can use different methods such as the trial of amplitudes corresponding to three eigenfrequencies of the system, etc. However, the method of generalized functions fully described in "Equations of Mathematical Physics" by V.S. Vladimirov seems to be the most convenient one. This method allows us to reduce the set of differential equations with defined initial conditions to an algebraic set of equations; its essence may be described as follows. The being functions sought are replaced by the functions which equal zero at $t < t_0$ and coincide with them at $t > t_0$, where t_0 is the initial time moment. The old equations with singular additional terms, which contain on the right-hand side the Dirac δ-functions depending on the initial conditions, turn out to hold for these new functions. As far as it concerns set (9.10) with the allowance for the initial conditions (9.9), this implies that an additional term $j\hbar\delta(t)$ appears on the right-hand side of the first of the (9.10) equations.

Now let us perform the Fourier transformation

$$\Psi(\varepsilon) = \int \Psi(t)e^{-j\varepsilon t}dt \qquad (9.11)$$

of the set of equations thus obtained. Then for the Fourier transforms of the ψ-functions being sought we have the set

$$\varepsilon\hbar\Psi_{1,0} - E_1\Psi_{1,0} - \frac{\mu_{12}\mathcal{E}_0}{2}\Psi_{2,-1} = j\hbar,$$

$$\varepsilon\hbar\Psi_{2,-1} - (E_2 - \hbar\omega)\Psi_{2,-1} - \frac{\mu_{21}\mathcal{E}_0}{2}\Psi_{1,0} - \frac{\mu_{23}\mathcal{E}_0}{2}\Psi_{3,-2} = 0, \qquad (9.12)$$

$$\varepsilon\hbar\Psi_{3,-2} - (E_3 - 2\hbar\omega)\Psi_{3,-2} - \frac{\mu_{32}\mathcal{E}_0}{2}\Psi_{2,-1} = 0.$$

The solution of this algebraic set of linear equations for the function of interest $\Psi_{3,-2}$ has the form

$$\Psi_{3,-2}(\varepsilon) = \cfrac{\begin{vmatrix} \varepsilon\hbar - E_1 & -\mu_{12}\mathcal{E}_0/2 & j\hbar \\ -\mu_{21}\mathcal{E}_0/2 & \varepsilon\hbar - E_2 + \hbar\omega & 0 \\ 0 & -\mu_{32}\mathcal{E}_0/2 & 0 \end{vmatrix}}{\begin{vmatrix} \varepsilon\hbar - E_1 & -\mu_{12}\mathcal{E}_0/2 & 0 \\ -\mu_{21}\mathcal{E}_0/2 & \varepsilon\hbar - E_2 + \hbar\omega & -\mu_{23}\mathcal{E}_0/2 \\ 0 & -\mu_{32}\mathcal{E}_0/2 & \varepsilon\hbar - E_3 - 2\hbar\omega \end{vmatrix}}. \qquad (9.13)$$

This expression written in terms of the transition frequencies $\omega_{ij} = (E_i - E_j)/\hbar$ and the Rabi frequencies $\Omega_{ij} = \mu_{ij}\mathcal{E}_0/\hbar$ (given energies E_2 and E_3 are counted off from level E_1) acquires the form

$$\Psi_{3,-2}(t) = \frac{j\Omega_{21}\Omega_{32}}{4\varepsilon(\varepsilon - \omega_{12} + \omega)(\varepsilon - \omega_{13} + 2\omega) - \varepsilon\Omega_{23}\Omega_{32} - (\varepsilon - \omega_{13} + 2\omega)\Omega_{12}\Omega_{21}}. \qquad (9.14)$$

Analyzing the expression obtained we confine ourselves to the case of moderate fields when the Rabi frequencies Ω_{ij} are small in comparison with the frequency detunings $\Delta_2 = \omega_{12} - \omega$, $\Delta_3 = \omega_{13} - 2\omega$ and with their difference $\Delta_{32} = \Delta_3 - \Delta_2$. Then, after the Fourier transformation inverse to (9.11), taking into account only the first term of (9.14), we get

$$\Psi_{3,-2}(t) = -\frac{\Omega_{21}\Omega_{32}}{\Delta_2\Delta_3} + \frac{\Omega_{21}\Omega_{32}}{\Delta_2\Delta_{32}}e^{j\Delta_2 t} - \frac{\Omega_{21}\Omega_{32}}{\Delta_3\Delta_{32}}e^{j\Delta_3 t}. \tag{9.15}$$

The expression obtained shows that the ψ-function and, therefore, the population of the third level is small unless one of the detuning frequencies becomes comparable with the Rabi frequencies, i.e., one of the transitions becomes resonant.

If $\Delta_2 \leq \Omega_{12}$ or $\Delta_{32} \leq \Omega_{32}$ single-photon resonance takes place at one of the transitions $1 \to 2$, $2 \to 3$. So, the three-level system in Fig. 9.1 becomes in fact equivalent to the two-level system fully described earlier. If $\Delta_3 \to 0$, the so-called "two-photon resonance" occurs. The population of the system begins to oscillate slowly between the first and the third levels (quasi-energy states $|1, 0\rangle$ and $|3, -2\rangle$). Indeed, in this case (9.14) may be represented in the form

$$\Psi_{3,-2}(\varepsilon) = \frac{-j\Omega_{21}\Omega_{32}/4\Delta_2}{\varepsilon(\varepsilon - \Delta_3) + \Omega_{23}\Omega_{32}\varepsilon/4\Delta_2 + \Omega_{12}\Omega_{21}(\varepsilon - \Delta_3)/4\varepsilon}. \tag{9.16}$$

Subjecting (9.16) to an inverse Fourier transformation, we find

$$\Psi_{3,-2}(t) = j\frac{\sin\Omega_{\text{eff}}t}{\Omega_{\text{eff}}}\frac{\Omega_{12}\Omega_{32}}{4\Delta_2}\exp\left[j\left(\frac{\frac{\Delta_3}{2} - \Omega_{12}^2 + \Omega_{32}^2}{8\Delta_2}\right)t\right], \tag{9.17}$$

where we use the notation

$$\Omega_{\text{eff}}^2 = \left(\frac{\Delta_3}{2} - \frac{\Omega_{12}^2 + \Omega_{32}^2}{8\Delta_2}\right)^2 + \frac{\Omega_{12}^2}{4\Delta_2}\Delta_3 \tag{9.18}$$

for a certain effective frequency of oscillations Ω_{eff}.

The population of the third level $\rho_{33} = |\Psi_3(t)|^2 \simeq |\Psi_{3,-2}(t)|^2$ amounts to

$$\rho_{33} = \frac{\Omega_{21}^2\Omega_{32}^2}{16\Delta_2^2}\frac{\sin^2\Omega_{\text{eff}}t}{\Omega_{\text{eff}}^2}. \tag{9.19}$$

This is a resonant-type dependence (see (9.18)). The resonant (minimum) value of the parameter Ω_{eff},

$$\Omega_{\text{eff}} = \Omega_{32}\Omega_{21}/4\Delta_2, \tag{9.20}$$

corresponds to the frequency detuning

$$\Delta_3 = (\Omega_{32}^2 - \Omega_{21}^2)/4\Delta_2. \tag{9.21}$$

Then we get

$$\rho_{33} = \sin^2\frac{\Omega_{32}\Omega_{21}}{4\Delta_2}t = \frac{1}{2}\left(1 - \cos\frac{\Omega_{32}\Omega_{21}}{2\Delta_2}t\right). \tag{9.22}$$

This expression is like that obtained in the second lecture (2.40) for the case of a two-level system. The difference between (9.22) and (2.40) is as follows.

In the case of a two-level system the transition matrix element, which is equal to the product of the matrix element of the dipole moment operator μ_{21} multiplied by the field strength $\mathcal{E}_0$, determines the value of the Rabi nutation frequency,

$$\Omega_0 = \mu_{21}\mathcal{E}_0/\hbar. \tag{2.36}$$

In the case of two-photon resonance in a three-level system, the nutation frequency is determined by the composite matrix element of the transition V_{13}, which from (9.22) and (9.14) is seen to be equal to

$$V_{13} = \frac{(\mu_{12}\mathcal{E}_0/2)(\mu_{23}\mathcal{E}_0/2)}{\hbar\Delta_2}. \tag{9.23}$$

The result obtained and the form in which it is written have deep physical meaning. Let us turn to Fig. 9.2 which shows the quasi-energy levels. The transition from state $|1, 0\rangle$ to state $|3, -2\rangle$ occurs through the intermediate level $|2, -1\rangle$ detuned from the resonance by the value of Δ_2. The probability amplitude of the occupation of this state is $\mu_{12}\mathcal{E}_0/2\pi\Delta_2$ times less than that of level $|1, 0\rangle$. Factor $\mu_{23}\mathcal{E}_0/2$ equals the matrix element of transition $|2, -1\rangle \rightarrow |3, -2\rangle$. The product of these two quantities determines the matrix element of transition $|1, 0\rangle \rightarrow |3, -2\rangle$ or transition $1 \rightarrow 3$ in the resonant approximation adopted.

We should point out another fundamental distinction between a two-photon resonance in a three-level system and a one-photon resonance in a two-level system. From (9.21) the resonant frequency of transition $1 \rightarrow 3$ is seen to depend on the field intensity. This dependence is determined by a mutual repulsion of levels $|1, 0\rangle$ and $|2, -1\rangle$ as well as $|2, -1\rangle$ and $|3, -2\rangle$ after the field has been turned on (see (2.29)). This is an instance of the well-known quantum-mechanical effect of the repulsion of the eigenenergy states on turning on the interaction between them. Let us take advantage of (2.29) for the pairs of levels considered, bearing in mind the smallness of the Rabi frequencies Ω_{12} and Ω_{23} as compared to the intermediate frequency detuning Δ_2. Then we obtain the changes of level energies $|1, 0\rangle$, $|2, -1\rangle$ and $|3, -2\rangle$ in the form

$$\Delta E_{1,0} = \frac{\hbar}{4}\frac{\Omega_{12}^2}{\Delta_2}, \quad \Delta E_{2,-1} = -\frac{\hbar}{4}\frac{\Omega_{12}^2}{\Delta_2} - \frac{\hbar}{4}\frac{\Omega_{23}^2}{\Delta_2}, \quad \Delta E_{3,-2} = \frac{\hbar}{4}\Omega_{23}^2\Delta_2, \tag{9.24}$$

respectively. These changes in fact represent the shift of energy levels due to the quadratic Stark effect. The difference $(\Delta E_{3,-2} - \Delta E_{1,0})/\hbar$ gives the resonant condition (9.21). Everything said above is illustrated in Fig. 9.3.

The peculiarities considered, including the appearance of the composite matrix element depending non linearly on the field and the presence of nonlinear resonance with the frequency depending on the field strength, are typical for the nonlinear oscillation processes in general and for the multiphoton processes in particular.

Everything said above remains valid in the case of a greater number of intermediate levels and higher degree of "multiphotonity" of the transition which we shall call photonity for short.

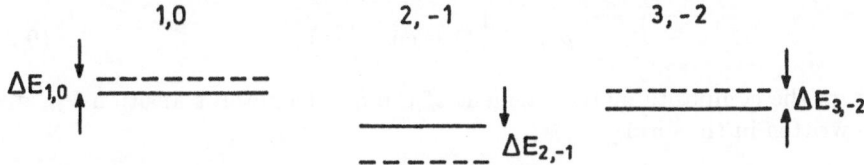

Fig. 9.3. The shift of the quasi-energy levels by the radiation field. The new positions of the levels are shown in dashed lines

In the case of n levels (the photonity equal to $n-1$) the quasi-energy method gives in resonant approximation

$$j\hbar\dot{\Psi}_{1,0} = \frac{\mu_{12}\mathcal{E}_0}{2}\Psi_{2,-1}\,,$$

$$j\hbar\dot{\Psi}_{2,-1} = (E_2 - \hbar\omega)\Psi_{2,-1} + \frac{\mu_{21}\mathcal{E}_0}{2}\Psi_{1,0} + \frac{\mu_{23}\mathcal{E}_0}{2}\Psi_{3,-2}\,,$$

$$\cdots\cdots\cdots\cdots\cdots\cdots\cdots\cdots\cdots\cdots\cdots\cdots$$

$$j\hbar\dot{\Psi}_{k,1-k} = (E_k - (k-1)\hbar\omega)\Psi_{k,1-k} + \frac{\mu_{k,k+1}\mathcal{E}_0}{2}\Psi_{k+1,-k} + \frac{\mu_{k,k-1}\mathcal{E}_0}{2}\Psi_{k-1,2-k}\,,$$

$$\cdots\cdots\cdots\cdots\cdots\cdots\cdots\cdots\cdots\cdots\cdots\cdots$$

$$j\hbar\dot{\Psi}_{n,1-n} = (E_n - (n-1)\hbar\omega)\Psi_{n,1-n} + \frac{\mu_{n,n-1}\mathcal{E}_0}{2}\Psi_{n-1,2-n}\,.$$

$$(9.25)$$

Here we assume $E_1 = 0$.

The method developed above for the three-level system allows us to represent the ψ-function of nth level in the form

$$\Psi_{n,1-n}(t) = \frac{j}{2\pi}\int\prod_{n=1}^{n-1}(\mu_{n,n-1}\mathcal{E}_0)\frac{\exp(j\varepsilon t)}{\text{Det}}d\varepsilon\,, \qquad (9.26)$$

where we denote by the symbol Det the determinant of the matrix

$$\begin{vmatrix} \varepsilon & -\frac{\mu_{12}\mathcal{E}_0}{2\hbar} & 0 & \cdots & 0 & 0 \\ -\frac{\mu_{21}\mathcal{E}_0}{2\hbar} & -\Delta_2 & -\frac{\mu_{23}\mathcal{E}_0}{2\hbar} & \cdots & 0 & 0 \\ 0 & -\frac{\mu_{32}\mathcal{E}_0}{2\hbar} & -\Delta_3 & \cdots & 0 & 0 \\ \cdot & \cdot & \cdot & \cdots & \cdot & \cdot \\ \cdot & \cdot & \cdot & \cdots & \cdot & \cdot \\ 0 & 0 & 0 & \cdots & -\Delta_{n-1} & -\frac{\mu_{n-1,n}\mathcal{E}_0}{2\hbar} \\ 0 & 0 & 0 & \cdots & -\frac{\mu_{n,n-1}\mathcal{E}_0}{2\hbar} & -\Delta_n + \varepsilon \end{vmatrix}$$

In this case, as in the case of two-photon resonance in a three-level system, all the intermediate frequency detunings from Δ_2 to Δ_{n-1} are considered to be small in comparison with ε. In this approximation (as in the case of (9.16)) the determinant in the denominator of (9.26) is a second-degree polynomial in ε. After we have found its roots, performed the Fourier transformation and evaluated the resonant value of Δ_n, we obtain for the population of upper state ρ_{nn} an expression resembling (9.22),

$$\rho_{nn} = \frac{1}{2}\left(1 - \cos\frac{V_{1n}}{\hbar}t\right), \tag{9.27}$$

where the composite matrix element of the $n - 1$-photon transition V_{1n} may be written in the form

$$V_{1n} = \frac{1}{2}\mu_{n-1,n}\mathcal{E}_0 \prod_{k=1}^{n-2}\left(\frac{\mu_{k,k+1}\mathcal{E}_0}{2\hbar\Delta_{k+1}}\right). \tag{9.28}$$

The interpretation of this expression based on the assumption of the rare populations of each of the $n - 2$ intermediate levels is quite analogous to that given in the case of two-photon resonance in a three-level system. The field shift of the resonant detuning Δ_n is determined in exactly the same way as in the three-level case by the repulsion of levels caused by the quadratic Stark effect.

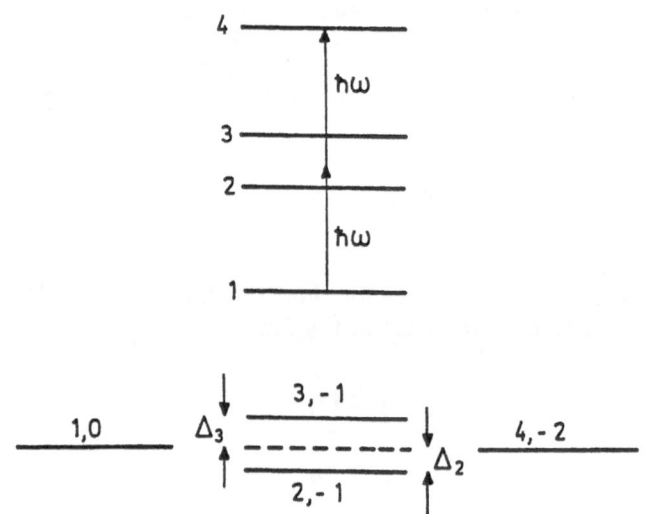

Fig. 9.4. A two-photon resonance in a system with two intermediate levels

The cases considered above of two and more photon resonances have a specific feature. The transition from the initial to the final state occurred along a single path ignoring a certain sequence of levels. However, situations topologically more complicated are possible. In these situations many channels lead from the initial state to the final one. Each channel has the appropriate composite matrix element. The resulting matrix element is the sum of the composite matrix elements of transitions via each channel. The simplest example of such a situation is a four-level system which has two closely adjacent levels in the vicinity of intermediate resonance. This system along with the corresponding scheme of quasi-energy levels is given in Fig. 9.4. The composite matrix element of such a system equals

$$V_{14} = \frac{\mu_{12}\mathcal{E}_0\mu_{24}\mathcal{E}_0}{4\hbar\Delta_2} + \frac{\mu_{13}\mathcal{E}_0\mu_{34}\mathcal{E}_0}{4\hbar\Delta_3}. \tag{9.29}$$

In the case presented in Fig. 9.4, one of the intermediate frequency detunings is positive while another is negative. In a fully symmetric case $\mu_{12} = \mu_{13}$, $\mu_{24} = \mu_{34}$, $\Delta_2 = -\Delta_3$ and, hence, $V_{14} = 0$, i.e., due to the interference of two channels the transition is impossible.

We have considered the cases of multiphoton excitation when each absorbed quantum can be attributed to one or even to several possibly detuned but real quantum states of the system. The vibrational spectra of molecules have this property, which is atypical of atoms. Nonetheless, multiphoton processes are also possible in atoms without intermediate atomic levels.

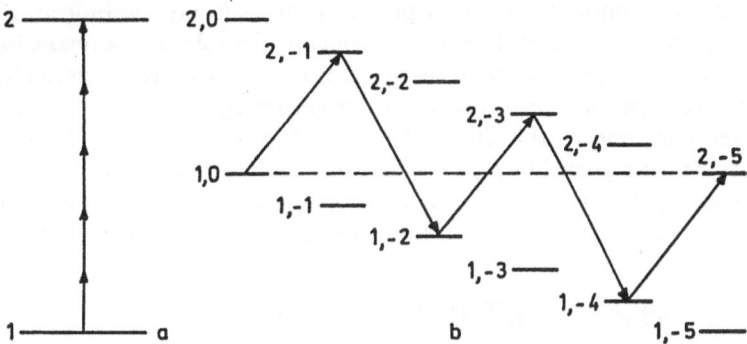

Fig. 9.5. A five-photon resonance in a two-level system: (a) the scheme of transition between the energy levels; (b) the scheme of transition between the quasi-energy levels

Let us consider a two-level system placed in an external field alternating with the frequency, which is a multiple of the interlevel distance, i.e., the conditions of multiphoton resonance are provided. An example of a five-photon resonance is shown in Fig. 9.5 a. The spectrum of the quasi-energy states of this system is given in Fig. 9.5 b. The shortest path of a five-photon transition $1 \to 5$ is represented in this figure as a sequence of transitions $|1, 0\rangle \to |2, -1\rangle$, $|2, -1\rangle \to |1, -2\rangle$, $|1, -2\rangle \to |2, -3\rangle$, $|2, -3\rangle \to |1, -4\rangle$, $|1, -4\rangle \to |2, -5\rangle$. Each successive transition corresponds to a dipole-allowed transition $1 \to 2$ or $2 \to 1$ and to the allowed transition in the space of quasi-energy indices $k \rightleftharpoons k \pm 1$. Such rules of selection over quasi-energy variables are caused by the type of interaction that has only one harmonic: $\hat{\mu}\mathcal{E}_0 \cos \omega t$. The main quota to the transition process is contributed by the share of the path mentioned. The following composite matrix element corresponds to this path:

$$V_{12} = \frac{\mu_{12}\mathcal{E}_0/2}{4\hbar\omega}\frac{\mu_{21}\mathcal{E}_0/2}{-2\hbar\omega}\frac{\mu_{21}\mathcal{E}_0/2}{2\hbar\omega}\frac{\mu_{21}\mathcal{E}_0/2}{-4\hbar\omega}\frac{\mu_{21}\mathcal{E}_0}{2}. \tag{9.30}$$

The meaning of this formula is that, thanks to the absence of intermediate energy levels in the system, the quasi-energetic states turn out to be considerably detuned. The appropriate detunings are equal to the field harmonics. This fact is illustrated in Fig. 9.5 b.

This simplified model demonstrates the main typical feature of the process: the probability amplitude of the transition is proportional to the field strength raised to the power of the resonance photonity. This, in turn, implies that under conditions of relaxation decay of the upper state, say, at the rate of Γ, the rate of its population is proportional to $|V_{12}|^2\Gamma^{-1}$ while the population is proportional to $|V_{12}^2\Gamma^{-2}$, i.e., the number of atoms in the upper state is proportional to the radiation intensity raised to the power of the photonity degree. The strong power dependence (at $\hbar\omega \ll E_2 - E - 1$) can be readily observed experimentally. It is used to determine the degree of transition photonity.

However, we have to stress that the model shown in fig.9.5b and formula (9.30) oversimplifies the multiphoton process in atoms. Indeed, two isolated energy levels are slightly shifted by a considerably nonresonant external field. However, the point is that there are no pure two-level systems in practice. After the field has been turned on, other energy levels really existing in the system can, due to the Stark effect, affect the position of the pair considered more significantly than the mutual interaction of these levels. The consequence of such an influence is the dependence of the energy position of each level on the external field. This dependence may be represented in the form

$$E_{1,2} = \alpha_{1,2}^{(1)}\mathcal{E}(t) + \alpha_{1,2}^{(2)}\mathcal{E}^2(t) + \alpha_{1,2}^{(3)}\mathcal{E}^3(t) + \alpha_{1,2}^4\mathcal{E}^4(t) + \cdots. \tag{9.31}$$

The substitution of this expression into the Schrödinger equation and the following passage to quasi-energy representation result in additional channels which connect the initial and final states. These additional channels bind the quasi-energy levels originating from the same level of the system under consideration. Due to the power-type dependence (9.31) the composite matrix element of the transition $1 \rightarrow 2$ remains proportional to the field amplitude raised to the photonity power. Its value can turn out to be considerably greater than (9.30) owing to the decrease of the energetic denominators which determine the coefficients of the expansion into the perturbation theory series (9.31). These denominators equal by the order-of-magnitude the distance of the closest level adjacent to the one considered. They may be small mainly for the upper state. The latter circumstance becomes especially apparent when the upper level corresponds to the continuum.

Figure 9.6 illustrates the modification of the scheme of transitions between the quasi-energy levels due to the effect just considered in the case of five-photon resonance (the same as in Fig. 9.5).

Multiphoton processes exist not only in a monochromatic external field but in the case of a polychromatic one as well. The quasi-energy method can

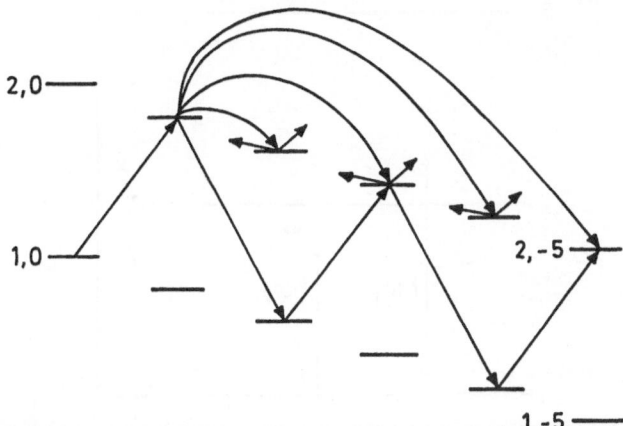

Fig. 9.6. Additional channels of transition in a five-photon resonance in a two-level system

be used in the latter case, too, even though the quasi-energy spectrum turns out to be much richer. However, it is very expedient to build this spectrum up since it allows us to visualize the possible transition paths and to readily evaluate the composite matrix element in accordance with the procedures given above.

In the case of the two-frequency action the initial formulae (9.1) and (9.2) are written in the form

$$j\hbar\dot{\Psi}_n = E_n\Psi_n + \sum_{n'} V_{nn'}^{(1)}\Psi_{n'}\cos\omega_1 t + \sum_{n'} Vnn'^{(2)}\Psi_{n'}\cos\omega_2 t, \qquad (9.32)$$

$$\Psi_n(t) = \sum_{k,l=-\infty}^{\infty} \Psi_{nkl}\exp(j\omega_1 tk + j\omega_2 tl). \qquad (9.33)$$

Then the equation corresponding to (9.3) takes on the form

$$j\hbar\dot{\Psi}_{nkl} = (E_n - k\hbar\omega_1 - l\hbar\omega_2)\Psi_{nkl}$$
$$+ \sum_{n'} V_{nn'}^{(1)}\left(\frac{1}{2}\Psi_{n',k+1,l} + \frac{1}{2}\Psi_{n',k-1,l}\right) + \sum_{n'} V_{nn'}^{(2)}$$
$$\times \left(\frac{1}{2}\Psi_{n',k,l+1} + \frac{1}{2}\Psi_{n'k,l-1}\right). \qquad (9.34)$$

Now let us consider the simplest case of a three-level system. A two-frequency action may be represented by the scheme in Fig. 9.7. As we have already mentioned the corresponding quasi-energy spectrum is much richer than the one represented in Fig. 9.2 for the case of a single-frequency action. Even after the extraction of the closest levels from this spectrum by analogy with Fig. 9.3, in a monochromatic case, we still obtain the very complex spectrum shown in Fig. 9.8.

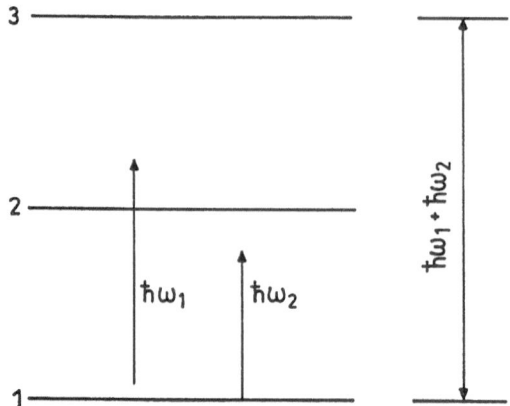

Fig. 9.7. A two-frequency two-photon resonant transition in a three-level system

The main contribution to the total composite matrix element is rendered by the transition paths shown in Fig. 9.8 by the arrows with the corresponding matrix elements marked in them. This situation turns out to be similar to the above-discussed case of single-frequency action onto a four-level system with two intermediate resonances (Fig. 9.4). As a result, by analogy with (9.29) we have

$$V_{13} = \frac{1}{4} \frac{V_{12}^{(1)} V_{23}^{(2)}}{E_2 - \hbar\omega_1 - E_1} + \frac{1}{4} \frac{V_{23}^{(1)} V_{12}^{(2)}}{E_2 - \hbar\omega_2 - E_1} . \tag{9.35}$$

In this expression we assume the two-frequency two-photon resonance condition to be satisfied,

$$\hbar\omega_1 + \hbar\omega_2 \approx E_3 - E_1 . \tag{9.36}$$

In other words, we assume quasi-energy levels $E_{1,0,0}$ and $E_{3,-1,-1}$ to coincide, as shown in Fig. 9.8. Here we ignore the Stark shift of the levels. Note, that by analogy with the symmetric case discussed above for the single-frequency two-photon transition in the system of two intermediate levels (see (9.29)), the resonant two-frequency two-photon transition in the equidistant three-level system,

$$E_3 - E_2 = E_2 - E_1 , \tag{9.37}$$

does not occur in accord with (9.36) and (9.35) due to the destructive interference of the two channels of transition considered.

The quasi-energy method also applies in the case of a high degree of poly-chromaticity of the field, i.e., in the case of complicated time dependence of the field. However, the number of quasi-energy levels that should be taken into consideration in the latter case turns out to be enormous. So, although the quasi-energy method is the most general one, it is not necessarily the most efficient one for the solution of such problems. Sometimes a method based on the consideration of the Schrödinger equations with variable coefficients is better suited for the solution of multiphoton problems.

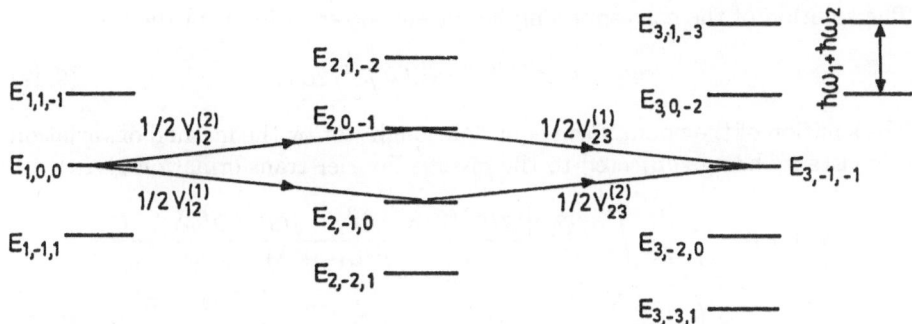

Fig. 9.8. The main transition paths in a two-frequency two-photon resonant transition in a three-level system

One of the simplest problems of this type which presents a certain interest is the problem of the excitation of the quantum system with a field frequency that linearly depends on time (the linear sweep of frequency),

$$E(t) = \mathcal{E}_0 \cos[(\omega_0 + \alpha t)t]. \qquad (9.38)$$

This problem refers, in fact, to the class of multiphoton problems since the spectrum of the oscillation of the (9.38) type contains many frequencies. In other words, the photon spectrum of radiation is initially assumed to be very rich.

Let us consider the excitation of a two-level system with the field (9.38). Note, that this problem is not only of interest in itself but its solution is of practical importance, as the consideration of multilevel systems may be reduced to it if only the field amplitude is not too large, i.e., the Stark shift of levels does not exceed the typical values of intermediate frequency detunings.

Selecting frequency ω_0 from the condition for the radiation to be in resonance at $t = 0$, $\hbar\omega_0 = E_2 - E_1$ we can write

$$j\hbar\dot{\Psi}_1 = V\Psi_2, \quad j\hbar\dot{\Psi}_2 = \hbar\alpha t\Psi_2 + V\Psi_1,$$
$$\Psi_1(t = 0) = 1, \quad \Psi_2(t = t_0) = 0, \qquad (9.39)$$

where $V = \mu_{21}\mathcal{E}_0/2$. Note that the given equation also describes the electron transition in the case of crossing molecular terms which is usually referred to as the Landau-Zener transition.

The Fourier transformation, with allowance for the initial conditions from (9.39), gives

$$\hbar\varepsilon\Psi_1 = V\Psi_2 + \exp(-jt_0\varepsilon),$$
$$\hbar\varepsilon\Psi_2 = j\hbar\alpha\partial\Psi_2/\partial\varepsilon + V\Psi_1. \qquad (9.40)$$

Substituting the first of these equations into the second we obtain a first-order linear differential equation with variable coefficients,

$$\frac{\partial}{\partial\varepsilon}\Psi_2 - \frac{\varepsilon}{j\alpha}\Psi_2 + \frac{V^2}{j\alpha\varepsilon\hbar^2}\Psi_2 = -\frac{\exp(-jt_0\varepsilon)}{j\alpha\varepsilon\hbar^2}. \qquad (9.41)$$

The solution of the corresponding homogeneous equation is of the form

$$\Psi_2 = (\varepsilon)^{V^2/j\hbar^2\alpha} \exp(-j\varepsilon^2/2\alpha)\,. \tag{9.42}$$

The solution of the nonuniform equation sought for by the method of variation of constants being subjected to the inverse Fourier transformation gives

$$\Psi_2(t) = -\int \frac{d\varepsilon}{2\pi} \int_{-\infty}^{0} \frac{\exp\big[(j/\alpha)(2\varepsilon\lambda + \lambda^2) - jt_0\varepsilon - jt_0\lambda + jt\varepsilon\big]}{j\hbar^2\alpha(\varepsilon + \lambda)}$$
$$\times \left(\frac{\varepsilon}{\varepsilon + \lambda}\right)^{V^2/j\hbar^2\alpha} d\lambda\,. \tag{9.43}$$

The obtained general solution (without expressing it in terms of hypergeometric functions) can be analyzed in two extremes.

For a start let us consider the case of the fast sweep of radiation frequency when the Rabi oscillation period is much longer than the typical time interval during which the levels are resonant

$$\hbar/V \gg V\alpha\,. \tag{9.44}$$

In this case factor $[\varepsilon/(\varepsilon + \lambda)]^{V^2/j\alpha\hbar^2}$ in the denominator of the term under the integral sign in (9.43) is close to unity and the integration over $d\varepsilon$ due to the presence of a large parameter $t_0 - t$ may be performed by the steepest descent method in the vicinity of point $\varepsilon = -\lambda$. Then the integration over $d\lambda$ can be performed. The resulting wavefunction Ψ_2 at $t_0 \to -\infty$ and $t \to +\infty$ turns out to equal by order-of-magnitude to

$$\Psi_2 \approx V^2/\alpha\hbar^2\,, \tag{9.45}$$

i.e., to be small. In the case of the fast sweep the resonance is being passed quickly. So, only a small fraction of the population of level 1 has time enough to undergo a transition to the upper level 2. This fraction equals by order of magnitude the product of the matrix element of the transition probability amplitude $V/\hbar$ multiplied by the time interval $V/\alpha\hbar$ when the system is in resonance.

In the opposite extreme (the case of the slow sweep) the value of parameter $V^2/\alpha\hbar^2$ happens to be large. So, if we apply the steepest descent method to integral (9.43) in the vicinities of points $\varepsilon = 0$ and $\varepsilon = -\lambda$, we obtain that at $t \to \infty$ the upper state wave function $\Psi_2 \to 1$ up to $\exp(-V^2/\alpha\hbar^2)$. This means that during the slow sweep levels 1 and 2 interchange their populations. This is the situation quite analogous to that which in the physics of molecules is referred to as the repulsion of molecular terms in the region of their formal intersection. In our case this effect can be explained as follows.

The adiabatically slow variation of the field frequency is equivalent to the parametric dependence of the quasi-energy level positions on time. These dependencies are presented in Fig. 9.9. Note that the quasi-energy method becomes convenient again in the adiabatically slow process. In the region of

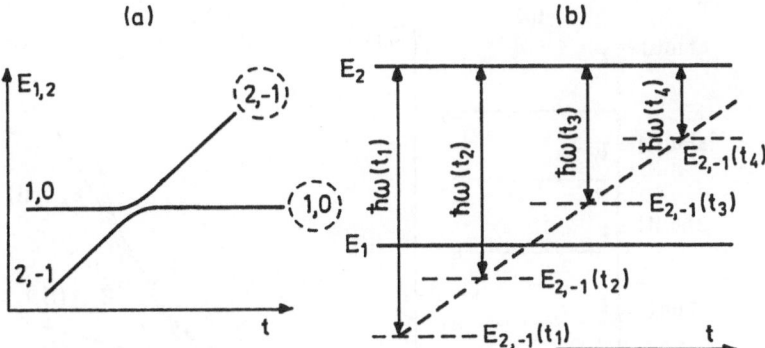

Fig. 9.9. The change in the position of the quasi-energy level $(2, -1)$ at adiabatically slow sweep of the field frequency (**b**) and the repulsion of the quasi-energy levels in resonance (**a**)

strict resonance the intersection of levels does not occur. So, after the first level has passed this region its position starts depending on time, as happened before with the second-level position. The position of the second level becomes independent of time as in the case of the first-level position. The process of the repulsion of levels could be called a swap of level numbers. In case the level energy changes adiabatically slowly, the level population remains practically constant (as we know, the process is adiabatic up to an exponentially small term). Therefore, this implies that levels 1 and 2 swap their populations.

The adiabatic sweep of frequency can be used for a successive excitation of ever higher states of the multilevel system in which the transition frequency diminishes as the system gains energy. This is illustrated in Fig. 9.10. A noticeable feature of the curve presented in this figure is the successive steepening of the time dependence of the quasi-energy level position with the level number. At the points of intersection of quasi-energy levels the Landau-Zener transitions analogous to the one considered above in the case of a two-level system take place successively. As a result, the populations are successively transferred from the lower levels to the upper ones, so that the number of the populated levels increases with time.

So far our consideration has concerned only two-level systems and those which can be reduced to them. However, if the field strength is sufficient for the interaction with the field to exceed the intermediate detuning frequencies, the situation changes radically and for the system dynamics to be described it is necessary to solve the complete set of Schrödinger equations containing Ψ-functions of all levels. Finding the solution of such a set is far from always possible. However, there is a whole class of problems which deal with the systems of nondegenerate, almost equidistant levels that sometimes admit an exact solution. In such systems the interaction of the levels is of the relay-race type, i.e., only adjacent levels are connected by transitions. The excitation is transferred from one level to another in succession and the population distribution at a certain moment is the superposition of the amplitudes of the

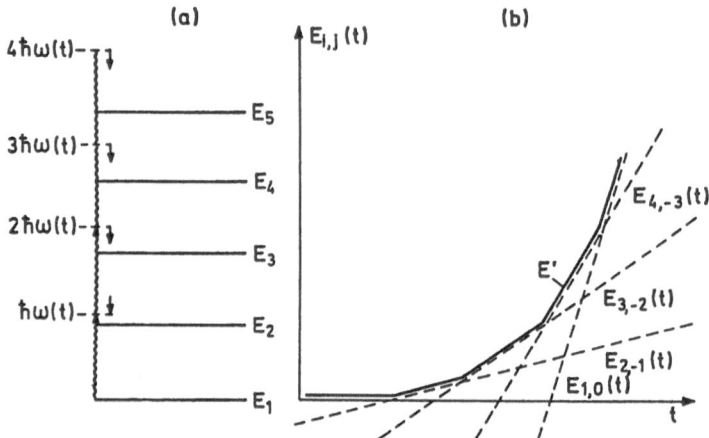

Fig. 9.10. A successive excitation of isolated levels of multilevel systems at an adiabatic frequency sweep: **(a)** the scheme of the energy levels; **(b)** the dependence of the positions of the quasi-energy levels $E_{i,j}(t)$ (dashed line) and of the populated quasi-energy level E' (solid line) in time

probability for the system to undergo such multiphoton transition (upward, upward, down, upward again, etc.). The following recurrent relations hold for the Fourier transforms of the quasi-energy wavefunctions

$$(\Delta_n - \varepsilon\hbar)\Psi_{n,1-k} + \frac{\mu_{n,n+1}\mathcal{E}_0}{2}\Psi_{n+1,k} - \frac{\mu_{n+1,k}\mathcal{E}_0}{2}\Psi_{n-1,2-k} = 0. \qquad (9.46)$$

Such three-term recurrent relations have a solution in the form of orthogonal polynomials. If these recurrent relations correspond to the classical orthogonal polynomials or hypergeometric functions, (9.46) admits analytical solutions. However, this is the case only for certain dependencies Δ_n and $\mu_{n,n+1}$ on n. So, for example, in the case of the system of levels infinite in both directions at $\Delta_n = 0$ and $\mu_{n,n=1} = \mu$, i.e., when all the detuning frequencies equal zero and the matrix elements $\mu_{n,n+1}$ do not depend on n, (9.46) is satisfied with the Chebyshev polynomials. After the inverse Fourier transformation this brings about a solution in the form of the Bessel function:

$$\Psi_{n,1-n} = j^n J_n\left(\frac{\mu\mathcal{E}_0}{\hbar}t\right). \qquad (9.47)$$

It is worth noting that in this simplest particular case the time-dependent Schrödinger Eq. (9.25) of the form

$$j\hbar\dot{\Psi}_{n,1-n} = \frac{\mu\mathcal{E}_0}{2}(\Psi_{n+1,-n} + \Psi_{n-1,2-n}), \qquad (9.48)$$

after the replacement $\Psi_{n,1-n} = j^n\phi_n$, exactly coincides with the tabular recurrent relation that binds the Bessel functions and their derivatives.

From (9.47) for the population of the n-th level we obtain

$$\rho_{nn} = J_n^2 \left(\frac{\mu \mathcal{E}_0}{\hbar} t \right) .$$
(9.49)

This implies that the population initially concentrated on the zero level diffuses with time over the adjacent levels. As long as the numbers of the levels being populated are not too great, i.e., while the Bessel function index is much less than its argument value,

$$n \ll \frac{\mu \mathcal{E}_0}{\hbar} t ,$$
(9.50)

we can use its asymptotic representation,

$$J_n(x) = (2/\pi)^{1/2} (x)^{-1/2} \cos(x - n\pi/2 - n/4) .$$
(9.51)

This gives a practically uniform population distribution among the levels whose numbers are less than the argument value. For the levels with numbers exceeding the Bessel function argument,

$$n > \frac{\mu \mathcal{E}_0}{\hbar} t ,$$
(9.52)

the population exponentially decreases due to the asymptotic-over-the-index behavior of the Bessel function.

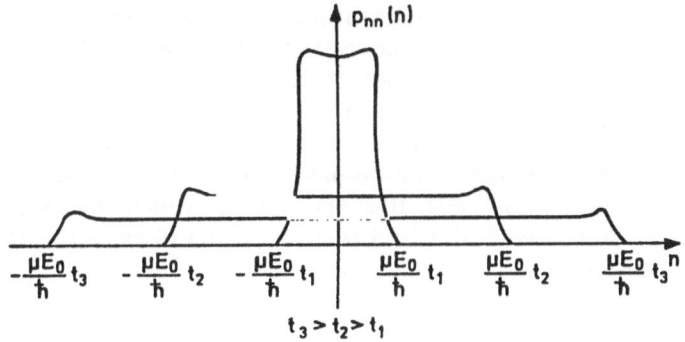

Fig. 9.11. The distribution of averaged populations among the levels at relay-type occupation of a system of exactly equidistant levels for three different time moments $t_3 > t_2 > t_1$

As a result, the process of relay-type population of the system of exactly equidistant levels with equal matrix elements of the dipole moment operator can be schematically represented as in Fig. 9.11. This figure shows the averaged population distributions at three different moments in time.

Note that there are other problems which can be reduced to orthogonal polynomials: the excitation of a simple oscillator, the excitation of the system with a decreasing matrix element and some others.

Lecture Ten

Coherent Processes
in Complicated Multilevel Systems

Statistical approach to spectrally complicated systems. Ensemble average. "Isolated level-band" system. Dense band. Decay of lower level into the band. Quasicontinuum. Distribution of populations in the band. Nondense band, nonsmooth distribution of matrix elements. Number of levels in the region of Stark withdrawal. "Band-band"-type system. Degenerate bands.

With an increase of laser light intensity, more and more eigenenergy states get involved in the process of the excitation of quantum particles. In fact, from the quantum-mechanical point of view, all contemporary problems of the intense resonant interaction of radiation with matter concern the spectrally complicated multilevel systems. There are two types of such problems. The problems of the first type can be simplified by the reduction of the number of levels involved owing to either the restrictions imposed on the time interval or the absence of relaxation processes, or the classical character of the interaction over certain degrees of freedom. The second class contains the problems in which the multilevelness of the system is of principal importance. These problems require a separate quantum-mechanical consideration with the use of specially developed methods. It is the ways in which the second type of problems are formulated and solved that are to be discussed in the present and subsequent lectures.

In the previous discussion we have been repeatedly convinced of the enormous mathematical complexity associated with the transition to multilevel systems, making an analytical solution of the Schrödinger equation practically impossible. However, in case we deal with the interaction of laser light with real quantum objects such as polyatomic molecules, the spectrum of the object's eigenstates turns out to be so intricate that the multilevelness of the system cannot be ignored. Fortunately, detailed information on the moduli and the phases of the ψ-functions of all the states, as a rule, is not necessary to obtain the values of physically meaningful quantities. Only the mean populations of the resonant groups of levels are often of real interest. These rather coarse-grain characteristics do not depend on the fine details of the problem, i.e., on the spectrum microstructure which affect the moduli and the phases of isolated state ψ-functions, but are determined by certain mean spectral parameters of the system such as the mean level density, the average interaction matrix element, etc.

In the theory of spectrally complicated systems the procedure of the ensemble average is used to find out the coarse-grain characteristics in case the spectrum microstructure is of no interest. This procedure may be used in our case as well.

We replace a single complicated system by an ensemble of similar systems with like properties. Each system of the ensemble has its own fine spectral characteristics while the coarse-grain macroscopic parameters are assumed to be the same. Solutions describing the evolution in time of the population distribution among all the levels of each system exist and depend strongly on the microstructure of the individual spectrum. These solutions may be written in the form of either an involved series or cumbersome determinants and, in fact, hardly admit any practical analysis. However, the ensemble average, which eliminates the dependence on the spectrum microstructure and retains only the dependence on coarse-grain system characteristics, significantly simplifies the solution. Such a procedure allows us to obtain quite "digestible" analytical expressions which are convenient for practical analysis. In such a way we obtain solutions averaged over the ensemble of systems. If the dispersion of the solutions, obtained after the ensemble average, is small, the mean solutions can be attributed to a single isolated system.

So, this implies, generally speaking, that after the ensemble average we should check that the dispersion is small. Should this be the case it would mean that we have correctly chosen not only the coarse-grain characteristics of the system considered but also the ensemble of systems itself. Nevertheless, one should not forget that due to its statistical nature this method of ensemble averaging cannot assure the exact coincidence of the coarse-grain characteristics of a certain given system with the averaged characteristics thus obtained. It only states that although such a discrepancy is generally feasible, its probability is small.

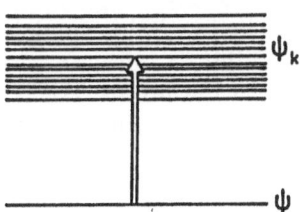

Fig. 10.1. The resonant transition in an "isolated level-band of levels" system

Let us consider for a start the simplest problem of how the quantum system of the "isolated level-band" type behaves under the action of resonant radiation (Fig. 10.1). The band levels are identified with index k and have wavefunctions ψ_k. We shall denote the wavefunction of an isolated level by ψ. Then the resonant-approximation equations in the case of a field of constant amplitude $\mathcal{E}_0$ may be written in the form

$$j\hbar\dot\psi = \sum_k \frac{\mu_k \mathcal{E}_0}{2}\psi_k\,,\quad jh\dot\psi_k = \hbar\Delta_k\psi_k + \frac{\mu_k\mathcal{E}_0}{2}\psi\,. \tag{10.1}$$

Let us take the following initial conditions

$$\psi(t=0)=1\,,\quad \psi_k(t=0)=0\,. \tag{10.2}$$

Then after the Fourier transformation we have

$$\varepsilon\hbar\psi = \sum_k \frac{\mu_k\mathcal{E}_0}{2}\psi_k - jh\,,\quad \varepsilon\hbar\psi_k = \hbar\Delta_k\psi_k + \frac{\mu_k\mathcal{E}_0}{2}\psi\,. \tag{10.3}$$

Substituting the second of these equations into the first one we find

$$\psi = -j\left[\varepsilon - \sum_k \frac{\mu_k^2\mathcal{E}_0^2}{4\hbar^2(\varepsilon-\Delta_k)}\right]^{-1}\,. \tag{10.4}$$

The following consideration is determined by the type of state spectrum in the band. The simplest case is a continuous spectrum or a spectrum so dense that it can be treated as continuous with the matrix element continuously varying from state to state. Then the summation over k may be replaced by integration over frequency detuning $d\Delta$. So we have

$$\psi = -j\left[\varepsilon - \int \frac{\mu^2(\Delta)\mathcal{E}_0^2}{4\hbar^2(\varepsilon-\Delta)}g(\Delta)d\Delta\right]^{-1}\,, \tag{10.5}$$

where $g(\Delta)$ is the density of band levels per unit interval of frequency detuning Δ[*]) The bracketed integral in (10.5) is a complex number with a small real part. The latter arises from the finite widths of distributions $\mu(\Delta)$ and $g(\Delta)$ and equals the Stark shift of the level in the vicinity of the resonance caused by its interaction with the band states. In the case of a broad band this shift is negligible.

The imaginary part of the integral is caused by the presence of the pole at $\Delta = \varepsilon$ and equals $j2\pi\mu^2(\varepsilon)g(\varepsilon)\mathcal{E}_0^2/4\hbar^2$.

The main contribution to the inverse Fourier transformation integral is rendered by the poles of (10.5) determined by the equation

$$\varepsilon - j2\pi\mu^2(\varepsilon)g(\varepsilon)\mathcal{E}_0^2/4\hbar^2 = 0\,. \tag{10.6}$$

Should the field $\mathcal{E}_0$ equal zero, the pole would be at $\varepsilon = 0$. The presence of the field shifts the pole, but due to the smooth dependence of $\mu(\varepsilon)$ and $g(\varepsilon)$ on energy, their values in the pole remain practically unchanged. Therefore, the inverse Fourier integral is written in this case in the form

$$\psi(t) = -\frac{1}{2\pi}\int \frac{j\exp(j\varepsilon t)d\varepsilon}{\varepsilon - j\pi\mu^2(0)g(0)\mathcal{E}_0^2/2\hbar^2} = \exp\left[-\frac{\pi\mu^2(0)g(0)\mathcal{E}_0^2}{2\hbar^2}t\right]\,. \tag{10.7}$$

---

[*]) Note that density g is sometimes expressed in energetic terms, i.e., represents the number of levels per unit energy interval. Then $g(E) = g(\Delta)\hbar^{-1}$.

Accordingly, the lower-level population exponentially decreases with time:

$$\rho(t) = \exp\left[-\pi\mu^2(0)g(0)\hbar^2 t\right] ,\qquad (10.8)$$

or, in other words, the lower level exponentially decays into the upper band with the typical time determined by the field strength

$$\tau = \hbar^2/pi\mu^2(0)g(0)\mathcal{E}_0^2 .\qquad (10.9)$$

The reciprocal with respect to τ quantity W, $W = 1/\tau$, is naturally called the transition probability while the corresponding expression $W = \pi\mu^2\mathcal{E}_0^2 g/\hbar^2$ is referred to as Fermi's golden rule.

In fact this result is obtained for the continuous band spectrum. It also holds for a sufficiently dense band, but only if the finite detuning between adjacent band levels δ does not bring about a noticeable change in the phase of the wavefunctions of these adjacent states within the typical decay time (10.9)

$$\tau\delta \ll \pi .\qquad (10.10)$$

Since there is an evident relation between level density g and the typical interlevel distance

$$g\delta \approx 1 ,\qquad (10.11)$$

the criterion of sufficient level density in the band where the band spectrum may be considered continuous can be written in the form

$$\pi g^2(0)\mu^2(0)\mathcal{E}_0^2/\hbar^2 \gg 1 .\qquad (10.12)$$

In other words the width of Stark draw $\mu(0)\mathcal{E}_0/\hbar$ should contain many levels.

If condition (10.12) is satisfied, the band with discrete levels behaves as a continuum from the point of view of an external exciting field. Such a spectral formation is called a quasicontinuum. Inequality (10.12) represents the condition for this formation to exist.

Now let us determine to which region in the band the population transferred by the radiation from the lower level falls. It turns out to be localized in the narrow vicinity of resonance. Let us prove the above statement. We substitute (10.5) into the second of Eqs. (10.3). Taking into account (10.6) and (10.9) we obtain

$$\psi_k = -\frac{j}{\varepsilon - j/2\tau}\frac{\mu_k\mathcal{E}_0}{2\hbar(\varepsilon - \Delta_k)} .\qquad (10.13)$$

This expression has two poles. At large times the main contribution into the inverse Fourier transformation integral is rendered by the pole on the real axis $\varepsilon = \Delta_k$. As a result the asymptotic value of $\psi_k(t)$ at large times becomes

$$\psi_k(t) = \frac{\mu_k\mathcal{E}_0}{2\hbar(\Delta_k - j/2\tau)}\exp(j\Delta_k t) .\qquad (10.14)$$

With allowance for the earlier-assumed smooth dependence of the dipole moment matrix element on the level number, $\mu_k = \mu$, this leads to

$$\rho_k = \frac{\mu^2 \mathcal{E}_0^2}{4\hbar^2 (\Delta_k^2 + 1/4\tau^2)},\tag{10.15}$$

which represents the stationary population distribution in the band. This distribution is seen to be Lorentzian, its width being determined in accordance with the uncertainly principle by the time of decay of the lower level into the band. Quantity $W = 1/\tau = \pi\mu^2\mathcal{E}_0^2 g/\hbar$ may also be called the Stark width of resonance in a band system. It differs from the Stark width of an isolated level $\mu\mathcal{E}_0/\hbar$ in factor $\sqrt{N}$ which is responsible for the cooperative behavior of all N levels, $N = \pi g/\tau = \pi^3\mu^2\mathcal{E}_0^2 g^2/\hbar^2$ resonating with the field. Note the interesting relation between the number of levels in the level Stark width, N_0, and the number of levels in the resonance Stark width, $N_0 = \sqrt{N}$.

Note that in the consideration just completed we do not explicitly perform the ensemble average. This procedure has been replaced by the assumption of continuous and smooth dependencies $\mu(\Delta)$ and $g(\Delta)$. Eventually this assumption was sufficient for the spectrum microstructure of the considered system not to manifest itself in rather strong fields.

In case the field is so weak that the condition for the level density (10.12) is not satisfied, i.e., the region of the Stark withdraw is comparable with or less than some typical mean distance between the band levels, the character of the system excitation is determined by the presence of at least one band level resonant with the field. In other words, the dynamics of the system excitation becomes sensitive to the band spectrum microstructure. In this case the probability of the system excitation is seen to depend on the position of the single band level closest to resonance. So, the two-level approximation is applicable. It yields (see (2.37)) the time-average population of the upper level,

$$\rho_k = \frac{1}{2}\frac{(\mu\mathcal{E}_0/\hbar)^2}{4\Delta_k^2 + (\mu\mathcal{E}_0/\hbar)^2}.\tag{10.16}$$

If we now perform the ensemble-average operation assuming that the position of the level with respect to the field frequency, i.e., the frequency detuning, is a random, uniformly distributed quantity, we obtain the following ensemble-average value

$$\langle\rho\rangle = \int_{-\infty}^{\infty} g\rho_k d\Delta_k = \frac{\pi}{2}\frac{\mu\mathcal{E}_0}{\hbar}g.\tag{10.17}$$

This quantity which in the case considered is much less than unity would be a suitable parameter of the problem, if the dispersion of random quantity ρ_k (10.16) were small. However, this dispersion,

$$\langle\rho^2\rangle - \langle\rho\rangle^2 \approx g\frac{\mu\mathcal{E}_0}{\hbar} \gg \langle\rho\rangle^2 \approx \left(g\frac{\mu\mathcal{E}_0}{\hbar}\right)^2,\tag{10.18}$$

is relatively large. Hence, the ensemble average does not provide a meaningful quantity in case the probability of resonance is small. The population of the upper band in this case is sometimes referred to as a non-self-averaging quantity.

Now let us consider the band in which the matrix elements vary from level to level considerably, i.e. by several orders of magnitude. The density of levels with significantly different matrix elements may also noticeably vary. Then (10.6) has many roots dispersed in an involved way over the complex plane and inequality (10.12), though holding for some group of levels with the same order-of-magnitude or greater values of μ, may be unsatisfied for the groups of levels with considerably smaller values of the matrix element. The point is that in such a system one cannot predict how many levels resonate with the field, whether this number is large finite or infinite, or how the ground state population decreases. For these questions to be clarified the method of ensemble averaging is necessary. The final result is as follows. If the number of resonating levels is infinite, then exponential decay analogous to (10.8) takes place. In case the number of levels is finite, the decay is incomplete and the mean population approaches its stationary asymptotic value according to the power law. Naturally the question arises as to what the number of levels resonating with the field is. We shall answer this question below (see 10.27).

Of course, the Schrödinger equation which describes the system considered may be solved in the form of (10.4) even under these conditions. However, the ψ-function obtained in such a way is not a self-averaging quantity, due to a relatively large dispersion of its phase sensitive to the values of matrix elements μ_k. Nevertheless, the population of the ground state whose Fourier transform is expressed as the product of the Fourier transform (10.4) of the ψ-function of this state multiplied by the Fourier image of the quantity conjugated with respect to it,

$$\psi(\varepsilon)\psi(\xi) = \left(\varepsilon - \sum_k \frac{\mu_k^2 \mathcal{E}_0^2}{4\hbar^2(\varepsilon - \Delta_k)}\right)^{-1} \left(\xi - \sum_k \frac{\mu_k^2 \mathcal{E}_0^2}{4\hbar^2(\xi - \Delta_k)}\right)^{-1}, \quad (10.19)$$

turns out to be a self-averaging quantity. Let us dwell on this point. The corresponding averaging over Δ is a rather lengthy exercise in computational mathematics which, in the case of uniformly distributed frequency detunings, may be described as follows.

Each of the factors in (10.19) can be represented in the integral form:

$$\psi(\varepsilon) = \int_0^\infty \frac{\exp(-\tau)}{\varepsilon} \prod_k \exp\left(-\tau \frac{\mu_k^2 \mathcal{E}_0^2}{4\hbar^2(\varepsilon - \Delta_k)\varepsilon}\right) d\tau, \quad (10.20)$$

$$\psi(\xi) = \int_0^\infty \frac{\exp(-\theta)}{\xi} \prod_k \exp\left(-\theta \frac{\mu_k^2 \mathcal{E}_0^2}{4\hbar^2(\xi - \Delta_k)\xi}\right) d\theta. \quad (10.21)$$

If we now expand the exponents under the product sign into a power series and then multiply the series extracted from expressions (10.20) and (10.21) for the same values of k, the common term of the series thus obtained turns out to contain a factor of the form $(\varepsilon - \Delta_k)^{-n_k}(\xi - \Delta_k)^{-m_k}$. Averaging of this factor over the frequency detunings yields

$$\int \frac{d\Delta_k}{(\varepsilon - \Delta_k)^{n_k}(\xi - \Delta_k)^{m_k}} = \frac{(-1)^{n_k}(m_k + n_k - 2)!}{(n_k - 1)!(m_k - 1)!}(\xi - \varepsilon)^{1-n_k-m_k}. \quad (10.22)$$

It is at this stage that we perform the ensemble averaging. In the process of the following calculations which include series summation, integration over one of the variables (10.20) or (10.21), and integration over $\eta = \xi + \varepsilon$, we obtain the following expression for the Fourier transform of the lower level population:

$$\rho(\zeta) = j\zeta^{-1} \int_0^\infty \exp(-\theta - 2\pi j\zeta I(\theta))d\theta,$$

$$I(\theta) = \int g(\mu)[I_0(-2z^2\theta) + I_1(-2z^2\theta)]z^2\theta \exp(2z^2\theta)d\mu, \quad (10.23)$$

where $\zeta = \xi - \varepsilon$, I_0 and I_1 are the modified Bessel functions. For brevity we have used the notation $z^2 = \mu^2 \mathcal{E}_0^2/\hbar^2\zeta^2$. In addition in (10.23) we use the spectral density of levels whose matrix elements lie within the unit interval $\delta\mu$ in the vicinity of μ, which is determined by the relation

$$g(\mu) = \int_{-\infty}^\infty g(\Delta)\delta(\mu - \mu(\Delta))d\Delta. \quad (10.24)$$

At small times, i.e., at large values of ζ, when the arguments and exponents of the Bessel functions are small, the averaged expression after integration over θ and after the Fourier transformation over ζ can be reduced to the one obtained previously for the "isolated level-continuum" system (see (10.8)). This is natural since at small times, in accord with the uncertainty principle, the microstructure of the band spectrum cannot be resolved.

At large times the situation is principally different. In this case ζ is small, i.e., the arguments and the exponents of the Bessel functions are large. Using asymptotic expressions for I_0 and I_1 yields:

$$\rho(\zeta) = \frac{j}{\zeta} \int_0^\infty \exp\left[\theta - 2(\pi\theta)^{1/2} \int g(\mu)\frac{\mu\mathcal{E}_0}{\hbar}d\mu\right]d\theta. \quad (10.25)$$

This expression after the inverse Fourier transformation and after integration over θ leads to

$$\rho(\infty) = 1 - \pi N_0 \exp(\pi N_0^2)[1 - \Phi(\pi^{1/2}N_0], \quad (10.26)$$

where Φ is the probability integral and we have used the notation

$$N_0 = \int_0^\infty g(\mu)\frac{\mu\mathcal{E}_0}{\hbar}d\mu. \quad (10.27)$$

In case N_0 is small the lower level population is close to unity:

$$\rho(\infty) = 1 - \pi N_0. \quad (10.28)$$

At large values of N_0 this population is small:

$$\rho(\infty) = 1/2\pi N_0^2. \quad (10.29)$$

Parameter N_0 has an obvious physical meaning. In a two-level system the criterion for the system to be resonant as we have already mentioned many times is that the transition detuning should not exceed the Rabi frequency. If we employ this criterion for "isolated level-band"-type systems in the form of $\mu_k \mathcal{E}_0 / \hbar > \Delta_k$, then under the assumption of uniform distribution of frequency detunings Δ_k, quantity N_0 is the mathematical expectation for the number of levels resonating with the radiation field, since the probability of such an event for the level with matrix element $\mu \pm d\mu/2$ equals in this case $(g(\mu)\mu\mathcal{E}_0/\hbar)d\mu$. In case N_0 is small, the probability of resonance is small and the criterion based on the two-level approximation is valid. Besides, it is evident that in this case the fraction of excited particles is small in accordance with (10.28). This extreme of a band with the matrix elements and the frequency detunings nonsmoothly varying from level to level in fact coincides with the previously considered case of a nondense band (see (10.17) and (10.18)). The band population turns out to be a non-self-averaging quantity.

In the opposite extreme, when N_0 is large, the band population is a self-averaging quantity. In this case the number of levels involved is large and the resonant criterion based on the two-level representation is not valid. The collective effects become essential and bring about the fact that the number of levels really resonating with the field is $N = N_0^2$. This particular circumstance is reflected in formula (10.29).

Let us elucidate all that has been said above as follows. Let the "level-band" system originate from the "level–M-fold-degenerate level" system in the process of removal of degeneracy. Bandwidth Γ is related to the level density g and number M by

$$\Gamma = M/g. \tag{10.30}$$

Given that in the "level-degenerate level" system the transition has the matrix element W, then after the removal of degeneracy in accord with the well-known optical sum rule, $W^2 = \sum_M (\mu_k \mathcal{E}_0/\hbar)^2$. If all μ_k are quantities of the same order, then $W = \mu \mathcal{E}_0 M^{1/2}/\hbar$. All M levels resonate with the field if $W > \Gamma$ and, hence, $\mu\mathcal{E}_0/\hbar > M^{1/2}/g$. This implies that all M levels resonate with the field if the band of the Stark draw $\mu\mathcal{E}_0/\hbar$ contains $M^{1/2}$ levels. In other words this cooperative effect results in the fact that if N_0 levels lie within the Stark draw band it turns out that $N = N_0^2$ levels resonate with the field.

Therefore, at small N_0, (10.27), the probability of excitation, linearly depends on N_0, while at large N_0, the probability for the particle to remain in the ground state is quadratically small over N_0.

Now let us consider a still more involved system of the "band-band" type. Let these bands contain a large number of levels; say, N levels in the lower band and K levels in the upper band. Then the total number of transition matrix elements equals NK, while the number of ψ-functions of different levels equals $N + K$. If the phases of transition matrix elements are essentially different, these differences cannot be compensated for by the appropriate choice of the

ψ-function phases[*]. Then it is reasonable to perform the average over the phases of all the matrix elements of the band-band transition operator.

The behavior of the "band-band"-type systems may significantly differ for narrow and wide bands. The wide bands reveal the relaxation dynamics and equalize their populations exactly in the same way as a single level relaxing into the band does. The degenerate (narrow) bands demonstrate quite another behavior, the process of incoherent nonexponential damping analogous to that described in lecture 3 is more characteristic of the tese bands. The analysis of the problem is rather cumbersome, but, however, typical of spectrally complicated systems. Its essence is the correct description of the interference of all possible excitation channels analogous to that considered in the previous lecture when studying the composite matrix elements. The main difficulty arising in this description apart from the great number of channels is that we are dealing with the excitation of multiphoton transitions via intermediate resonant levels $\Delta_k = 0$ (see (9.28)). So we should also make allowance for the width of the levels, i.e., for their finite lifetime caused by the same radiative transitions (see (10.9)). Due to this fact the frequency detunings acquire the imaginary additions accounted for by the difference between the eigenenergy states and the eigenstates of the quasi-energy operator. These additions could be called polarization additions by analogy with those arising in the many-particle problem.

The Schrödinger equation for the "band-band" system is written in the form

$$j\hbar\dot{\psi}_n = \hbar\Delta_n\psi_n + \sum_k \frac{\mu_{nk}\mathcal{E}_0}{2}\psi_k \,,$$

$$j\hbar\dot{\psi}_k = \hbar\Delta_k\psi_k + \sum_n \frac{\mu_{kn}\mathcal{E}_0}{2}\psi_n \,. \tag{10.31}$$

The complex conjugated ψ-functions satisfy the system

$$-j\hbar\dot{\psi}_n^* = \hbar\Delta_n\psi_n^* + \sum_k \frac{1}{2}\mu_{kn}\mathcal{E}_0\psi_k^* \,,$$

$$-j\hbar\dot{\psi}_k^* = \hbar\Delta_k\psi_k^* + \sum_k \frac{1}{2}\mu_{nk}\mathcal{E}_0\psi_n^* \,. \tag{10.32}$$

The initial conditions are

$$\psi_n(t=0) = \psi_n^*(t=0) = \delta_{n,0} \,, \quad \psi_k(t=0) = \psi_k^*(t=0) = 0 \,. \tag{10.33}$$

Indices n and k run the values

$$n = 0, \pm 1, \pm 2, \ldots, \pm N \gg 1 \,,$$
$$k = 0, \pm 1, \pm 2, \ldots, \pm K \gg 1 \,. \tag{10.34}$$

The Fourier transformation of these quantities along with the initial conditions yields the system

---

[*] Such a situation is typical of spectra corresponding to stochastic motion.

$$\varepsilon\psi_n(\varepsilon) = \Delta_n\psi_n(\varepsilon) + \frac{\mathcal{E}_0}{2\hbar}\sum_k \mu_{nk}\psi_k(\varepsilon) + j\delta_{n,0},$$

$$\varepsilon\psi_k(\varepsilon) = \Delta_k\psi_k(\varepsilon) + \frac{\mathcal{E}_0}{2\hbar}\sum_n \mu_{kn}\psi_n(\varepsilon),$$
(10.35)

$$\xi\psi_n(\xi) = \Delta_n\psi_n(\xi) + \frac{\mathcal{E}_0}{2\hbar}\sum_k \mu_{nk}\psi_k(\xi) - j\delta_{n,0},$$

$$\xi\psi_k(\xi) = \Delta_k\psi_k(\xi) + \frac{\mathcal{E}_0}{2\hbar}\sum_n \mu_{kn}\psi_n(\xi),$$
(10.36)

The solutions of these equations can be represented as a power series over $\frac{\mu_{nk}\mathcal{E}_0}{2\hbar}$:

$$\psi_n(\varepsilon) = \frac{j}{\varepsilon - \Delta_0}\delta_{n,0} + \sum_k \frac{j}{\varepsilon - \Delta_n}\frac{\mu_{nk}\mathcal{E}_0}{2\hbar(\varepsilon - \Delta_k)}\frac{\mu_{n0}\mathcal{E}_0}{2\hbar(\varepsilon - \Delta_0)} + \cdots,$$

$$\psi_k(\varepsilon) = \frac{j}{\varepsilon - \Delta_0}\frac{\mu_{k0}\mathcal{E}_0}{2\hbar(\varepsilon - \Delta_k)}$$
(10.37)

$$+ \sum_{k',n} \frac{j}{\varepsilon - \Delta_0}\frac{\mu_{0k'}\mathcal{E}_0}{2\hbar(\varepsilon - \Delta_{k'})}\frac{\mu_{k'n}\mathcal{E}_0}{2\hbar(\varepsilon - \Delta_n)}\frac{\mu_{nk}\mathcal{E}_0}{2\hbar(\varepsilon - \Delta_k)} + \cdots.$$

$$\psi_n(\xi) = \frac{-j}{\xi - \Delta_0}\delta_{n,0} + \sum_k \frac{-j}{\xi - \Delta_n}\frac{\mu_{kn}\mathcal{E}_0}{2\hbar(\xi - \Delta_k)}\frac{\mu_{0k}\mathcal{E}_0}{2\hbar(\xi - \Delta_0)} + \cdots,$$

$$\psi_k(\xi) = \frac{-j}{\xi - \Delta_0}\frac{\mu_{k0}\mathcal{E}_0}{2\hbar(\xi - \Delta_k)}$$
(10.38)

$$+ \sum_{k',n} \frac{-j}{\xi - \Delta_0}\frac{\mu_{k'0}\mathcal{E}_0}{2\hbar(\xi - \Delta_{k'})}\frac{\mu_{nk'}\mathcal{E}_0}{2\hbar(\xi - \Delta_n)}\frac{\mu_{kn}\mathcal{E}_0}{2\hbar(\xi - \Delta_k)} + \cdots.$$

It is convenient to use the diagram technique in the following analysis. For this purpose let us represent each term of the series for $\psi(\varepsilon)$ graphically in accordance with the following scheme (Fig. 10.2 a). We shall represent each factor $\mu_{nk}\mathcal{E}_0/2\pi$ entering the terms of the series (10.37) by a straight line drawn between the quasi-energy levels n and k which lie in the bands N and K, respectively. Let us denote factors $(\varepsilon - \Delta_n)^{-1}$ and $(\varepsilon - \Delta_k)^{-1}$, respectively, by points on these lines. In Fig. 10.2 a we have shown as an example one of the terms corresponding to the second term of the second series of (10.37). The summation over all possible trajectories of different length which pass through all the possible levels yields $\psi(\varepsilon)$.

Note that in fact the same procedure was used earlier when calculating the composite matrix element of a multiphoton transition. The distinction from the case considered in the previous lecture is that in the band-band transitions the frequency detuning Δ_k must not always be considered large and so ε cannot be neglected against them.

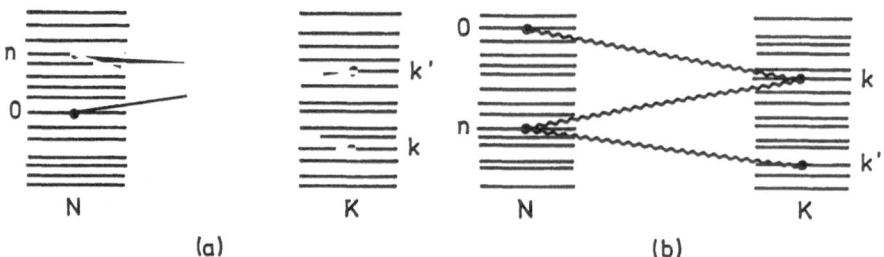

Fig. 10.2. The diagrams of the transition composite matrix elements for $\psi(\varepsilon)$ and $\psi(\xi)$

It is evident that the graphic technique presented in Fig. 10.2 a can be applied for the determination of function $\psi(\xi)$. To distinguish between the diagrams corresponding to functions $\psi(\varepsilon)$ and $\psi(\xi)$ we shall in the latter case denote the transition matrix elements with wavy lines (Fig. 10.2 b).

The population being sought is

$$\rho_n = \frac{1}{(2\pi)^2}\int \psi(\varepsilon)\psi(\xi)e^{-j(\varepsilon-\xi)}d\varepsilon d\xi \equiv \frac{1}{(2\pi)^2}\int_{-\infty}^{\infty}\rho_{nn}(\varepsilon,\xi)e^{-j(\varepsilon-\xi)}d\varepsilon d\xi .$$

(10.39)

Hence, for its determination we should perform the multiplication of series (10.37) and (10.38). The product obtained is the series in terms of the type of the product of some item in the series for $\psi(\varepsilon)$ by some item in that for $\psi(\xi)$. It is convenient to illustrate this operation by the superposition of a diagrams for $\psi(\xi)$ and $\psi(\varepsilon)$, i.e., in the form of diagram with straight and wavy lines as shown, for instance, in Fig. 10.3. For the population to be determined it is necessary to sum all the possible diagrams of that type.

We have seen in the previous lecture when considering the two-photon transition in the system with two intermediate levels (Fig. 9.4) that the destructive interference of the transition channels may lead to their mutual cancellation. Such effects are essential in the multilevel problems where, due to the complexity of the structure of the transition operator matrix elements, the contributions of the majority of channels are compensated for. The net contribution of these channels is sensitive to the microstructure of the spectrum. Side by side with these there are trajectories whose contributions are not compensated for at all. Although there are only a few such trajectories, their total contribution is large, being independent of the spectrum details. It is these trajectories that determine the result of the ensemble average. The contribution of the trajectories which compensate for each other vanishes in the process of ensemble averaging.

Let us illustrate all that has been said above with the following example. Let the positions of the levels be fixed and the matrix elements of the dipole moment operator μ_{nk} corresponding to the transitions between the levels of the upper and lower bands strongly differ for different systems. Let the ensemble averages be $\langle \mu_{nk} \rangle = 0$, $\langle \mu_{nk}^2 \rangle = \mu^2$, and the matrix elements be normally

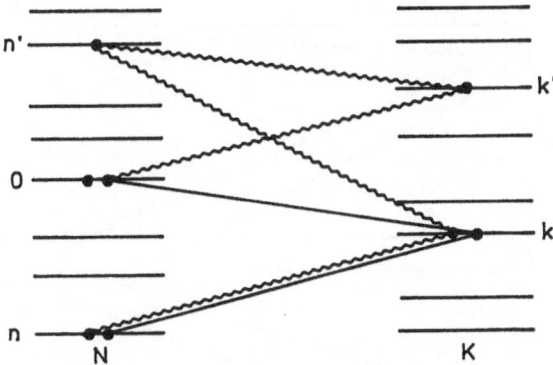

Fig. 10.3. The superposition of the composite matrix element diagrams of the transitions for $\psi(\varepsilon)$ and $\psi(\xi)$

distributed with parameter μ^2. Such an ensemble in which all the matrix elements μ_{nk} are normally distributed with the same dispersion μ^2 is referred to as Wigner's ensemble. Let us average the series for the population (10.39) over Wigner's ensemble. Then each series term in which some transition matrix element occurs only once (or an odd number of times), vanishes after averaging. Such a series term corresponds to the diagrams with single lines of the type $0 \to k$, $0 \to k'$, $k \to n$, $k' \to n'$ (Fig. 10.3). So, all the diagrams analogous to that presented in Fig. 10.3 result in zero contribution. The nonzero contribution after averaging corresponds to the diagrams in which all the transition lines go in pairs, i.e., which result from the transitions with an even number of matrix elements. An example of such a diagram is given in Fig. 10.4. This figure pictures some particular realization of the possible transition $0 \to n$. Much more involved trajectories may correspond to other realizations of this transition. So, it seems expedient to draw such diagrams overlooking the specific level scheme and retaining only the topological structure of the diagram. In such a representation Fig. 10.4 resembles Fig. 10.5. This diagram explicitly shows one of the trajectories of nonzero contribution to transition $0 \to n$.

Calculating the population being sought (10.39), we shall confine ourselves to the diagrams of a simplest topology. First, we shall discount the self-interactions. Then we shall assume that only coupled trajectories are of importance, so the even trajectories of higher order (4,6,...) will be neglected. In addition, the factor $\mu^2 \mathcal{E}_0^2$ will correspond to each section of the coupled trajectory.

Everything said above turns out to come true for large systems (i.e., for the bands containing a large number of levels) if the system is considered within a limited time interval. The point is that at the time moment t, of importance are only those trajectories where the number of sections does not exceed that of the Rabi oscillation periods which can be nested in the time interval t. This corresponds to the rather simple fact that each transition takes the time of the order of the Rabi oscillation period. Then the total number of transitions

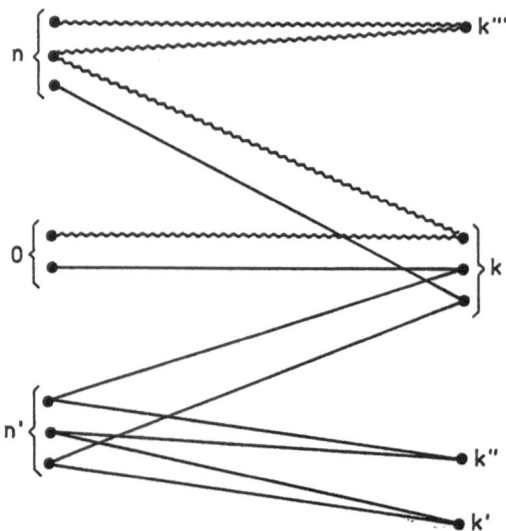

Fig. 10.4. Diagram of $0 \to n$ transition with even number of composite matrix elements. One of the possible implementations of the transition is shown

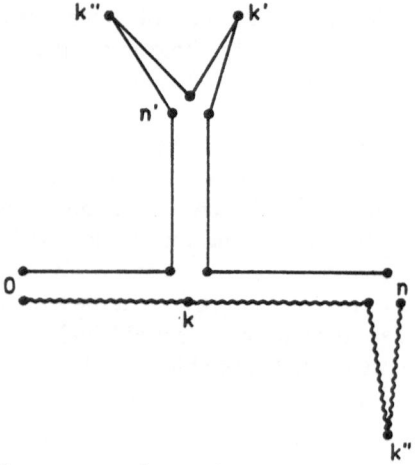

Fig. 10.5. Topological structure of one of the trajectories of the $0 \to n$ transition

within the time interval t, i.e., the number of trajectory sections, will not exceed $t\mu\mathcal{E}_0/\hbar$. However, in case this number is much less than that of the levels in the lower and upper bands which are resonant with respect to the field (because of the cooperative effect, the latter number for dense bands equals $(g\mu\mathcal{E}_0/\hbar)^2$), then the self-intersecting trajectories and the even trajectories of higher order (4, 6 and so on) represent only a small fraction of all possible trajectories and may be disregarded. Such a consideration is, of course, only an approximation. It fails at long time intervals when the trajectory length becomes so large that the self-intersections cannot be ignored. In this situation the returns of the trajectory to its initial point become essential, so one says that the time reaches the so-called Poincaré return time.

The summation of the perturbation theory series with the terms being determined by the diagrams of the above described topology may be reduced to a summation of geometric progressions and therefore can be easily performed. Note that the diagram that contains several straight or wavy lines starting from the same point, for example, point k in Fig. 10.5, looks like a tree. Such a tree can branch endlessly. The number of trees growing from a particular point is also unlimited (Fig. 10.6).

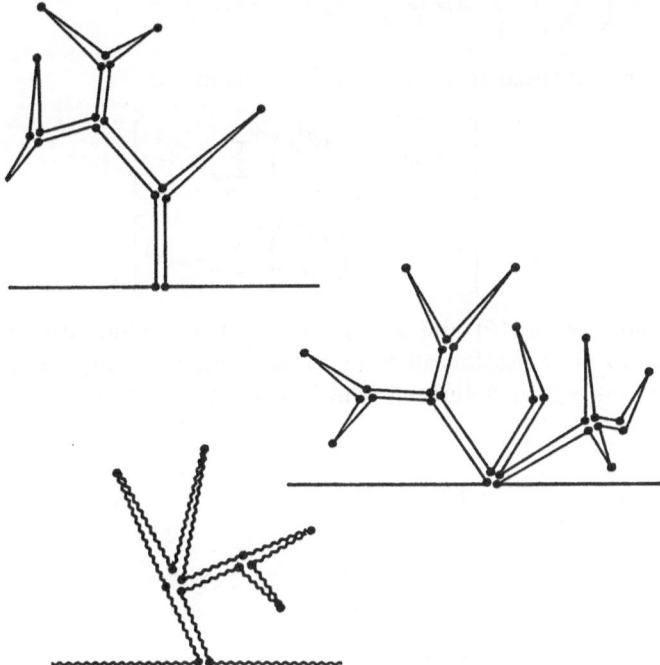

Fig. 10.6. Branching structures of diagrams

Let us denote the sums of all the trees corresponding to levels n or k, i.e., originating at points n or k, by the symbols $x_n(\varepsilon)$ or $x_k(\varepsilon)$ (for wavy lines, $x_n(\xi)$ or $x_k(\xi)$). These operators x will be graphically represented by a circle currounding the given point. Note, that the branching of a tree can be replaced by the graphic image of operator x. Then we have

$$x_n(\varepsilon) = \frac{1}{\varepsilon - \Delta_n} + \frac{1}{\varepsilon - \Delta_n}\left(\frac{\mu\mathcal{E}_0}{\hbar}\right)^2 \sum_k x_k(\varepsilon)\frac{1}{\varepsilon - \Delta_n}$$

$$+ \frac{1}{\varepsilon - \Delta_n}\left(\frac{\mu\mathcal{E}_0}{\hbar}\right)^2 \sum_k x_k(\varepsilon)\frac{1}{\varepsilon - \Delta_n}\left(\frac{\mu\mathcal{E}_0}{\hbar}\right)^2 \sum_{k'} x_{k'}(\varepsilon)\frac{1}{\varepsilon - \Delta_n}$$

$$+ \frac{1}{\varepsilon - \Delta_n}\left(\frac{\mu\mathcal{E}_0}{\hbar}\right)^2 \sum_k x_k(\varepsilon)\frac{1}{\varepsilon - \Delta_n}\left(\frac{\mu\mathcal{E}_0}{\hbar}\right)^2 \sum_{k'} x_{k'}(\varepsilon)\frac{1}{\varepsilon - \Delta_n}$$

$$\times \left(\frac{\mu\mathcal{E}_0}{\hbar}\right)^2 \sum_{k''} x_{k''}(\varepsilon)\frac{1}{\varepsilon - \Delta_n} + \cdots . \tag{10.40}$$

The summation of these geometric progressions gives

$$x_k(\varepsilon) = \left[\varepsilon - \Delta_k - \left(\frac{\mu\mathcal{E}_0}{\hbar}\right)^2 \sum_n x_n(\varepsilon)\right]^{-1},$$

$$x_n(\varepsilon) = \left[\varepsilon - \Delta_n - \left(\frac{\mu\mathcal{E}_0}{\hbar}\right)^2 \sum_k x_k(\varepsilon)\right]^{-1}. \tag{10.41}$$

The expressions for $x_n(\xi)$ and $x_k(\xi)$ are analogous. Quantities x_k, x_n and their pictorial representation allow us to essentially simplify the structure of the series terms which determine the integrand in (10.39). Indeed, due to (10.40),

$$\rho_{kk} = x_0(\varepsilon)x_0(\xi)\left(\frac{\mu\mathcal{E}_0}{\hbar}\right)^2 x_k(\varepsilon)x_k(\xi) + x_0(\varepsilon)x_0(\xi)\left(\frac{\mu\mathcal{E}_0}{\hbar}\right)^2$$

$$\times x_k(\varepsilon)x_k(\xi)\left(\frac{\mu\mathcal{E}_0}{\hbar}\right)^2 \sum_{k'} x_{k'}(\varepsilon)x_k(\xi)\left(\frac{\mu\mathcal{E}_0}{\hbar}\right)^2 \sum_{n'} x_{n'}(\varepsilon)x_{n'}(\xi) + \cdots ,$$

$$\rho_{nn} = x_0(\varepsilon)x_0(\xi)\delta_{0n} + x_0(\varepsilon)x_0(\xi)\left(\frac{\mu\mathcal{E}_0}{\hbar}\right)^2 x_n(\varepsilon)x_n(\xi)\left(\frac{\mu\mathcal{E}_0}{\hbar}\right)^2$$

$$\times \sum_{k'} x_{k'}(\varepsilon)x_{k'}(\xi) + \cdots . \tag{10.42}$$

The summation of these geometric progressions yields

$$\rho_{kk} = x_0(\varepsilon)x_0(\xi)\frac{\left(\frac{\mu\mathcal{E}_0}{\hbar}\right)^2 x_k(\varepsilon)x_k(\xi)}{1 - \left(\frac{\mu\mathcal{E}_0}{\hbar}\right)^2 \sum_{k'} x_{k'}(\varepsilon)x_{k'}(\xi)\left(\frac{\mu\mathcal{E}_0}{\hbar}\right)^2 \sum_{n'} x_{n'}(\varepsilon)x_{n'}(\xi)}$$

$$\rho_{nn} = x_0(\varepsilon)x_0(\xi)\delta_{0n} \tag{10.43}$$

$$+ x_0(\varepsilon)x_0(\xi)\frac{\left(\frac{\mu\mathcal{E}_0}{\hbar}\right)^2 x_n(\varepsilon)x_n(\xi)\left(\frac{\mu\mathcal{E}_0}{\hbar}\right)^2 \sum_k x_k(\varepsilon)x_k(\xi)}{1 - \left(\frac{\mu\mathcal{E}_0}{\hbar}\right)^2 \sum_{k'} x_{k'}(\varepsilon)x_{k'}(\xi)\left(\frac{\mu\mathcal{E}_0}{\hbar}\right)^2 \sum_{n'} x_{n'}(\varepsilon)x_{n'}(\xi)}.$$

Expressions (10.43), (10.41) and (10.39) determine the dynamics of the "band-band" system under consideration, averaged over the ensemble of such systems. The dispersion of quantity ρ_n obtained by such an ensemble-average operation is determined by the self-intersecting diagrams and by the even diagrams of higher order which were neglected. In the framework of the assumptions made above this dispersion is unessential.

In order to progress, (10.41) must be solved. This problem is rather difficult. The solution of (10.41) is possible under quite definite assumptions of the band structure.

Let us consider the case of degenerate bands $\Delta_n = \Delta_k = 0$. Note that it resembles in general the case of coherent damping in a two-level system already discussed in lecture 2. If the multiplicity of the degeneration of the lower band equals that of the upper one, i.e., both bands contain the same number of levels, $N = K$, Eqs. (10.41) may be reduced to a single equation,

$$x = \left[\varepsilon - \left(\frac{\mu\mathcal{E}_0}{\hbar}\right)^2 Nx\right]^{-1}. \tag{10.44}$$

The solution of this equation is

$$x(\varepsilon) = \frac{\varepsilon - [\varepsilon^2 - 4N(\mu\mathcal{E}_0/\hbar)^2]^{1/2}}{2N(\mu\mathcal{E}_0/\hbar)^2}. \tag{10.45}$$

The sign of the bracketed expression corresponds to the requirement $x \to 1/\varepsilon$ at $\mu\mathcal{E}_0 \to 0$ in accordance with (10.41). It is evident that $x(\xi)$ may be written by analogy. Now we should substitute $x(\varepsilon)$ and $x(\xi)$ in (10.43) and then in (10.39).

Let us insert new variables with the relations

$$\varepsilon = 2\frac{\mu\mathcal{E}_0}{\hbar}N^{1/2}\cosh(\chi), \quad \xi = 2\frac{\mu\mathcal{E}_0}{\hbar}N^{1/2}\cosh\phi, \quad t = \frac{\hbar}{2\mu\mathcal{E}_0 N^{1/2}}\tau. \tag{10.46}$$

Then $x(\varepsilon) = e^\chi$ and $x(\xi) = e^\phi$. The following substitution and the summation over degenerate states give the expressions for the upper- and lower-band populations. The sum of these populations turns out to be unity, as one would anticipate, while their difference equals

$$\delta\rho = \int_C (e^{\phi+\chi} + 1)^{-1} \sinh\phi \sinh\chi \exp[j\tau(\cosh\chi - \cosh\phi)]d\chi d\phi. \quad (10.47)$$

The integration path is determined by the paths of integration over ε and ξ, bearing in mind the substitution (10.46). Integral (10.47) is reduced to one of the integral representations of the Bessel functions and we obtain as a result

$$\delta\rho = 2\frac{J_1(\tau)}{\tau} = 2\frac{J_1((\mu\mathcal{E}_0/\hbar)\sqrt{N}\,t)}{(\mu\mathcal{E}_0/\hbar)\sqrt{N}\,t}. \quad (10.48)$$

The type of time dependence obtained is similar to that discussed in lecture 2 for the case of the three-dimensional isotropic distribution of identical dipole moments. This proximity is natural since the problem of the dynamics of the occupation of N-fold degenerate bands can be reduced to the problem of the excitation of N two-level systems by means of the transformation diagonalizing the matrix of the transition operator. The diagonal elements thus obtained are, in fact, the Rabi frequencies of these two-level systems. However, the Rabi frequency distribution in the Wigner ensemble is distinct from (2.68) and has a semicircular form $g(\Omega) \propto (\Omega_{max}^2 - \Omega^2)^{1/2}$. Using such a distribution function we can obtain (10.48) by averaging the two-level population difference $\delta\rho = \cos\Omega t$. The distinction between (10.48) and (2.70) is caused by the difference in the dipole moment (i.e., the Rabi frequency) distribution functions.

Coherent Processes
in Complicated Multilevel Systems (continued)

Two nondegenerate bands. Kinetic equations. Degenerate level-nondegenerate band. Multiband system. Diffusion of populations in bands. Occupation of resonance vicinities. Kinetic coefficients of diffusion. Decay of multiband system in continuum. Competition of the decay and excitation rates. The switching of decay channels by the field. Relaxation of the simple oscillator energy into a thermostat.

This lecture is devoted to a consideration of the excitation dynamics of some particular multilevel systems, the methods used to obtain kinetic equations of the system with wide bands and the methods to solve these equations. We also pay attention to coherence in spectrally involved systems which becomes apparent in both the Lorentz type of population distribution in the vicinity of resonance and the nonexponential type of coherent damping processes, should these systems contain degenerate bands of levels.

In the previous lecture when considering systems of the "single level – dense band" type we have seen that the interference of different excitation paths results in an imaginary additional term to the denominator of the ψ-functions (10.4) and (10.5) of the initial level. This addition is explicitly described by (10.6). In other words, the eigenenergy of this state acquires an imaginary additional term which brings about the exponential decay of the population of the level (10.8). This is just an instance of a common situation typical of involved multilevel systems when the excitation process may, in a certain sense, be taken to be in its initial stage, i.e., when the initial state loses its population to many other unoccupied states, while the reverse processes have not begun yet. An exact description of such an initial stage based on completely reversible equations can be substituted for quite adequately by an approximate one. The latter is based on irreversible equations which lead to the decay of the initial states. The validity of this approximation is, of course, lost at times approaching the so-called Poincaré return time where the interference of the ψ-functions of all the system states can give rise to the inverse flux of populations, which brings the system back to its initial state. As a rule, this time is exponentially large, the exponent index being the number of the levels involved in the process. This resembles the nature of a large fluctuation event in a system in thermodynamic equilibrium

Throughout this book our considerations are, as a rule, restricted to time intervals small as compared to Poincaré time unless special arrangements are made to shorten this time, as for example, in the case of the photon echo (Lecture 3). Note that the self-intersecting diagrams and the even diagrams of high order (4, 6 and so on) become essential only approaching the Poincaré time. These diagrams were discarded when summing the series for the populations in the previous lecture.

The imaginary additions to the eigenenergy values leading to the irreversible decay of states within the time intervals considered are taken into account by $x(\varepsilon)$ and $x(\xi)$ (10.40). Values $x(\varepsilon)$ and $x(\xi)$ have been obtained as a result of the already mentioned summation of diagrams. This becomes apparent most clearly in the system of two nondegenerate bands with equal numbers of levels and with equal densities of levels.

Indeed, let us turn to Eqs. (10.41). Since the bands are identical, $\sum_k x_k = \sum_n x_n = Q$. From (10.41) quantity Q thus constructed seen to equal

$$Q = \sum_n \left[\varepsilon - \Delta_n - \frac{\mu^2 \mathcal{E}_0^2}{2\hbar^2} Q \right]^{-1}. \tag{11.1}$$

Assuming for simplicity $\Delta_n = n\pi\beta$, i.e., assuming the levels in the band to be equidistant, we obtain for Q (the assumption made is not restrictive in the case of rather dense bands in which many levels lie in the Stark draw region)

$$Q\beta = \sum_n \left[\frac{\varepsilon}{\beta} - \left(\frac{\mu\mathcal{E}_0}{2\hbar} \right)^2 \frac{Q}{\beta} - n\pi \right]^{-1} = \cot \left[\frac{\varepsilon}{\beta} - \left(\frac{\mu\mathcal{E}_0}{2\beta\hbar} \right)^2 Q\beta \right]. \tag{11.2}$$

Let us solve this transcendental equation with respect to $Q\beta$ by taking advantage of the large number of levels in the Stark draw band:

$$(\mu\mathcal{E}_0/2\hbar\beta)^2 \gg 1. \tag{11.3}$$

Note that $\beta = 1/\pi g$ and, hence, (11.3) is equivalent to condition (10.12) for the band to be dense.

We now search for solutions independent of the spectrum microstructure. If condition (11.3) holds, such solutions are

$$Q = \pm j/\beta = \pm j\pi g, \tag{11.4}$$

since the cotangent of the large imaginary quantity equals $\pm j$. This solution is accurate up to a factor exponential with respect to parameter $(\mu\mathcal{E}_0/\beta\hbar)^2$, i.e., within a factor inversely proportional to the Poincaré return time. In accordance with the causality principle the sign, either "+" or "−", should be chosen depending on which of the Fourier variables, ε or ξ, (11.2) corresponds to; $Q_\varepsilon = -j\pi g$ while $Q_\xi = +j\pi g$. Then

$$x_n(\varepsilon) = \left[\varepsilon - \Delta_n - j \left(\frac{\mu\mathcal{E}_0}{2\hbar} \right)^2 \pi g \right]^{-1}, \tag{11.5}$$

$$x_n(\xi) = \left[\xi - \Delta_n + j \left(\frac{\mu\mathcal{E}_0}{2\hbar} \right)^2 \pi g \right]^{-1}, \tag{11.6}$$

The expressions obtained within an accuracy up to $\Delta_n \neq 0$ are similar to the ψ-function of the level decaying under the action of an external field into the upper band in the "isolated level-band" system (see (10.4)–(10.6)). This analogy is not just casual. It reflects the fact that functions $x_n(\varepsilon)$ and $x_n(\xi)$ would be the ψ-functions of the nth level, should we take into consideration only the decay of this level without allowance for its population on account of other levels. It is the last circumstance that we have taken into account when summing series (10.42). The substitution of (11.5) and (11.6) into (10.43) at $\Delta_0 = 0$, along with identity

$$\sum_m (a - m\pi)^{-1}(b - m\pi)^{-1} = (a - b)^{-1} \sum_m [(b - m\pi)^{-1} - (a - m\pi)^{-1}]$$

$$= (a - b)^{-1}(\cot b - \cot a) \tag{11.7}$$

and the fact that the cotangent of a large positive imaginary quantity equals $-j$ and that of a negative one equals $+j$, give

$$\rho_{kk} \left(\frac{\mu \mathcal{E}_0}{2\hbar} \right)^2 ABC, \tag{11.8}$$

$$A = \left[\varepsilon - j \left(\frac{\mu \mathcal{E}_0}{2\hbar} \right)^2 \pi g \right]^{-1} \left[\xi + j \left(\frac{\mu \mathcal{E}_0}{2\hbar} \right)^2 \pi g \right]^{-1},$$

$$B = \left[\varepsilon - \Delta_k - j \left(\frac{\mu \mathcal{E}_0}{2\hbar} \right)^2 \pi g \right]^{-1} \left[\xi - \Delta_k + j \left(\frac{\mu \mathcal{E}_0}{2\hbar} \right)^2 \pi g \right]^{-1},$$

$$C = \left[\varepsilon - \xi - 2j \left(\frac{\mu \mathcal{E}_0}{2\hbar} \right)^2 \pi g \right]^2 (\varepsilon - \xi)^{-1} \left[\varepsilon - \xi - 4j \left(\frac{\mu \mathcal{E}_0}{2\hbar} \right)^2 \pi g \right]^{-1}.$$

When summing over k we have, due to (11.7),

$$\sum_k \rho_{kk} = 2j\pi g \left(\frac{\mu \mathcal{E}_0}{2\hbar} \right)^2 D,$$

$$D = \frac{\left[\varepsilon - \xi - 2j \left(\frac{\mu \mathcal{E}_0}{2\hbar} \right)^2 \pi g \right] (\varepsilon - \xi)^{-1}}{\left[\varepsilon - \xi - 4j \left(\frac{\mu \mathcal{E}_0}{2\hbar} \right)^2 \pi g \right] \left[\varepsilon - j \left(\frac{\mu \mathcal{E}_0}{2\hbar} \right)^2 \pi g \right] \left[\xi + j \left(\frac{\mu \mathcal{E}_0}{2\hbar} \right)^2 \pi g \right]}. \tag{11.9}$$

After replacing $\varepsilon - \xi = \zeta$ and the following inverse Fourier transformation (10.39) by means of integration over ε and then over ζ, we obtain

$$\sum_k \rho_k = \frac{1}{2} \left\{ 1 - \exp \left[-\pi g \left(\frac{\mu \mathcal{E}_0}{2\hbar} \right)^2 t \right] \right\}. \tag{11.10}$$

By analogy,

$$\sum_n \rho_n = \frac{1}{2} \left\{ 1 + \exp \left[-\pi g \left(\frac{\mu \mathcal{E}_0}{2\hbar} \right)^2 t \right] \right\}. \tag{11.11}$$

Note that if, without summing over k, we expand (10.8) into simple fractions and then perform the inverse Fourier transformation, we can obtain the distribution of populations among the levels of the kth band. As one would anticipate (compare with (10.15)) this is the Lorentz distribution of width $\pi g (\mu \mathcal{E}_0 / 2\hbar)^2$.

Relations (11.10) and (11.11) show that the expressions for total populations of the upper and lower bands $\rho_2 = \sum_k \rho_k$ and $\rho_1 = \sum_n \rho_n$ are, respectively, the solutions of the following kinetic equations:

$$\dot{\rho}_1 = - \left(\frac{\mu \mathcal{E}_0}{\hbar} \right)^2 \pi g \rho_1 + \left(\frac{\mu \mathcal{E}_0}{\hbar} \right)^2 \pi g \rho_2 + \delta(t),$$
$$\dot{\rho}_2 = \left(\frac{\mu \mathcal{E}_0}{\hbar} \right)^2 \pi g \rho_1 - \left(\frac{\mu \mathcal{E}_0}{\hbar} \right)^2 \pi g \rho_2. \tag{11.12}$$

The reason is the linear topology of diagrams (10.42), which allows us to express the populations of the upper and lower bands in terms of each other as follows:

$$\sum_n \rho_{nn}(\varepsilon, \xi) = x_0(\varepsilon) x_0(\xi) + \sum_n x_n(\varepsilon) x_n(\xi) \left(\frac{\mu \mathcal{E}_0}{2\hbar} \right)^2 \sum_k \rho_{kk}(\varepsilon, \xi),$$
$$\sum_k \rho_{kk}(\varepsilon, \xi) = \sum_k x_k(\varepsilon) x_k(\xi) \left(\frac{\mu \mathcal{E}_0}{2\hbar} \right)^2 \sum_n \rho_{nn}(\varepsilon, \xi). \tag{11.13}$$

Substituting (11.5) and (11.6), bearing in mind (11.7) and inverting the Fourier transformation by analogy with (11.10) and (11.11) we obtain the system of kinetic Eqs. (11.12) irreversible in time. These equations are written for the bands of the same level density g. However, they can be generalized to cover the case of different level densities: g_1, for the lower band and g_2, for the upper one. In this case the same procedure yields the system

$$\dot{\rho}_1 = - \left(\frac{\mu \mathcal{E}_0}{\hbar} \right)^2 \pi g_2 \rho_1 + \left(\frac{\mu \mathcal{E}_0}{\hbar} \right)^2 \pi g_1 \rho_2 + \delta(t),$$
$$\dot{\rho}_2 = \left(\frac{\mu \mathcal{E}_0}{\hbar} \right)^2 \pi g_2 \rho_1 - \left(\frac{\mu \mathcal{E}_0}{\hbar} \right)^2 \pi g_1 \rho_2. \tag{11.14}$$

In other words, the dynamics of the populations of narrow resonance vicinities is described by balance Eqs. (11.14). The transition probabilities or else the kinetic equations are, in fact, the products of the mean-square matrix element of the transition between the levels belonging to these vicinities by the density of states in the vicinities the transitions are directed to. So, the kinetic coefficient for the transition between bands 1 and 2 is

$$D_{12} = \pi \left\langle \left(\frac{\mathcal{E}_0 \mu_n^m}{\hbar} \right)^2 \right\rangle_{n,m} g_2 , \qquad (11.14\,a)$$

where index n corresponds to the first band and m, to the second one; g_2 is the level density in the second band m in the vicinity of resonance and the averaging $\langle \rangle_{n,m}$ is performed over the levels belonging to the resonance vicinities.

We have considered two opposite extremes of the interaction of resonant radiation with a two-band system. In the case of two degenerate bands the problem has been reduced to the solution of the Schrödinger equation for the ensemble of two-level systems with subsequent averaging over Rabi's frequencies (the coherent damping). In the case of two nondegenerate bands we have arrived at the kinetic (rate) equation, i.e., at the equation of the population balance.

However, the reduction of the problem to such simple equations is far from always possible. An example of where it is possible is the case of the. "degenerate level – nondegenerate band of levels" system.

Let the lower level be degenerate and the band above this level consist of equidistant levels. Then (10.41) leads to

$$x_n(\varepsilon) = \left[\varepsilon - j \left(\frac{\mu \mathcal{E}_0}{2\hbar} \right)^2 \frac{1}{\beta} \right]^{-1} , \quad x_n(\xi) = \left[\xi + j \left(\frac{\mu \mathcal{E}_0}{2\hbar} \right)^2 \frac{1}{\beta} \right]^{-1} ,$$

$$x_k(\varepsilon) = \left\{ \varepsilon - \pi\beta k - N \left(\frac{\mu \mathcal{E}_0}{2\hbar} \right)^2 \left[\varepsilon - j \left(\frac{\mu \mathcal{E}_0}{2\hbar} \right)^2 \frac{1}{\beta} \right]^{-1} \right\}^{-1} , \qquad (11.15)$$

$$x_k(\xi) = \left\{ \xi - \pi\beta k - N \left(\frac{\mu \mathcal{E}_0}{2\hbar} \right)^2 \left[\xi + j \left(\frac{\mu \mathcal{E}_0}{2\hbar} \right)^2 \frac{1}{\beta} \right]^{-1} \right\}^{-1} ,$$

where N is the degeneracy of the lower level. The substitution of (11.15) into (10.43), the summation with the help of (11.7) and oft-mentioned inverse Fourier transformation all give rise to

$$\rho_2 = \sum_k \rho_k(t) = \pi g \left(\frac{\mu \mathcal{E}_0}{2\hbar} \right)^2 \int_0^t J_0 \left(N^{1/2} \frac{\mu \mathcal{E}_0}{2\hbar} t \right) \exp\left[-\pi g \left(\frac{\mu \mathcal{E}_0}{2\hbar} \right)^2 t \right] dt .$$
$$(11.16)$$

The expression obtained is seen to bear the typical features of both the coherent damping (10.48) in the form of the Bessel function and the irreversible decay of the lower band in the form of the exponential term. In the asymptotic limit of large times, (11.16) brings about

$$\rho_2(t \to \infty) = \pi g \frac{\mu \mathcal{E}_0}{\hbar} \left[\left(\pi g \frac{\mu \mathcal{E}_0}{\hbar} \right)^2 + N \right]^{-1/2} . \qquad (11.17)$$

Deriving (11.17) we take into account the condition for the band to be dense (10.12). However, this condition alone is insufficient for $\rho_2(t \to \infty) \to 1$. It should be supplemented by the condition that the number of levels belonging to the band and resonating with the field (which, due to the cooperative effect, equals $(\pi g \mu \mathcal{E}_0/\hbar)^2$) must be well in excess of the degeneracy of the lower level N.

In the opposite extreme, $N \gg (\pi g \mu \mathcal{E}_0/\hbar)^2 \gg 1$,

$$\rho_2(t \to \infty) = \pi g \frac{\mu \mathcal{E}_0}{\hbar} N^{-1/2} . \tag{11.18}$$

This extreme may be illustrated as follows. Suppose we were successful transforming the "N-fold degenerate level-dense band"-system into N systems of the "isolated level-band" type. This makes the root-mean-square value of the matrix element μ in each of the systems increase by the factor of $N^{1/2}$ due to the spectroscopic sum rule and the fact that the level density in each of the bands decreases by factor N. Then the width of the Stark draw region in each of these systems is comparable to, or less than, the mean interlevel distance. Hence, the mathematical expectation of the resonance (compare with (10.17)) comprises $(\pi g/N)(\mu N^{1/2})\mathcal{E}_0/\hbar = \pi g \mu \mathcal{E}_0 \hbar^{-1} N^{-1/2}$. As distinct from the "level-band" system (see (10.16)–(10.18)) the question of self-averaging does not arise here due to the large ($N \gg 1$) number of systems which the average is performed over.

Now let us consider the system of a large number of bands coupled via successive (cascade) transitions. We may describe the population dynamics in such a system by the balance equations

$$\dot{\rho}_n = -\pi g_{n+1} \frac{\mu_{n+1,n}^2 \mathcal{E}_0^2}{\hbar^2} \rho_n - \pi g_{n-1} \frac{\mu_{n-1,n}^2 \mathcal{E}_0^2}{\hbar^2} \rho_n$$
$$+ \pi g_n \frac{\mu_{n,n+1}^2 \mathcal{E}_0^2}{\hbar^2} \rho_{n+1} + \pi g_n \frac{\mu_{n,n-1}^2 \mathcal{E}_0^2}{\hbar^2} \rho_{n-1} . \tag{11.19}$$

These equations can be derived from (11.13) similarly to (11.14) and (11.12).

As in the case of a two-band system, (11.10) and (11.11), the populations in the bands are concentrated in the vicinities of the resonances. Their distribution is of the Lorentz type, the distribution width being determined by the values of g and μ near the resonance region. The resonant position of the band region being occupied is determined by the initial starting level of the first band and by the energy of the radiation quantum $\hbar \omega$ (Fig. 11.1). The relative sharpness of the resulting distribution permits us to use the resonant values of g and μ in case the band is rather broad and nonuniform, i.e., in case the matrix element μ and the level density g significantly differ for different band regions.

Therefore, the dynamics of the excitation of the systems with many broad bands can be analyzed with the help of the kinetic equation, if the condition of quasi-continuum existence is valid and the Stark width of the resonance encompasses many levels. This along with the determination of the applicability region of such a consideration (the time interval is less than the Poincaré

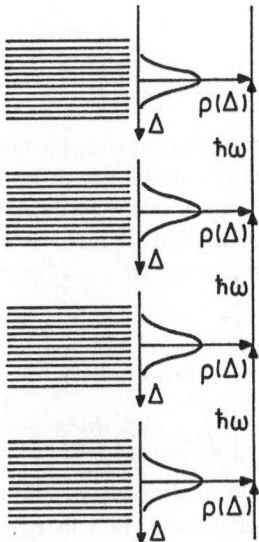

Fig. 11.1. The cascade excitation of a multiband system. The concentration of populations in the resonant vicinities $\rho(\Delta)$ is shown schematically

return time) is the most essential result of the examination of multilevel systems.

Let us look into some particular cases.

Let all $\mu_{ik} = \mu$ and $g_n = g$ be the same for all the bands. Let us insert the dimensionless time τ by the replacement $\tau = \pi g(\mu \mathcal{E}_0/\hbar)^2 t$. Then (11.19) turns to be

$$\dot{\rho}_n = -2\rho_n + \rho_{n+1} + \rho_{n-1} , \qquad (11.20)$$

where the dot above the symbol stands for the time derivative. Let us try the solution in the form $\rho_n = \widehat{\rho}_n \exp(-2\tau)$. Then we have

$$\dot{\tilde{\rho}}_n = \tilde{\rho}_{n+1} + \tilde{\rho}_{n-1} . \qquad (11.21)$$

This is the recurrent relation which defines the modified Bessel function $I_n(2\tau)$, $\rho_n = I_n(2\tau)$. As a result,

$$\rho_n = e^{-2\tau} I_n(2\tau) . \qquad (11.22)$$

Using the well-known integral representation of the modified Bessel function

$$I_n(z) = \frac{1}{2\pi} \int_{-\pi}^{\pi} \exp((z \cos \theta - jn\theta)d\theta , \qquad (11.23)$$

we rewrite (11.22) in the form

$$\rho_n(\tau) = \frac{1}{2\pi} \int_{-\pi}^{\pi} \exp[-2\tau(1 - \cos \theta) - jn\theta]d\theta . \qquad (11.24)$$

let us consider the case of large τ:

$$\tau = \pi g \frac{\mu^2 \mathcal{E}_0^2}{\hbar^2} t \gg 1 . \tag{11.25}$$

This formula says that we are considering times greater than the time required for one transition. At large τ, only small θ are of importance in the expression under the integral sign in (11.24), i.e., we may expand $1 - \cos\theta \approx \theta^2/2$. Then, complementing $\theta^2 + jn\theta/\tau$ to the full square $(\theta + jn/2\tau)^2$ by adding and subtracting term $n^2/4\tau^2$ and replacing $(\theta + jn/2\tau)^2 = u^2$, we obtain

$$\rho_n(\tau) \approx \frac{1}{2\pi} \exp\left[-\frac{n^2}{4\pi}\right] \int_{-\pi-jn/2\tau}^{\pi-jn/2\tau} e^{-u^2} \frac{du}{\sqrt{\tau}}$$

$$\approx \frac{1}{2\pi\sqrt{\tau}} \exp\left[-\frac{n^2}{4\tau}\right] \int_{-\infty}^{\infty} e^{-u^2} \, du = \frac{1}{\sqrt{4\pi\tau}} e^{-n^2/4\tau} . \tag{11.26}$$

The solution obtained is of the diffusion type. The population blurs over the bands with time, remaining, of course, in the resonant domain. This is of no surprise since the initial equation (11.20) is the discrete analogue of the diffusion equation,

$$\partial\rho/\partial t = D \partial^2 \rho/\partial x^2 , \tag{11.27}$$

where D is the diffusion coefficient and the continuous variable stands for the discrete variable n. Such a replacement of (11.20) by (11.27) is possible owing to the fact that the population ρ being sought is real and everywhere positive.

At $\rho(x, t = 0) = \delta(x)$ and $\rho(x, t < 0) = 0$, the solution of (11.27) is well known

$$\rho = \exp\left[-\frac{x^2}{4Dt}\right] (4\pi Dt)^{-1/2} . \tag{11.28}$$

The comparison of (11.25)–(11.28) leads to the value of the diffusion coefficient of the population over the bands

$$D = \pi g \mu^2 \mathcal{E}_0^2/\hbar^2 . \tag{11.29}$$

Reformulating, this means that within the time interval t the radiation occupies all the bands with the number up to

$$n = \left(4\pi g \frac{\mu^2 \mathcal{E}_0^2}{\hbar^2} t\right)^{1/2} . \tag{11.30}$$

Now let the level density be the same in all the bands, and the matrix element μ increase with the band number in proportion to the square root of this number $\mu_{n-1,n}^2 = n\mu_{0,1}^2$. Then (11.19) takes on the form

$$\dot\rho_n = -(2n+1)\rho_n + n\rho_{n-1} + (n+1)\rho_{n+1} + \delta_{n,0}\delta(t) . \tag{11.31}$$

Here, as in the previous example, the derivative is being taken with respect to dimensionless time (11.25) with the evident distinction that μ^2 should be replaced by $\mu_{0,1}^2$. The initial condition in (11.31) is $\rho(0) = \delta_{n,0}$. This equation is the diffusion equation with the diffusion coefficient linearly increasing with the spatial coordinate. We give one of the methods to solve this equation (see (11.42)).

Let us consider the difference

$$\rho_n - \rho_{n-1} = \mathcal{P}_n \,. \tag{11.32}$$

Then (11.31) can be rewritten in the form

$$\sum_0^n \dot{\mathcal{P}}_n = -n\mathcal{P}_n + (n+1)\mathcal{P}_{n+1} + \delta_{n,0}\delta(t) \,. \tag{11.33}$$

Let us take advantage of the generating function method. For this purpose we introduce

$$\Phi(x,\tau) = \sum_n x^n \mathcal{P}_n \,. \tag{11.34}$$

Let us multiply (11.33) by x^n and perform the summation over all n. Then, noting that

$$\sum_{n=0}^{\infty} x^n \sum_0^n \dot{\mathcal{P}}_k = \sum_{m=0}^{\infty} \sum_{k=0}^{\infty} \dot{\mathcal{P}}_k x^{k+m} = \sum_{m=0}^{\infty} x^m \sum_{k=0}^{\infty} \dot{\mathcal{P}}_k x^k = \frac{1}{1-x}\dot{\Phi}(x,\tau) \,,$$

$$\sum_n n\mathcal{P}_n x^n = x\frac{\partial}{\partial x}\Phi(x,\tau), \quad \sum_n (n+1)x^n \mathcal{P}_{n+1} = \frac{\partial}{\partial x}\Phi(x,\tau) \,,$$

$$\sum_n \delta_{n,0}\delta(\tau) = \delta(\tau) \,,$$

we obtain the following equation for the generating function

$$(1-x)^{-1}\dot{\Phi}(x,\tau) = -x\frac{\partial}{\partial x}\Phi(x,\tau) + \frac{\partial}{\partial x}\Phi(x,\tau) + \delta(\tau) \,. \tag{11.35}$$

We replace the variable

$$u = (1-x)^{-1} \,. \tag{11.36}$$

Then (11.35) takes on the form

$$\dot{\Phi} = \Phi'_u + u^{-1}\delta(\tau) \,. \tag{11.37}$$

The solution of this equation is of the form

$$\Phi(x,\tau) = \theta(\tau)[\tau + (1-x)^{-1}]^{-1} \,, \tag{11.38}$$

where $\theta(\tau)$ is the Heaviside function, which turns to zero at $\tau < 0$ and equals unity at $\tau \geq 0$.

Quantities $\dot{\mathcal{P}}_k$ can be obtained from the generating function $\Phi(x,\tau)$ with the help of an inverse transformation of the type

$$\mathcal{P}_k(\tau)\frac{1}{2\pi j}\int_C x^{-(k+1)}\Phi(x,\tau)dx \,, \tag{11.39}$$

where the integration path C rounds point $x = 0$ in the positive direction. Hence:

$$\rho_n = \sum_{k=0}^{n} \mathcal{P}_k = \frac{1}{2\pi j} \int_C \frac{1 - x^{-(n+1)}}{x - 1} \Phi(x, \tau) dx. \tag{11.40}$$

The substitution of (11.38) in (11.40) gives

$$\rho_n = \frac{1}{2\pi j} \frac{\theta(\tau)}{\tau} \int_C \frac{1 - x^{-(n+1)}}{x - 1 - 1/\tau} dx. \tag{11.41}$$

The integral of the first term in the numerator of the integrand (11.41) is regular inside the integration contour and, therefore, vanishes. The second term yields

$$\rho_n(\tau) \frac{\theta(\tau)}{\tau} \left(1 + \frac{1}{\tau}\right)^{-(n+1)}. \tag{11.42}$$

In the limit of large times, $\tau \gg 1$

$$\rho_n(\tau) = \frac{\theta(\tau)}{\tau} \exp\left(-\frac{n+1}{\tau}\right). \tag{11.43}$$

If we come back to dimensional time $t = \tau \hbar^2 / \pi g \mu_{01} \mathcal{E}_0^2$ we obtain

$$\rho_n(t) = \frac{\theta(\tau)}{\tau} \frac{\hbar^2}{\pi g \mu_{01}^2 \mathcal{E}_0^2} \exp\left(-\frac{n+1}{\tau} \frac{\hbar^2}{\pi g \mu_{01}^2 \mathcal{E}_0^2}\right). \tag{11.44}$$

As in the previous case (11.31) correlates with its diffusion analogue

$$\frac{\partial}{\partial t} \rho = \frac{\partial}{\partial x} D(x) \frac{\partial}{\partial x} \rho. \tag{11.45}$$

The solution of this equation coincides with (11.44) if D is given by formula (11.29) with allowance made for $\mu^2 = \mu_{01}^2$. In fact, we are dealing with the diffusion of a population in bands with the linearly growing diffusion coefficient $D(x)$.

The examples considered are only instances of a more general rule that states that the diffusion of particles in discrete bands can be replaced by continuous diffusion, if during the time of the interaction with radiation a large number of transitions occur. This is rather convenient since the continuous diffusion equations can often be essentially easier to solve. By the way, note that the irreversibility of the equations we use, considering the spectrally involved multilevel systems within the time intervals much shorter than the Poincaré return time essentially restricts the character of coherence of the processes considered. The residual manifestation of the interaction coherence is the sharp distribution of particles among the levels of the bands being excited in the resonance vicinities. The dynamics of the population of the resonance vicinities is determined by the irreversible balance equations with kinetic coefficients of the form (11.29). In other words, coherence becomes apparent in involved systems only as the requirement of the resonance.

Among the problems of coherent interaction of radiation with spectrally involved multilevel systems is the case of a multiband system in which each level is coupled with the continuum (Fig. 11.2). Even though the coupling of the level n, m, where n is the band number and m is the number of the level in the band , with the continuum of the density $g_{n,m}$ may not be of a radiative type , this level decays into continuum similarly to what has been shown in the previous lecture for the case of radiative interaction of the isolated level with the dense band (see (10.8) and (10.15)). Hence, in this situation the total populations of all the bands are not conserved in time. The populations of the narrow vicinities of the resonances in the bands flow down into the corresponding narrow regions of the continuum. The rate of this flow is determined by the decay rate of the level coupled with the continuum, the rate being averaged over the populated resonance vicinity

$$\Gamma_n = \langle \gamma_{n,m}^2 g_{n,m} \rangle_m . \tag{11.46}$$

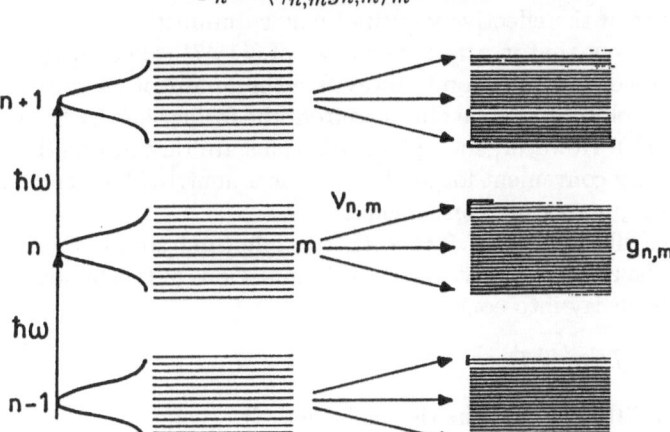

Fig. 11.2. Scheme of the excitation multiband system in which each level of every band is coupled with the continuum. Level m of band n decays at the rate of $\gamma_{n,m}$ into the continuum, the level density of which comprises $g_{n,m}$ in the vicinity of n, m. The width of the continuum draw region equals $\gamma_{n,m}^2 g_{n,m}$

The dynamics of the population of the nth band is then determined by (11.19) with the additional term $-\Gamma_n \rho_n$. If during the typical decay time of the band levels Γ^{-1} there occur many radiative transitions between the bands, the equation in finite differences thus obtained may be replaced with the continuous diffusion equation

$$\frac{\partial}{\partial t} \rho = \frac{\partial}{\partial x} D(x) \frac{\partial}{\partial x} \rho - \Gamma(x)\rho . \tag{11.47}$$

In the particular case of $D(x) = D$ and $\Gamma(x) = \Gamma$ the solution of this equation may be written in the form

$$\rho(t)(4\pi D t)^{-1/2} \exp(-x^2/4Dt - \Gamma t) . \tag{11.48}$$

As one would expect, we have established that the diffusive distribution of populations among bands damps with time because it is transferred into the continuum. The stationary distribution of the population in the continuum can be found with the help of (11.48). This distribution

$$n(x) = \int_0^\infty \Gamma(x)\rho(x,t)dt \qquad (11.49)$$

is in our case

$$n(x)\Gamma \int_0^\infty \frac{\exp(-x^2/4dt - \Gamma t)}{(4\pi Dt)^{1/2}}dt = (\Gamma/D)^{1/2}\exp[-x(\Gamma/d)^{1/2}]. \qquad (11.50)$$

The extent of the continuum excitation is seen to increase with the diffusion coefficient D (11.29). An increase of the withdrawal rate from the band system hinders the growth of the extent of continuum excitation, since this flow makes the number of the effectively excited bands diminish.

It is evident that in case the solution of (11.47) is known at $\Gamma(x) = 0$, then the solution of this equation is always known for any other constant $\Gamma(x) = \Gamma$. The latter solution may be obtained from the former one by the multiplication by $\exp(-\Gamma t)$. However, such plain situations are rare in practice even though they are very convenient for analysis, while a nontrivial dependence $\Gamma(x)$ may bring about interesting consequences.

Let us illustrate this fact with the following simplest example. Let us consider the balance equations for the populations of two bands, whose specific rate of the decay into continuum is

$$\dot{\rho}_1 = -D_1\rho_1 + D_1\rho^2 - \Gamma_1\rho_1, \quad \dot{\rho}_2 = -D_1\rho_2 + D_1\rho_1 - \Gamma_2\rho_2, \qquad (11.51)$$

where (11.29) holds for kinetic coefficients D.

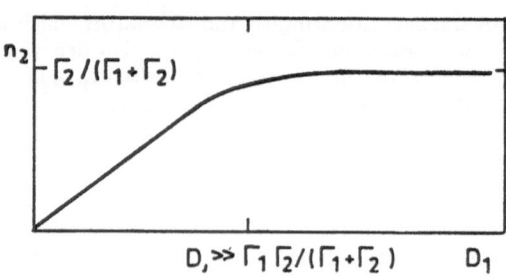

Fig. 11.3. Saturation effect in the process of population transfer into the continuum through the second band of a two-band system

Solving Eqs. (11.51) with standard methods for population n_2 transferred into the continuum through the decay of the second band, we find

$$n_2 = \int_0^\infty \Gamma_2\rho_2(t)dt, \qquad (11.52)$$

which yields

$$n_2 = \frac{D_1 \Gamma_2}{D_1(\Gamma_1 + \Gamma_2) + \Gamma_1 \Gamma_2}. \tag{11.53}$$

In this expression the saturation effect may be observed at $D_1 \gg \Gamma_1\Gamma_2/(\Gamma_1 + \Gamma_2$ (Fig. 11.3)). Let $\Gamma_2 \gg \Gamma_1$. Then $n_2 = D_1/(D_1 + \Gamma_1)$. At small D_1, $D_1 < \Gamma_1$ the second band transfers only a small portion of particles into the continuum. If the radiative transition from the second band into some third one with kinetic coefficient D_2 is possible and if the third band decays into the continuum at the rate $\Gamma_3 \gg \Gamma_2 \gg \Gamma_1$, then while $D_2 < \Gamma_2$ and $D_1 \gg \Gamma_1$, the main channel of the decay is the one through the second band. The increase in the field intensity, i.e., the increase of kinetic coefficients toward the region where $D_2 > \Gamma_2$, switches the decay channel over to the third band. Indeed, let us write down the balance equation system corresponding to the three-band model:

$$\begin{aligned}
\dot{\rho}_1 &= -D_1\rho_1 + D_1\rho_2 - \Gamma_1\rho_1, \\
\dot{\rho}_2 &= -D_1\rho_2 - D_2\rho_2 + D_1\rho_1 + D_2\rho_3 - \Gamma_2\rho_2, \\
\dot{\rho}_3 &= -D_2\rho_3 + D_2\rho_2 - \Gamma_3\rho_3.
\end{aligned} \tag{11.54}$$

We now find the energy the system gives off into the continuum. In terms of $\hbar\omega$ this energy equals

$$W = \int_0^\infty (2\rho_3\Gamma_3 + \rho_2\Gamma_2)dt. \tag{11.55}$$

Solving system (11.54) with standard methods and substituting the values of ρ_2 and ρ_3 thus obtained into (11.55) we find

$$W = \frac{D_1(D_2 + \Gamma_3)\Gamma_2 + 2D_1D_2\Gamma_3}{(D_1 + \Gamma_1)(D_1 + D_2 + \Gamma_2)(D_2 + \Gamma_3) - D_2^2(D_1 + \Gamma_1) - D_1^2(D_2 + \Gamma_3)}. \tag{11.56}$$

Figure 11.4 shows the dependence of $W(\mathcal{E}_0^2)$ which illustrates everything said above.

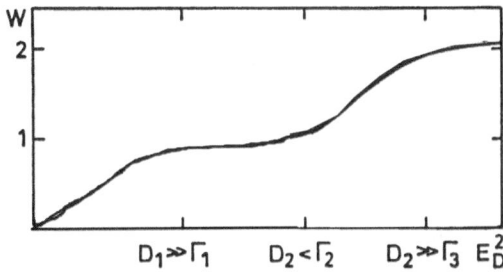

Fig. 11.4. Energy (in terms of $\hbar\omega$) transferred by a three-band system into the continuum depending on the intensity of radiation for the case of $\Gamma_3 > \Gamma_2 > \Gamma_1$

We have already considered the situations when the populations of the band system under investigation flow out into other, statistically more capacious states, thus falling out from the process of the interaction with the field. In fact, we have been dealing with the kind of irreversible relaxation which makes those particles resonant with respect to the field disappear. However, there are other types of relaxation which result in the dissipation of energy, while the total population of the states of the system interacting with the field remains unchanged. Such a relaxation does not lead to the disappearance of resonant particles. These particles are only transferred to energetically lower states.

We shall illustrate this with the example of a harmonic oscillator interacting with both the external field and the thermostat. Such an interaction provides the mechanism of relaxation.

The question of the relaxation into a thermostat has been discussed in a general form in Lecture 4, where we have given the matrix elements (4.41) of the relaxation matrix $\widehat{R}$ entering the equation for the density matrix (4.7). In the simplest possible case the energy of the interaction of the harmonic oscillator with the thermostat is proportional to the displacement. Then, among all the matrix elements J_{iklj} (see(4.41)), only the elements $J_{i,i\pm1,i\pm1,i}$ are nonzero. Hence, in the order of perturbation theory considered, the relaxation of nondiagonal elements of the density matrix develops with the longitudinal relaxation time. Besides, if the thermostat temperature is much less than the quantum energy, $kT \ll \hbar\omega$, only the downward relaxation transitions corresponding to the energy flow from the oscillator to the thermostat are essential. The rate of this relaxation is proportional to the number n of the relaxation level.

This is easily understandable.

The relaxation matrix is proportional to the square of the interaction and, therefore, to the square of the displacement. In turn, the energy accumulated by the oscillator equal to $n\hbar\omega$ is proportional to n. This fact may be strictly proven by the calculation of integrals (4.41) under the assumptions adopted above.

Then, in the situation considered, (4.7) can be represented in the form

$$j\hbar\dot{\rho}_{n,n} = \frac{\mu\mathcal{E}_0}{2}(n+1)^{1/2}\rho_{n+1,n} + \frac{\mu\mathcal{E}_0}{2}n^{1/2}\rho_{n-1,n} - \frac{\mu\mathcal{E}_0}{2}n^{1/2}\rho_{n,n-1}$$

$$- \frac{\mu\mathcal{E}_0}{2}(n+1)^{1/2}\rho_{n,n+1} - j\hbar Rn\rho_{n,n} - j\hbar R(n+1)\rho_{n+1,n+1},$$

$$(11.57)$$

$$j\hbar\dot{\rho}_{k,m} = \frac{\mu\mathcal{E}_0}{2}(k+1)^{1/2}\rho_{k+1,m} + \frac{\mu\mathcal{E}_0}{2}k^{1/2}\rho_{k-1,m} - \frac{\mu\mathcal{E}_0}{2}m^{1/2}\rho_{k,m-1}$$

$$- \frac{\mu\mathcal{E}_0}{2}(m+1)^{1/2}\rho_{k,m+1} - j\hbar Rn\rho_{n,n} - j\hbar R(k+m)\rho_{k,m},$$

$$(11.58)$$

where the matrix elements μ and R refer to the transition $1 \to 0$. If the radiation is intense, $\mu\mathcal{E}_0/\hbar \gg R$, the levels with large number n, $n \gg 1$, have noticeable populations. In this case we may confine ourselves to the diagonal elements and to the closest nondiagonal elements of the density matrix when considering Eqs. (11.57) and (11.58). This allows us, using (11.58), to express the nondiagonal elements, with the index difference ± 1, in terms of the diagonal ones and to substitute them in (11.57). As a result we have

$$
\dot{\rho}_{nn} = \frac{1}{2R}\left(\frac{\mu\mathcal{E}_0}{2\hbar}\right)^2 (2\rho_{nn} - \rho_{n+1,n+1} - \rho_{n-1,n-1})
$$
$$
- Rn\rho_{nn} + R(n+1)\rho_{n+1,n+1}. \tag{11.59}
$$

But by analogy with the cases of the diffusion in the band structure considered earlier, the recurrent relation obtained is equivalent to the diffusion equation,

$$
\frac{\partial}{\partial t}\rho = \frac{1}{2R}\left(\frac{\mu\mathcal{E}_0}{2\hbar}\right)^2 \frac{\partial^2}{\partial x^2}\rho + R\frac{\partial}{\partial x}(x\rho). \tag{11.60}
$$

The stationary solution of (11.60) is of the form

$$
\rho = 2\pi^{-1/2}R\left(\frac{\mu\mathcal{E}_0}{2\hbar}\right)^{-1}\exp\left[-x^2R^2\left(\frac{\mu\mathcal{E}_0}{2\hbar}\right)^{-2}\right]. \tag{11.61}
$$

Therefore we have obtained the distribution for the total population being conserved as distinct from the distributions for the cases of level decay (see, for example, (11.48)). This stationary distribution has another important property: it differs from the uniform distributions in levels which are typical in the absence of any relaxation (see, for example, (11.28) and (11.44) at $t \to \infty$). The principle difference of distribution (11.61) from those obtained earlier is that in the case in which the energy relaxation is not accompanied by the disappearance of the particles interacting with the field, they only lose energy in the relaxation process.

The energy flow from the oscillator to the thermostat equals

$$
W = \hbar\omega\int_0^\infty Rx\rho(x)dx = \pi^{-1/2}\omega\frac{\mu\mathcal{E}_0}{2} = \pi^{-1/2}\hbar\omega\frac{\mu\mathcal{E}_0}{2\hbar}. \tag{11.62}
$$

This last expression admits an evident interpretation. The energy transferred from the oscillator to the thermostat per unit time equals, up to a factor close to unity, the quantum energy multiplied by the Rabi frequency, i.e., the energy of the interaction of the field with the oscillator multiplied by the field frequency. Note also that the number of an efficiently populated level is seen from (11.61) to be determined by the ratio of the Rabi frequency to the relaxation rate.

Lecture Twelve

Collisional Relaxation

Strong and weak collisions. The Weisskopf radius. Strict resonance, V–T-relaxation, V–V-exchange, V–V'-exchange. Collisional dynamics of the single-particle density matrix. Weak collisions of simple oscillators, the Boltzmann distribution. Strong collisions, τ-approximation. Weak collisions of anharmonic oscillators, the Rich-Treanor and Gordiez-Osipov-Shelepin distributions.

In the previous lectures the relaxation processes were considered as processes of energy transfer via collisions of resonantly excited particles with particles of a buffer gas or via interactions with some thermostat. In either case the energy was transferred to qualitatively different particles distinct from resonant ones and not interacting immediately with the radiation field. Nonetheless, relaxation may also occur in a gas containing only the particles resonant with respect to the radiation due to their mutual collisions. In the process of such collisions the change in the quantum state of one participant is accompanied by a corresponding change in the state of another. Processes of this kind evidently take place not only in pure resonant gases, but are also observed in gaseous mixtures along with relaxation caused by collisions with buffer gas particles.

The processes of energy exchange in collisions of identical particles are rather diverse and do not ever admit a general-form analysis. So, it appears convenient to distinguish two extremes: the cases of strong and weak collisions. Having the same net effect as strong collisions, weak collisions are more frequent, though each one only slightly affects the initial states of colliding particles; strong collisions are rather rare and change the initial particle states considerably. Note that so far we have considered only weak collisions, mainly owing to the fact that the theory of such collisions is better developed. However, one should not overlook the possibility of the existence of strong collisions, so in the following consideration we shall pay appropriate attention to them. For now let us just emphasize that the answer to the question of whether the collisions are strong or weak is determined by the type of potential of the colliding particles: the long-range potential leads to weak collisions while the short-range potential leads to strong ones.

Let us start by considering the simplest case of collisional interaction – the resonant exchange of excitation energy in the course of weak binary collisions in the ensemble of identical two-level particles.

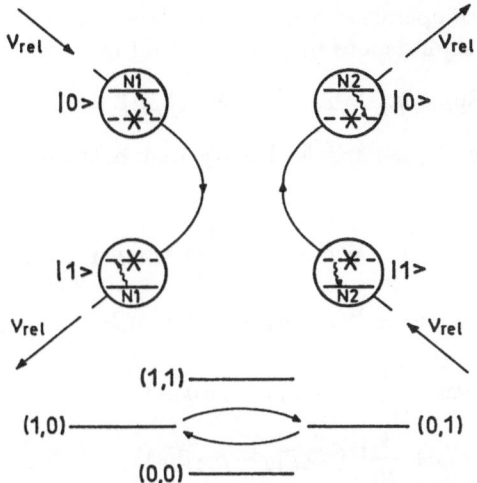

Fig. 12.1. Schematic representation of a collisional exchange

Figure 12.1 schematically shows excited and unexcited particles flying toward each other and exchanging their excitation as the result of a collision. The same figure presents the energy spectrum of the system of two two-level particles. It is obvious that this spectrum should contain four energy levels, and two indices corresponding to different particles are to be attributed to each level. The arrows show the exchange of populations between the levels (1,0) and (0,1) corresponding to the resonant exchange by excitation. Such a process may occur due to the dipole-dipole interaction of colliding particles. This case is referred to as the exchange by a photon, i.e., a particle losing a quantum emits a photon which is absorbed by another particle (see note 2 the Appendix). The interaction operator corresponding to this process is

$$\widehat{\widehat{V}} = -\frac{\widehat{\mu}_1^+ \widehat{\mu}_2^- + \widehat{\mu}_1^- \widehat{\mu}_2^+}{R^3(t)}, \tag{12.1}$$

where $R(t)$ is the interparticle distance, and indices 1 and 2 correspond to the first and second colliding particles, respectively. Operators $\widehat{\mu}^+$ and $\widehat{\mu}^-$ may be written in the form

$$\widehat{\mu}^+ = \begin{pmatrix} 0 & \mu \\ 0 & 0 \end{pmatrix}, \quad \widehat{\mu}^- = \begin{pmatrix} 0 & 0 \\ \mu & 0 \end{pmatrix}. \tag{12.2}$$

Operator $\widehat{\widehat{V}}$ of the form (12.1) describes the process in the course of which the transition of one particle from the upper level to the lower one is accompanied by the opposite transition of the other particle.

In order to obtain an equation for the single-particle density matrix we should substitute (12.2) into (4.16) and (4.17), bearing in mind that

$$\widehat{\mu}_{1,2}^+ \widehat{\mu}_{1,2}^+ = \widehat{\mu}_{1,2}^- \widehat{\mu}_{1,2}^- = 0. \tag{12.3}$$

Performing the trace operation over the variables of the second particle whose density matrix we shall denote by $\tilde{\rho}$, one should take into account that

$$\mathrm{Sp}\widehat{\tilde{\mu}_2^-}\,\tilde{\rho}\widehat{\tilde{\mu}_2^+} = \tilde{\rho}_{00}\mu^2\,, \quad \mathrm{Sp}\widehat{\tilde{\mu}_1^+}\,\tilde{\rho}\widehat{\tilde{\mu}_1^-} = \tilde{\rho}_{11}\mu^2\,. \quad (12.4)$$

As a result for the single-particle density matrix ρ we obtain the following equations

$$\dot{\rho}_{00} = -W\rho_{00}\tilde{\rho}_{11} + W\tilde{\rho}_{00}\rho_{11} + \frac{1}{j\hbar}V(\rho_{10}\tilde{\rho}_{01} - \rho_{01}\tilde{\rho}_{10})\,,$$

$$\dot{\rho}_{11} = -W\tilde{\rho}_{00}\rho_{11} + W\rho_{00}\tilde{\rho}_{11} + \frac{1}{j\hbar}V(\rho_{01}\tilde{\rho}_{10} - \rho_{10}\tilde{\rho}_{01})\,,$$

$$\dot{\rho}_{01} = -W\rho_{01} + \frac{1}{j\hbar}V(\rho_{11}\tilde{\rho}_{01} - \rho_{00}\tilde{\rho}_{01})\,, \quad (12.5)$$

$$\dot{\rho}_{10} = -W\rho_{10} + \frac{1}{j\hbar}V(\rho_{00}\tilde{\rho}_{10} - \rho_{11}\tilde{\rho}_{10})\,,$$

where

$$W = 2\int\frac{w d\Gamma}{\hbar^2}\int_{-\infty}^{\infty}\frac{\mu^2 dt}{R^3(t)}\int_{\infty}^{t}\frac{\mu^2 d\tau}{R^3(\tau)}\,, \quad (12.6)$$

$$V = \int w d\Gamma\int_{-\infty}^{\infty}\frac{\mu^2 dt}{R^3(t)}\,. \quad (12.7)$$

The structure of system (12.5) resembles in some aspects the equation for the variation of the distribution function in the collisions described by the Boltzmann integral. This is not surprising since both are based on the same *ad hoc* assumptions about the binary character of the collisions. The main distinction consists in the fact that the Boltzmann integral describes the comings and goings of the particles into the states with definite momenta while system (12.5) corresponds to the change of the intrinsic eigenenergy quantum states. Besides, this system encompasses the equations for the off-diagonal density matrix elements responsible for the transverse relaxation.

If, as was assumed before, we consider the collisions of identical particles the single-particle density matrices ρ and $\tilde{\rho}$ are equal. Then (12.5) takes on the form

$$\dot{\rho}_{00} = \dot{\rho}_{11} = 0\,,$$

$$\dot{\rho}_{01} = -W\rho_{01} + \frac{1}{j\hbar}V(\rho_{11} - \rho_{00})\rho_{01}\,,$$

$$\dot{\rho}_{10} = -W\rho_{10} + \frac{1}{j\hbar}V(\rho_{00} - \rho_{11})\rho_{10}\,, \quad (12.8)$$

It may be seen that in the collisions of identical particles the energy of the system is conserved, and the longitudinal relaxation time turns out to be infinite. The equations for the off-diagonal matrix elements in first-order perturbation theory acquire, as is usually the case, an imaginary additional term which brings about the frequency shift. The second order of the perturbation theory gives the relaxation of the off-diagonal matrix elements. Therefore, an exchange of quanta in the system of identical particles does not lead to the longitudinal relaxation, but just to the transverse one. Note that the appearance of the frequency shift caused by the interaction between two two-level systems is in complete agreement with the well-known predictions of the vibration theory concerning the splitting of the frequencies of coupled oscillators.

Now let us turn to the calculation of integrals (12.6) and (12.7). If the trajectory is supposed to change only slightly in the course of collision, then the interparticle distance is

$$R(t) = [r^2 + (vt)^2]^{1/2},\qquad(12.9)$$

where v is the relative velocity and r is the impact parameter. Velocity v obeys the Maxwell distribution for relative velocities. Then

$$wd\Gamma = nv\left[\pi^{-1}\left(\frac{m}{4kT}\right)^{1/2}\right]^3 \exp\left(-\frac{mv^2}{4kT}\right)2\pi r dr d^{(3)}v,\qquad(12.10)$$

where n is the density of particles. The substitution of (12.10) into (12.6) and (12.7) gives

$$W = 2n\int dvdrv\pi^{1/2}\left(\frac{m}{kT}\right)^{3/2}\exp\left(-\frac{mv^2}{4kT}\right)\mu^4 r^{-3}\hbar^{-2}I_1 I_2,\qquad(12.11)$$

$$I_1 = \int_{-\infty}^{\infty}\frac{dx}{(1+x^2)^{3/2}},\quad I_2 = \int_{-\infty}^{x}\frac{du}{(1+u^2)^{3/2}}.$$

$$V = n\hbar\int dvdrv^2\pi^{1/2}\left(\frac{m}{kT}\right)^{3/2}\exp\left(-\frac{mv^2}{4kT}\right)\mu^2 r^{-1}\hbar^{-1}\int_{-\infty}^{\infty}\frac{dx}{(1+x^2)^{3/2}}.$$
$$(12.12)$$

The value of integral (12.12) is determined generally by the large impact parameters where the interaction is small and, hence, the model of weak collisions applies. In the case of integral (12.11) diverging as r^{-2} in the region of small impact parameters, the situation is significantly more involved. The point is that in the region of small impact parameters the model of weak collisions may not be applicable.

The margin of applicability of this model may be found from the following considerations. The typical interaction time during a collision is about r/v_T, where v_T is the mean thermal velocity of relative motion. The interaction energy at a distance of the order of r is about μ^2/r^3. This interaction corresponds to some effective Rabi frequency of the transition $(1,0)\rightarrow(0,1)$ (Fig. 12.1) equal to $\mu^2/r^3\hbar$. Then the interaction is weak if

$$\frac{\mu^2}{r^3\hbar}\frac{r}{v_t}\ll 1.\qquad(12.13)$$

In other words, the model of weak collisions applies if the impact parameter value exceeds

$$r_W = (\mu^2/\hbar v_T)^{1/2}.\qquad(12.14)$$

Quantity r_W is referred to as the Weisskopf radius. This is the impact parameter value at which the action of the interaction forces coincides by order of magnitude with the Planck constant $\hbar$, $\int V(t)dt \sim \hbar$. If the typical geometric size of the particle exceeds the Weisskopf radius, $r_0 > r_W$, then all the collisions are weak. For thermal velocities in the range 10^4–10^5 cm/c and dipole moments 1–0.1 D (i.e., $(1$–$0.1)\cdot 10^{-18}$ CGSE), r_W is confined to the region of from 30 to 1 Å.

In case the geometric size of the particle is less than the Weisskopf radius, the contribution to the process of excitation transfer rendered by the collisions with impact parameters between the particle size and the Weisskopf radius r_W should be taken into account according to the theory of strong collisions.

In the frame of the weak-collision model, integrals (12.11) and (12.12) yield

$$W = 4\pi^{1/2}n(m/kT)^{1/2}\mu^4\hbar^{-2}r_0^{-2}\,, \tag{12.15}$$

$$V = 4\pi n\mu^2 \ln(\lambda/r_0)\,. \tag{12.16}$$

Formula (12.16) for the quantity V that characterizes the strength of the averaged dipole-dipole interaction contains wavelength λ of the photon, corresponding to the frequency of transition $0 \to 1$ in each of the identical two-level particles considered. This is caused by the fact that at distances greater than λ the wave retardation starts to affect the interparticle interaction, and the interference effects caused by this retardation eliminate the logarithmic divergence. If the typical interparticle distance $n^{-1/3}$ is less than wavelength λ, it is this distance that is substituted instead of λ in (12.16).

At $v_T = 10^5$ cm·s^{-1}, $\mu = 0.1$ D, $n = 10^{16}$ cm^{-3}, $r_0 = 10^{-8}$ cm, $\lambda = 10^{-3}$ cm we obtain $W \approx 1 \cdot 10^5$ s^{-1} and $V/\hbar \approx 10^7$ s^{-1}.

As we have already mentioned, the consideration just discussed is valid only for weak collisions when the impact parameter exceeds the Weisskopf radius. However, the above evaluations show that this radius may be significantly greater than the gas-kinetic size of the particles. Then for the description of an elementary collision event one should use the model of strong collisions and retain not only two first orders of the perturbation theory series as in Lecture 4 when calculating the relaxation matrix, but sum up the whole series. In other words, in the process of collision the particles have enough time to interchange their quanta many times. In fact, a coherent process of population exchange occurs. In order to determine the final result of such an exchange, we have to take into account at least the number of terms of the perturbation theory series equal to the number of transitions which take place during the collision event.

However there is no need to perform such a complex procedure. In the case considered of the excitation exchange at strict resonance, such a problem admits an exact solution. Here we take into account that the population dynamics of levels (1,0) and (0,1) (Fig. 12.1) completely coincides with that of a two-level system in strict resonance with a radiation field, if only the amplitude of the field varies slowly in time in the same way as the amplitude of an oscillating-dipole field in the static zone does during the collision event: $\mathcal{E}(t) = \mu/R^3(t)$. This dynamics was discussed in Lecture 2. If, as previously, we insert the area under the pulse $\theta(r, v)$ with relation

$$\theta(r, v) = \int_{-\infty}^{\infty} \frac{\mu^2 dt}{\hbar R^3(t)} = \int_{-\infty}^{\infty} \frac{\mu^2 dt}{\hbar (r^2 + v^2 t^2)^{3/2}}, \qquad (12.17)$$

the change in the density matrix of the ensemble of two particles which results from a single collision event may be written in the form

$$\widehat{\rho} - \widehat{\rho}_0 = \widetilde{\widehat{u}} \widetilde{\rho}_0 \widetilde{\widehat{u}}^* - \widehat{\rho}_0, \qquad (12.18)$$

where the evolution operator is given by

$$\widehat{u} = \begin{pmatrix} u_{11}^{11} & u_{11}^{10} & u_{11}^{01} & u_{11}^{00} \\ u_{10}^{11} & u_{10}^{10} & u_{10}^{01} & u_{10}^{00} \\ u_{01}^{11} & u_{01}^{10} & u_{01}^{01} & u_{01}^{00} \\ u_{00}^{11} & u_{00}^{10} & u_{00}^{01} & u_{00}^{00} \end{pmatrix} = \begin{pmatrix} 1 & 0 & 0 & 0 \\ 0 & \cos\theta & j\sin\theta & 0 \\ 0 & j\sin\theta & \cos\theta & 0 \\ 0 & 0 & 0 & 1 \end{pmatrix}.$$

Let us look into the physics of the approach proposed here. The particle collisions bring about coherent transitions between the quantum states of the entire system presented in Fig. 12.1. This process does not touch upon ψ-functions and, hence, the populations of levels (0,0) and (1,1), while a coherent process of population exchange (quantum nutations) develops between levels (0,1) and (1,0). The dynamics of the wavefunctions of these states is described by Eqs. (2.49). In other words, the interaction pulse of area θ acts upon the system of two levels with initial wavefunctions ψ_{01} and ψ_{10} and after its termination, leaves the amplitudes of the probabilities for the system to be found in states (0,1) and (1,0) equal to $\psi_{01}\cos\theta + j\psi_{10}\sin\theta$ and $\psi_{10}\cos\theta - j\psi_{10}\sin\theta$, respectively. Such a transformation of the wavefunctions from their initial to final values is performed with matrix $\widehat{u}$, while the corresponding change in the density matrix $\widehat{\rho}$ is given as usual by the product $\widetilde{\widehat{u}} \widetilde{\rho} \widetilde{\widehat{u}}^*$.

Representing $\widehat{\rho}_0$ in the form $\widetilde{\rho}\widetilde{\rho}$, performing the trace summation over the second particle variables (with "tilde") and averaging over the collision parameters, from (12.18) we obtain an equation whose form completely coincides with (12.5) at

$$W = 2\int w d\Gamma \sin^2 \frac{\theta(r, v)}{2} = 2\int w d\Gamma \sin^2 \frac{\mu^2}{\hbar v r^2}, \qquad (12.19)$$

$$V = 2\hbar \int w d\Gamma \sin \frac{\theta(r, v)}{2} = 2\hbar \int w d\Gamma \sin \frac{\mu^2}{\hbar v r^2}. \qquad (12.20)$$

Let us mention that such a coincidence of equations describing the cases of weak and strong collisions is the consequence of the strict resonance of levels (1,0) and (0,1) and the dipole type of binary interaction. Note also that expressions (12.6) and (12.7) for W and V are instances of (12.19) and (12.20), respectively, in the limit of small θ.

Substituting expression (12.10) for $wd\Gamma$ into (12.19) and (12.20) we obtain after integration

$$W = \pi^2 n\mu^2/\hbar, \tag{12.21}$$
$$V = 4\pi n\mu^2 \ln(\lambda/r_W). \tag{12.22}$$

The comparison of (12.22) with (12.16) shows that the type of frequency shift dependence for weak and strong collisions is the same. However, the role of the geometric size of the particle in (12.22) is played now by the Weisskopf radius. Since in strong collisions $r_W > r_0$ the frequency shift in this case is a bit smaller than for weak ones.

The comparison of (12.21) with (12.15) points to a qualitatively different type of relaxation rate dependence which in the case of strong collisions is determined only by the density of the particles and the dipole moment. At $n = 10^{16}$ cm^{-3} and $\mu = 1$ D we obtain $W = 10^8$ s^{-1}. In other words, in the case of the resonant exchange between identical particles the time of the off-diagonal matrix element relaxation (the transverse relaxation time) $T_2 = W^{-1}$ is about 10^{-8} s or, in the customary representation of the experimental results, corresponds to the relaxation with time 2.5 T · ns.

When the collisional partners are similar but not quite identical the longitudinal relaxation takes place as well. The point is that in such a situation the transition from state (1,0) into state (0,1) implies the escape of an excitation quantum from the ensemble of particles considered. If we are interested in the energy transfer from a small resonant impurity to a buffer gas of much greater density, then it is easy to see that in this case the rate of the longitudinal relaxation equals W. Indeed, let at the initial time moment $\rho_{01} = \rho_{10} = \tilde{\rho}_{01} = \tilde{\rho}_{10} = 0$, $\rho_{00} = 0$, $\rho_{11} = 0$, $\tilde{\rho}_{00} = 1$, $\tilde{\rho}_{11} = 0$. Besides, let $\tilde{n} \gg n$. Then we may consider $\tilde{\rho}_{00}$ and $\tilde{\rho}_{11}$ unchanged in the course of the energy transfer from small impurity. Hence, Eqs. (12.5) take on the form

$$\dot{\rho}_{00} = W\rho_{11}, \quad \dot{\rho}_{11} = -W\rho_{11}, \tag{12.23}$$

wherefrom the statement of the longitudinal relaxation at rate W stems.

Everything said above refers to the case of the collisional interaction of resonant particles. The cross sections of the energy transfer and the relaxation rate were determined by the Weisskopf radius and were large, due to resonance. In the presence of a small frequency detuning the situation is much like the resonant one, the only difference being that the interaction cross section is determined by some effective radius of resonance, where the interaction energy becomes equal to the energy detuning, $\mu^2/R^3 = \Delta E$, rather than to the Weisskopf radius. (As previously, we remain in the limits of dipole-dipole interaction.) Let us mention that a small energy deficiency arising in such collisions may be easily compensated for by an appropriate change in the energy of translational motion.

Of course, a collisional energy exchange is feasible as well in the cases when the energy deficiency exceeds the energy of dipole-dipole interaction even at distances equal to the gas-kinetic size of a particle. The cross section of such an exchange is small and this is quite natural, since the process considered may be reduced to the X–T-relaxation (Lecture 8) between levels (0,1) and (1,0) of the compound system of two particles, i.e., in the case of the compound system presented for this case in Fig. 12.2. In this process the translational degrees of freedom serve as a thermostat. Let us consider such a process.

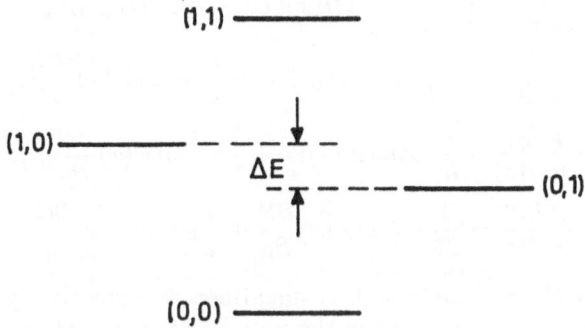

Fig. 12.2. Scheme of levels of compound system with energy deficiency ΔE

The probability of such a process turns out to be determined by the Fourier transform of the correlation function of the interaction energy. Indeed, the interaction energy depends on time and, hence, has a finite spectrum width. If from the time dependence of the interaction energy we extract the Fourier component corresponding to the frequency of the transition considered, then it is this component that the transition probability amplitude is proportional to. The probability itself is equal to the modulus squared of this quantity. This implies that it equals the Fourier transform of the correlator.

Indeed,

$$H^2(\omega) = \int e^{j\omega t} H(t)dt \int e^{-j\omega\tau} H(\tau)d\tau = \int e^{j\omega(t-\tau)} H(t)H(\tau)dtd\tau$$

$$= \int e^{2j\omega y} H(x+y)H(x-y)2dxdy = \int e^{2j\omega y} \int H(x-2y)H(x)2dxdy$$

$$= \int e^{j\omega y} \int H(x-y)H(x)dxdy \,.$$

If the change in the internal energy of the system is significantly less than the translational motion energy, then this process may be considered using the quasiclassical approach given in Lecture 7. Let us dwell on this approach in more detail.

An elementary act of the X–T-process in the simplest case of the relaxation of the two-level particle energy into the translational motion, when it collides with a heavy unstructured buffer particle, may be described using (7.3) and Pauli's matrices (see (4.8)). In the interaction representation for the Hamiltonian,

$$\begin{pmatrix} 0 & H(x)e^{j\omega t} \\ H(x)e^{-j\omega t} & 0 \end{pmatrix} = \hat{\sigma}_x \cos \omega t + \hat{\sigma}_y \sin \omega t, \qquad (12.24)$$

in terms of Bloch's variables N, P, Q and W, we have equations (compare with (7.12) and (17.21)):

$$\begin{aligned} \frac{\partial W}{\partial t} + \frac{p}{m}\frac{\partial W}{\partial x} &= -\frac{1}{\hbar}\cos \omega t H(x)Q + \frac{1}{\hbar}\sin \omega t H(x)P, \\ \frac{\partial P}{\partial t} + \frac{p}{m}\frac{\partial P}{\partial x} &= -\frac{1}{\hbar}\sin \omega t H(x)W + \frac{1}{2}\cos \omega t F(x)\frac{\partial N}{\partial p}, \\ \frac{\partial Q}{\partial t} + \frac{p}{m}\frac{\partial Q}{\partial x} &= \frac{1}{\hbar}\cos \omega t H(x)W + \frac{1}{2}\sin \omega t F(x)\frac{\partial N}{\partial p}, \\ \frac{\partial N}{\partial t} + \frac{p}{m}\frac{\partial N}{\partial x} &= \frac{1}{2\hbar}\cos \omega t F(x)\frac{\partial P}{\partial p} + \frac{1}{2}\sin \omega t F(x)\frac{\partial Q}{\partial p}, \end{aligned} \qquad (12.25)$$

where $F(x) = \partial H/\partial x$. The first three equations describe the dynamics of the populations and the polarization in the course of the relaxation process, while the fourth one is responsible for the description of translational motion.

In the case of weak collisions, this case being most frequently considered at V–T-relaxation, system (12.25) may be solved with perturbation theory methods. Assuming that in the zero approximation $P = Q = 0$ and W and N change slightly in the process of the collision, we find

$$\begin{aligned} P(t, x) = \frac{1}{\hbar}\int_{-\infty}^{\infty} &\left[\frac{1}{2}\cos \omega \left(t - \frac{mu}{p} \right) F(x - u)\hbar\frac{\partial N}{\partial p} \right. \\ &\left. - \sin \omega \left(t - \frac{mu}{p} \right) H(x - u)W \right] du, \\ Q(t, x) = \frac{1}{\hbar}\int_{-\infty}^{\infty} &\left[\cos \omega \left(t - \frac{mu}{p} \right) H(x - u)W \right. \\ &\left. + \frac{1}{2}\sin \omega \left(t - \frac{mu}{p} \right) F(x - u)\hbar\frac{\partial N}{\partial p} \right] du. \end{aligned} \qquad (12.26)$$

Substituting (12.26) into the first of Eqs. (12.25), we find that the correction to W per collision equals

$$\begin{aligned} \delta W = &-\frac{1}{\hbar^2}\int_{-\infty}^{0} \cos \frac{m\omega u}{p} H(x)H(x - u)W\,du \\ &+ \frac{1}{2\hbar}\int_{-\infty}^{0} \sin \frac{m\omega u}{p} H(x)F(x - u)\frac{\partial N}{\partial p}\,du. \end{aligned} \qquad (12.27)$$

Correction (12.27) refers to a single particle and a single collision event. An average over the collision parameters gives the rate of the population difference change in the form

$$\frac{\partial W}{\partial t} = -\frac{1}{\hbar^2}\int w\,d\Gamma \left[W + \frac{m\omega u}{2p}\hbar\frac{\partial N}{\partial p} \right] \int_{-\infty}^{0} \cos \frac{m\omega u}{p} H^{(2)}(u)\,du. \qquad (12.28)$$

Symbol $H^{(2)}(u)$ denotes here the correlation function $\int_{-\infty}^{\infty} H(x)H(x+u)dx$. We additionally use

$$\int_{-\infty}^{\infty} F(x+u)H(x)dx = \partial H^{(2)}(u)/du.$$

Expression (12.28) shows that the probability of V–T-relaxation is determined by the Fourier transform of the correlation function of the interaction energy. This transform should be taken at the frequency of transition and averaged over the collision parameters. A small value of the V–T-relaxation probability is caused by the fact that many spatial periods of oscillations $\cos(m\omega u/p)$ may be nested on the correlation radius length. This is to say that the time spent by a molecule flying through the typical interaction radius is much greater than the period of oscillations on frequency ω.

The last of Eqs. (12.25) describes the translational acceleration of the particles during the X–T-process, which may be considered in the framework of the same approach we used for the processes of light pressure.

The stationary population difference is determined by the condition of the vanishing expression $W + (\hbar\omega m/2p)\partial N/\partial p$ (see (12.28)). Should the distribution of particles over momenta be Boltzmannian, $N \propto \exp(-p^2/2mkT)$, then

$$W = N\hbar\omega/2kT,\qquad (12.29)$$

which coincides with the equilibrium population difference in a two-level system at the same temperature T but only in the limit of small frequencies,

$$W = N\frac{1 - \exp(-\hbar\omega/kT)}{1 + \exp(-\hbar\omega/kT)} \approx N\frac{\hbar\omega}{2kT}.\qquad (12.30)$$

This implies that analyzing the V–T-relaxation, a formula of the (12.28) type may be used only if the transferred quantum is small with regard to the typical energy of thermal motion so it only slightly affects the character of this motion (the quasi-classical approach holds).

When the vibrational quantum cannot be considered small, the cross section of the relaxation process is to be estimated with the help of the Born approximation mentioned at the end of Lecture 8 (see (8.33)).

Now let us turn to one of the possible applications where the relaxation processes and the processes of the excitation energy transfer circumscribed above may be used and consider the relaxation processes on the vibrational energy levels of molecules.

The processes of vibrational excitation energy transfer are diverse, especially in the case of polyatomic molecules. Among them the processes of V–V-exchange, V–T-relaxation, and V–V'-exchange, combining the features of the first two, have been studied in most detail, owing to their most striking experimental manifestation.

The V–V-exchange process is, in fact, an excitation exchange between oscillators of the same vibrational modes of colliding molecules. During such an exchange in the systems of molecular vibrational levels, the transitions go in opposite directions. So, for example, if a molecule loses vibrational energy as a result of a collision and transits from level n to level $n - 1$, its counterpart in collision gains vibrational energy and transits from level k to $k + 1$. In case the oscillators are truly harmonic the process of the $V-V'$-exchange is strictly resonant. Therefore, the $V-V$-exchange is the resonant vibrational-vibrational exchange by energies between colliding molecules. However, the term "$V-V'$-exchange" is often used when describing the collisional transfer of vibrational energy in the case of anharmonic oscillators. When the anharmonicity is small, then in accordance with the above consideration, it does not lead to appreciable complications.

The process of $V-T$-relaxation is, as we know, a relaxation process of the $X-T$-type where the relaxing excitation is vibrational. The process of $X-T$-relaxation has been discussed earlier and all its typical features mentioned above refer to $V-T$-relaxation as well. For this process, as for any process of interaction with a thermostat, the probabilities of upward and downward transitions differ by the Boltzmann factor $\exp(-\Delta E/kT)$.

The process of $V-V'$-exchange in the course of which the vibrational energy is transferred from one mode of a molecule into another mode (with different frequency) of some other molecule – its counterpart in collision – occurs with an energy deficiency. So, the probability of transferring a vibrational quantum in the process of collision from the mode of higher frequency into that of a lower one is $\exp(\Delta E/kT)$ times less than the probability of the inverse process. One should keep in mind that at a high extent of excitation of polyatomic molecules, the separation of the collision processes into a nonresonant $V-V$-exchange and a $V-V'$-exchange becomes somewhat conventional due to the increase of anharmonic detunings and the loss of the mode specificity of the vibrations.

In the general form the dynamics of the single-particle density matrix in the course of binary collisions is

$$\dot{\rho}_{nk} = \sum_{n',k',p',q',,p,q} \left(W_{n,k,p,q}^{n',k',p',q'} \rho_{p'q'} \rho_{n'k'} - W_{n',k',p',q'}^{n,k,p,q} \rho_{pq} \rho_{nk} \right). \qquad (12.31)$$

This complex nonlinear equation is general. It says that the change in the density matrix of a particle takes place due to its interaction with other particles and is accompanied by appropriate changes in the matrix element of its counterpart in collision. Quantities W are the probabilities of such processes averaged over velocities and impact parameters. The summation ensures averaging of the result over the states of the partner in collision. In the case of a collision of two identical two-level particles, this equation is equivalent to system (12.25) while the action of operator W upon the two-particle density matrix occurs in accordance with (12.18), the evolution matrix u corresponding to the collisional interaction of multilevel systems. Equation (12.31) as well as (12.5) possesses the property that its off-diagonal matrix elements equal zero at the initial time moment and remain zero unless there are some external reasons, such as an intense external field which make them change. In the latter case the collision is called radiative. Hence, for nonradiative collisions the description of collisional dynamics may be reduced to an investigation of the equation for the diagonal elements:

$$\dot{\rho}_n \equiv \dot{\rho}_{nn} = \sum_{n',p',q'} (W^{n',p'}_{n,p} \rho_{n'}\rho_{p'} - W^{n,p}_{n',p'}\rho_n\rho_p).$$ (12.32)

Quantities $W^{n',p'}_{n,p}$ entering (12.32) are the probabilities of transitions between the states n,p and n',p' averaged over the translational thermal motion of colliding particles and, hence, possessing the property

$$W^{n',p'}_{n,p} = W^{n,p}_{n',p'} \exp([-(E_n + E_p - E_{n'} - E_{p'})/kT].$$ (12.33)

Equation (12.32) is a straightforward generalization of the Boltzmann equation for the case of a collision of particles with internal quantum structure. It describes both the processes of resonant vibrational exchange and the nonresonant processes such as nonresonant V–V-, V–V'- and V–T-relaxations, etc.

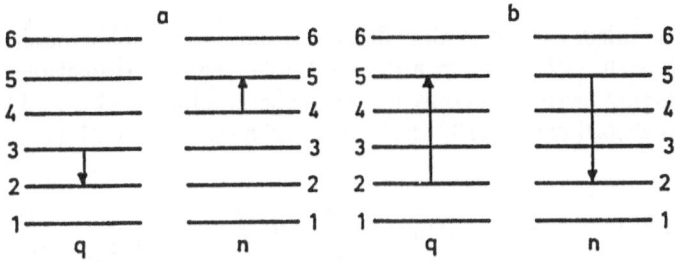

Fig. 12.3. One-quantum transitions in weak collisions: (a) transition between the closest adjacent levels, (b) jump over several levels

It is appropriate to mention here that the difference between strong and weak collisions, which has become apparent in the case of two-level particles only in the distinction between quantities W and V (compare (12.15) and (12.16) with (12.19) and (12.20)), affects the form of the kinetic equations in the case of multilevel systems. Indeed, in the case of weak collisions, (12.32) describes only one-quantum transitions in a compound multilevel system of colliding particles. Such a transition is shown, for example, in Fig. 12.3 a, where one particle acquires an energy quantum while the other losses one. In the case of strong interactions, several successive one-quantum transitions occur during a single collision event. Such a process is shown in Fig. 12.4, where one particle acquires several energy quanta, while the other loss several. The distinction between strong and weak collisions of multilevel particles is not in the changes of the quantum states of interacting particles. Should the selection rules permit it and the law of energy conservation be satisfied, the weak collisions may lead not only to transitions to the closest adjacent level but to the transitions over a level or even several levels (Fig. 12.3 b). In strong collisions the transition between the same levels occurs as a result of the coherent transfer of several quanta in succession. Such a distinction in the physics of interactions manifests itself in the difference of the coefficients W in (12.32). Let us also mention that the action in $\hbar$ units, which one particle exerts upon another in the course of strong collisions (the so-called Masey parameter), is large, while for weak collisions this quantity, $\hbar \int V(t)dt$, is small.

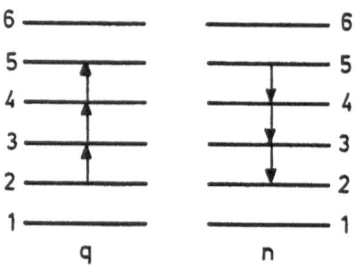

Fig. 12.4. Successive one-quantum transitions in a single strong collision event

In the simplest case of weak collisions in the system of harmonic oscillators interacting with each other in a dipole-dipole way, the transitions which are strictly resonant are allowed only between closely adjacent states. Their matrix elements in accord with (4.16) and (4.18) and by analogy with (12.6) are given by

$$W_{n,q}^{n\pm1,q\mp1} = 2|\mu_n^{n\pm1}|^2|\mu_q^{q\mp1}|^2 \int wd\Gamma \int_{-\infty}^{\infty} \frac{dt}{R^3(t)} \int_{-\infty}^{t} \frac{d\tau}{R^3(\tau)}$$

$$= \begin{cases} (n+1)qW & \text{for upper signs,} \\ n(q+1)W & \text{for lower signs,} \end{cases} \qquad (12.34)$$

where W is given by formula (12.6). In Fig. 12.3 a we show the transition $q = 3 \rightarrow q = 2$, $n = 4 \rightarrow n = 5$ with the matrix element $W_{5,2}^{4,3}$.

In the simplest case considered, (12.32) is written as

$$\dot{\rho}_n = W \sum_q [(n+1)q(\rho_{n+1}\rho_{q-1} - \rho_n\rho_q) + n(q+1)(\rho_{n-1}\rho_{q+1} - \rho_n\rho_q)]. \quad (12.35)$$

Taking into account the normalization $\sum \rho_q = 1$ and the conservation law for the total number of quanta in the system $\sum q\rho_q = N$, (12.28) may be rewritten in the form

$$\dot{\rho}_n = W[(n+1)((N+1)\rho_{n+1} - N\rho_n) + n(N\rho_{n-1} - (N+1)\rho_n)]. \quad (12.36)$$

Note that the dependence on the quantum state of the particle – partner in collision – has dropped from (12.36). This considerably facilitates its solution. Under stationary conditions when $\dot{\rho}_n = 0$, the solution of (12.36) is

$$\rho_{n+1} = \frac{N}{N+1}\rho_n. \qquad (12.37)$$

Bearing in mind the normalization condition $\sum \rho_n = 1$, this implies that

$$\rho_n = \frac{1}{N+1} \left[\frac{N}{N+1} \right]^n .$$ (12.38)

At $N \gg 1$ the former is equivalent to the formula

$$\rho_n = N^{-1} e^{-n/N} .$$ (12.39)

This equation shows that the stationary solution of (12.36) is the Boltzmann distribution, while the rate of approaching it is greater, the higher the energy storage, i.e., the mean number of quanta N present in the oscillators. The approach to the equilibrium is of a diffusive type, since, due to the weakness of collisions, (12.36) includes the populations of only three neighboring levels (compare with (11.20)).

For strong collisions the stationary distribution is also Boltzmannian, as it follows from the more general Eq. (12.32) under the assumption about the conservation of vibrational energy in the system. However, the process of approaching the equilibrium state in the case of strong collisions acquires an essentially nondiffusive character and differs from the process just considered. Each particular case requires a special treatment. However, one often encounters the situation in which the analysis of approaching the equilibrium distribution may be carried out in the so-called τ-approximation. In this approximation we may assume that all the levels relax to their equilibrium values with the same typical time, i.e., according to the law

$$\dot{\rho}_n = -(\rho_n - \rho_n^{(0)})/\tau .$$ (12.40)

For the τ-approximation to be valid, the number of quanta transferred during one collision should coincide by order of magnitude with the number of quanta stored in the system.

The previous consideration referred to the case in which strict resonances are present (for instance, in the case of a harmonic oscillator), where the stationary solutions could be either Boltzmannian or equal to zero. For physical reasons we always choose the former.

If transitions in a system are essentially nonresonant, as, for instance, for anharmonic oscillators, and should the energy deficiency have to be compensated for by the thermal motion ($X–T$-processes), there is a nonzero solution among the stationary solutions distinct from the Boltzmann distribution. In the case of anharmonic oscillators this distribution is called the Rich-Treanor distribution.

The point is that the probability of a quantum transfer from a highly to a slightly excited molecule in a gas of anharmonic oscillators differs by the Boltzmann factor from the probability of the inverse process. The exchange of a hard quantum for a softer one is statistically profitable. If the quantum energy decreases with the extent of excitation, then the energy transfer is mainly directed from the vibrationally cold molecules to those vibrationally hot, and a population inversion occurs. The decrease of the vibrational entropy is compensated for by a decrease in the translational one. If the anharmonicity is positive, the hot molecules get cold while the cold ones heat up, and we have no problems with entropy.

Although the Rich-Treanor distribution is in many aspects exotic since it assumes the complete absence of V–T-relaxation processes, its consideration seems to be expedient from the viewpoint of methodology.

Let us elucidate what has been said above as follows.

In weak collisions when a single collision event may result only in one-quantum transitions between neighboring levels, Eqs. (12.32) and (12.33) are written in the form

$$\dot{\rho}_n = \sum_i [W_{n,i}^{n+1,i-1} \rho_{n+1}\rho_{i-1} - W_{n+1,i-1}^{n,i}\rho_n\rho_i$$

$$+ W_{n,i}^{n-1,i+1} \rho_{n-1}\rho_{i+1} - W_{n-1,i+1}^{n,i}\rho_n\rho_i], \qquad (12.41)$$

$$W_{n,i}^{n+1,i-1} = W_{n+1,i-1}^{n,i} \exp\left[-\frac{E_n + E_i - E_{n+1} - E_{i-1}}{kT}\right] \frac{\rho_{i-1}}{\rho_i}, \qquad (12.42)$$

respectively. The condition for the distribution to be stationary is evident from (12.41). With allowance made for (12.42), it takes on the form

$$\frac{\rho_n}{\rho_{n+1}} = \frac{W_{n,i}^{n+1,i-1}}{W_{n+1,i-1}^{n,i}} \frac{\rho_{i-1}}{\rho_i} = \exp\left[\frac{E_n + E_i - E_{n+1} - E_{i-1}}{kT}\right] \frac{\rho_{i-1}}{\rho_i}. \qquad (12.43)$$

When the anharmonicity is small,

$$E_n = n\hbar\omega + \alpha n^2. \qquad (12.44)$$

Then $E_n + E_i - E_{n+1} - E_{i-1} = -2\alpha[n - (i-1)]$, and

$$\frac{\rho_n}{\rho_{n+1}} \exp\left[-\frac{2\alpha n}{kT}\right] = \frac{\rho_{i-1}}{\rho_i} \exp\left[-\frac{2\alpha(i-1)}{kT}\right]. \qquad (12.45)$$

It is seen that the right- and left-hand sides do not depend on each other's indices. Hence they are constant. Then

$$\rho_{n+1} = C\rho_n \exp\left[-\frac{2\alpha n}{kT}\right], \qquad (12.46)$$

where C is some constant factor. From the condition for the distribution to become uniform at $\alpha \to 0$, we obtain that $C = 1$. Then

$$\rho_n = A \prod_n \exp\left[-\frac{2\alpha n}{kT}\right] = A \exp\left[-\frac{\alpha n(n+1)}{kT}\right], \qquad (12.47)$$

where A is the normalization factor determined by the condition $\sum \rho_n = 1$. As a result, we obtain the Rich-Treanor distribution:

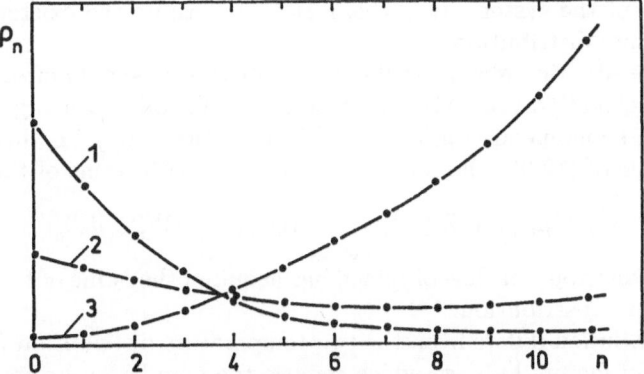

Fig. 12.5. The Boltzmann (1) and Rich-Treanor (2 and 3) distributions: 2 – positive anharmonicity, $\alpha > 0$; 3 – negative anharmonicity, $\alpha < 0$

$$\rho_n = \frac{\exp\left[-\frac{\alpha n(n+1)}{kT}\right]}{\sum_n \exp\left[-\frac{\alpha n(n+1)}{kT}\right]} \cdot \qquad (12.48)$$

The distribution obtained decreases more smoothly with level number then the Boltzmann one (12.39) does. These two distribution types are shown for comparison in Fig. 12.5. If the anharmonicity constant is negative as is usually the case, then the inverse population on the upper levels corresponds to the Rich-Treanor distribution. As we have said above, the point here is that in a system of anharmonic oscillators the process of energy transfer from the slightly excited molecules to the highly excited ones may take place. Indeed, at $\alpha < 0$ the vibrational quantum of a highly excited molecule is less than that of a slightly excited one. So the collisions result with greater probability in the process corresponding to a transfer of positive energy – the difference of the quanta of highly and slightly excited molecules – to the thermostat of translational motion. The inversion of populations in the Rich-Treanor distribution grows up until the anharmonic oscillators model holds. The type of steady-state distribution established in each specific case depends on the parameters which are external with respect to the problem under consideration – the boundary conditions and the type of the excitation source.

Note that the initial Eq. (12.41) has a stationary solution of a more general form leading to the so-called two-temperature distribution. This distribution is the combination of the Boltzmann and Rich-Treanor distributions and is written as

$$\rho_n = A \exp\left[-\frac{n\hbar\omega}{kT_1} - \frac{\alpha n(n+1)}{kT^2}\right]. \qquad (12.49)$$

One can say that constant C from (12.46) takes on the value $C = \exp(-\hbar\omega/kT_1)$. At $T_1 = T_2$ the distribution is, in fact, Boltzmannian, while at $T_1 \to \infty$, we have the Rich-Treanor distribution. In (12.41) we ignore the processes responsible for the equalization of vibrational and translational temperatures of the system. It is this fact that permits the existence of the two-temperature distribution.

In fact the situation may be still more involved. One should often take into account the relaxation processes in which the vibrational exchange is absent (for example, spontaneous radiation, V–T-relaxation, etc.). In this case the right-hand side of (12.36) should be supplemented with terms of the form

$$W_n^{n+1}\rho_{n+1} + W_n^{n-1}\rho_{n-1} - W_{n+1}^n\rho_n - W_{n-1}^n\rho_n \,.$$

The complex distribution thus obtained has acquired the name of the Gordiez-Osipov-Shelepin distribution.

This distribution differs from the two-temperature distribution (12.49) by the presence of factors $\prod_{i=0}^n \phi_i$ which remove the population inversion in the region of high energies. At high extents of excitation, where the rate of the V–V-process that tends to invert the population of the neighbouring levels becomes equal to the rate of the V–T-process which removes this inversion, the distribution function reaches its maximum. Under conditions of rapid growth of the probability of the V–T-process, it is possible that the region with the population inversion does not exist at all. Then the distribution reveals the so-called "plateau" where the population slowly (not according to the Boltzmann law) decreases in energy. Note that in the absence of any external excitation, the Gordiez-Osipov-Shelepin distribution, in which the number of vibrational quanta is not conserved, is not a steady-state distribution. We postpone a detailed analysis of the Rich-Treanor and Gordiez-Osipov-Shelepin distributions until later.

In an important case in which the population distribution becomes steady in the presence of radiation whose influence is significant at the times of the order of the molecule free-path time but only negligiblly affects the actual process of collision (the collision is nonradiative), the right-hand side of (12.36) should be supplemented with terms of the form

$$\frac{I}{\hbar\omega}(\sigma_n^{n+1}\rho_{n+1} + \sigma_n^{n-1}\rho_{n-1} - \sigma_{n+1}^n\rho_n - \sigma_{n-1}^n\rho_n)\,,$$

where σ denotes the cross section of the radiative transition and I, the radiation intensity. In this case, (12.36) describes the so-called radiative-collisional cascade process.

Lecture Thirteen
Parametric Phenomena
as Processes in Multilevel Systems

Linear and nonlinear coupling between oscillators, linear and nonlinear polarizabilities. Harmonics and undertones. Cubic nonlinearity. Compound level schemes, multilevel transitions in them. Intermediate resonances. IRS, the Stokes and anti-Stokes components. CARS. Spontaneous processes, probability of spontaneous decay and spontaneous scattering. Spontaneous processes in the presence of a strong field.

Many processes of interaction between radiation and matter are, in fact, parametric and well known in the theory of vibrations. The methods of parametric interaction analysis are well developed in classical theories. But in order to study the processes of resonant interaction we sometimes have to use a quantum approach. This proves inevitable when dealing with the emission of a small number of radiation quanta, spontaneous processes and quantum effects in matter.

Parametric processes in quantum mechanics are to be regarded as those in which the quantum states of both the matter and the field are changed. Generally speaking these are rather complex processes, their dynamics being determined by a number of factors. We shall consider the most simple ones, disregarding relaxation processes and ensemble averages. Neither shall we deal with wave propagation phenomena such as wave synchronism, the presence of resonators and specific modes, spatial coherence, etc., which, in fact, determine the conditions of relaxation for the quantum states of the field. The purpose of this lecture is to describe some nonlinear optical phenomena, with maximum emphasis on the quantum nature of the processes they are based upon.

Parametric processes may be both spontaneous and induced. In the latter case the number of quantum parameters (quantum numbers) required to describe the system is usually not large. Indeed, if many similar photons participate in the process, the induced transitions dominate and the radiation field state may be described by specifying the occupation numbers of a few field oscillators. On the other hand, when the spontaneous processes are of interest, it is necessary to take into account the presence of a large number of field oscillators. For all these reasons it is much more convenient to begin the study of parametric phenomena with the analysis of induced processes.

Let us consider two interacting oscillators. Quantum electrodynamics usually assigns a certain number of quanta of the corresponding field oscillators to each energy level of these oscillators. Parametric interaction of oscillators is a nonlinear interaction and so, as it turns out, is a multiphoton process. This distinguishes the case of parametric coupling of oscillators from the simplest case of their linear coupling.

Indeed, the classical Hamiltonian of linearly coupled oscillators of the same frequency taken to be unity, $\omega = 1$, is

$$H_l = \frac{1}{2}x_1^2 + \frac{1}{2}x_2^2 + \frac{1}{2}p_1^2 + \frac{1}{2}p_2^2 + \alpha x_1 x_2 , \qquad (13.1)$$

where x_1 and x_2 are the generalized coordinates; p_1, p_2, momenta; and α, the coupling coefficient. The motion equations which follow from it,

$$\ddot{x}_1 = -x_1 - \alpha x_2 , \quad \ddot{x}_2 = -x_2 - \alpha x_1 , \qquad (13.2)$$

may be assigned the following quantum Hamiltonian:

$$\widehat{H}_l = a_1^+ a_1 + a_2^+ a_2 + (1/2)\alpha(a_1^+ + a_1)(a_2^+ + a_2) , \qquad (13.3)$$

where symbols a^+ and a denote the creation and annihilation operators, respectively:

$$a^+ = (1/\sqrt{2})(x - jp), \quad a = (1/\sqrt{2})(x + jp) . \qquad (13.4)$$

The system motion (13.2) in the quantum region corresponds to the mutual exchange of quanta (Fig. 13.1 a) or, which is equivalent, to the transitions between the levels of the compound system which are shown in Fig. 13.1 b. For the description of such transitions it is sufficient to retain only the resonant terms in the Hamiltonian (13.3) (the resonant approximation):

$$\widehat{H}_l = a_1^+ a_1 + a_2^+ a_2 + (1/2)\alpha(a_1^+ a_2 + a_2^+ a_1) . \qquad (13.5)$$

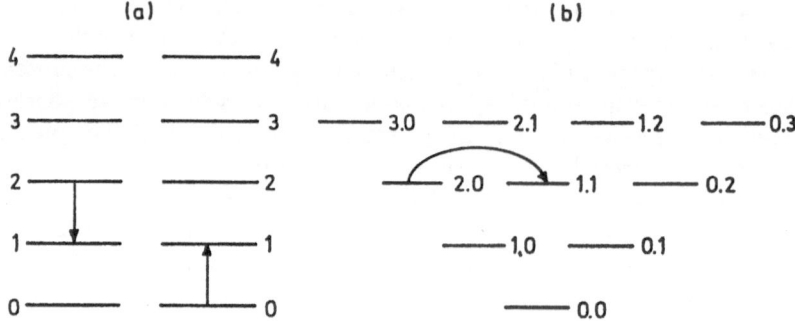

Fig. 13.1. Transition in the system of coupled-linear-oscillator levels (a) and its representation in the compound scheme of levels (b)

In the simplest nonlinear case (the parametric resonance on double frequency) the additional cubic term in the Hamiltonian

$$H_{n.-l.} = (1/2)x_1^2 + x_2^2 + (1/2)p_1^2 + p_2^2 + \alpha x_1^2 x_2 \qquad (13.6)$$

leads to the quadratic nonlinearity in the equations of motion:

$$\ddot{x}_1 = -x_1 - 2\alpha x_1 x_2, \quad \ddot{x}_2 = -4x_2 - 2\alpha x_1^2, \tag{13.7}$$

just as the quadratic addition $\alpha x_1 x_2$ in the Hamiltonian (13.1) gives the linear coupling between Eqs. (13.2). In the resonant approximation the corresponding quantum Hamiltonian has the form

$$\widehat{H}_{n.-l.} = a_1^+ a_1 + 2a_2^+ a_2 + \frac{1}{2\sqrt{2}}\alpha(a_1^+ a_1^+ a_2 + a_1 a_1 a_2^+), \tag{13.8}$$

and the population dynamics may be reduced to transitions in opposite directions, the one-quantum transition in one of the oscillators and the two-quantum transition in the other (Fig. 13.2 a). The corresponding transition in the compound system is shown in Fig. 13.2 b. The comparison of Figs. 13.1 and 13.2 reveals an analogy in the sequence of transitions for the cases of linear and parametric coupling of oscillators. However, a significant difference in the dynamics of the processes of linear and parametric interactions, evident in the classical case from the notation of Eqs. (13.2) and (13.7), becomes apparent in the quantum range in the distinction between matrix elements of the transition between the adjacent levels.

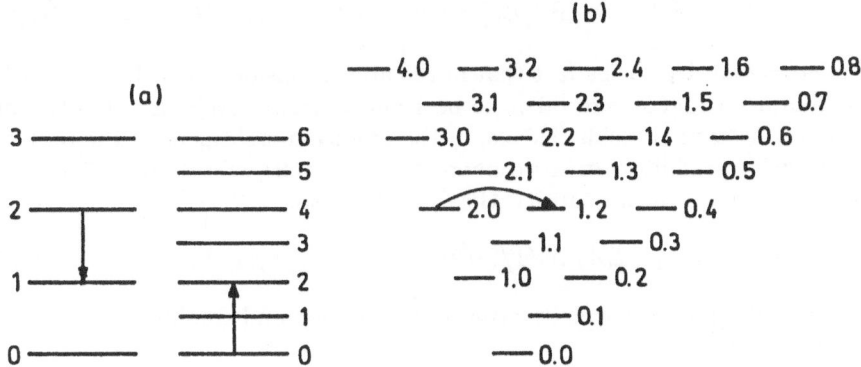

Fig. 13.2. Single- and two-quantum transitions in a system of coupled oscillators with quadratic nonlinearity (**a**) and its representation in the compound system of levels (**b**)

Everything said above refers to abstract oscillators. We are interested in oscillators of field. In a vacuum there is no coupling between them. Coupling arises only in the presence of a material medium as a result of its polarizability. The energy of the interaction between the field and the medium is determined by the product of the medium polarization by the field strength $\boldsymbol{PE}$. If the medium polarizability depends only on the field and may be represented as the sum of linear and nonlinear (say, quadratic) parts over field,

$$P = \kappa\mathcal{E} + \xi\mathcal{E}^2, \tag{13.9}$$

then the energy of the interaction contains a nonlinearity which is of the third order over the field and is responsible for the parametric interaction of the field oscillators. The part of the polarizability linear over the field only changes the propagation velocity of the radiation. These changes are to be taken into account when analyzing the conditions of the wave synchronism (phase matching) well known in nonlinear optics; we shall, however, disregard them in our further consideration.

Let us now consider the interaction of two spatial field harmonics with wave vectors k and $2k$[*)]; then

$$\mathcal{E} = \mathcal{E}_1 \cos kz + \mathcal{E}_2 \cos 2kz . \tag{13.10}$$

Substituting (13.10) in wave Eq. (3.17) and equating the terms corresponding to the same spatial frequency, we obtain

$$(1/c^2)\ddot{\mathcal{E}}_1 = -k^2 \mathcal{E}_1 - (4\pi/c^2)\xi(2\dot{\mathcal{E}}_1\dot{\mathcal{E}}_2 + \mathcal{E}_1\ddot{\mathcal{E}}_2 + \mathcal{E}_2\ddot{\mathcal{E}}_1) ,$$
$$(1/c^2)\ddot{\mathcal{E}}_2 = -4k^2 \mathcal{E}_2 - (4\pi/c^2)\xi(\ddot{\mathcal{E}}_1\mathcal{E}_1 + \dot{\mathcal{E}}_1\dot{\mathcal{E}}_1) . \tag{13.11}$$

This along with the smallness of nonlinearity ξ and the multiplicity of the vibration frequencies of oscillators $\mathcal{E}_1$ and $\mathcal{E}_2$ leads to

$$\ddot{\mathcal{E}}_1 = -\omega^2 \mathcal{E}_1 - 2\alpha\omega^2 \mathcal{E}_1\mathcal{E}_2 , \quad \ddot{\mathcal{E}}_2 = -4\omega^2 \mathcal{E}_2 - 2\alpha\omega^2 \mathcal{E}_1^2 , \tag{13.12}$$

where $\alpha = 4\pi\xi$. Up to the fact that in the above consideration the frequencies of oscillators 1 and 2 were taken to be 1 and 2, respectively, Eqs. (13.12) and (13.7) are seen to coincide. Hence, in the quantum treatment of the problem the following Hamiltonian corresponds to Eqs. (13.2), which are written with respect to the classical quantity, the field strength,

$$\widehat{H} = A_1^+ A_1 + 2A_2^+ A_2 (1/2\sqrt{2})(A_1^+ A_1^+ A_2 + A_1 A_1 A_2^+) . \tag{13.13}$$

This Hamiltonian is written in terms of the creation and annihilation operators in dimensionless time ωt.

Therefore, the quadratic term in the polarizability leads to a parametric interaction of the field oscillators described by the cubic nonlinearity in the quantum Hamiltonian. This reduces the problem of, say, parametric vibration swing by quadratic nonlinearity in the medium polarizability to a problem of population dynamics in a multilevel quantum system (Fig. 13.2 b). If the transition shown in this figure with an arrow occurs from right to left, then such a process is called a generation of the second harmonic or first overtone, while the transition from left to right is referred to as the generation of an undertone. In this case a single quantum of high energy is changed into two quanta of lower energy.

Now let us turn to the origin of the nonlinear polarizability of a substance. The energy of the interaction between the field and the medium is proportional to

$$\boldsymbol{PE} = \langle \Phi | e\widehat{\mu} | \Phi \rangle , \tag{13.14}$$

[*)] We neglect the higher harmonics for simplicity.

where $\varPhi$ is the ψ-functions describing the medium state. If the external field is absent, then, as a rule, $\varPhi = \varPhi_0$ is the ψ-function of the ground state and $\langle \varPhi_0 | \hat{\mu} | \varPhi_0 \rangle = 0$. When the outer field is turned on, the polarization becomes nonzero. The most evident example to illustrate the appearance of cubic nonlinearity is the three-level system shown in Fig. 13.2. When the external field is turned on, then, in accordance with perturbation theory, the ψ-function of the system which was previously in the ground state becomes mixed with the ψ-functions of levels 1 and 2 being

$$\varPhi = \varPhi_0 + (\mathcal{E}\mu_{01}/\hbar\omega_{01})\varPhi_1 + (\mathcal{E}\mu_{02}/\hbar\omega_{02})\varPhi_2 \,. \tag{13.15}$$

The calculation of the interaction energy in (13.14) yields the cubic expansion terms of the form

$$V_3 = 2\mathcal{E}^3 \mu_{01}\mu_{12}\mu_{20}/\hbar\omega_{01}\hbar\omega_{02} \,, \tag{13.16}$$

which ensure the transitions shown in Fig. 13.2 b. The form of these terms coincides with that of the composite matrix elements of multiphoton transitions (9.28). This fact is not just due to chance, since the matrix elements of the type of (13.16) correspond to multiphoton (three-photon, in our case) transitions in the total compound system of the energy levels of the field and the material medium.

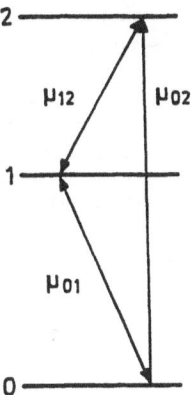

Fig. 13.3. Transitions in a three-level system

The matrix elements of the nonlinear interaction of the type of (13.16) are properly related with nonlinear susceptibilities. Substituting formula (13.9) into the left-hand side of (13.14), and formula (13.15) into its right-hand side, we obtain by the comparison of the terms with the same powers of $\mathcal{E}$

$$\chi = 2\mu_{01}\mu_{12}\mu_{20}/\hbar\omega_{01}\hbar\omega_{02} \,. \tag{13.17}$$

The transitions considered for the case in which the energy of the field quanta is much less than the typical spacing between the energy levels of the material medium are shown in Fig. 13.4. Factor 2 in (13.6) corresponds to two possible paths of transition $2.0.0 \rightarrow 1.2.0$ in the system of the field-oscillator energy levels. The transitions via the intermediate levels shown in Fig. 13.4 correspond to the transition from left to right in the system of the levels of field oscillators in Fig. 13.2 b.

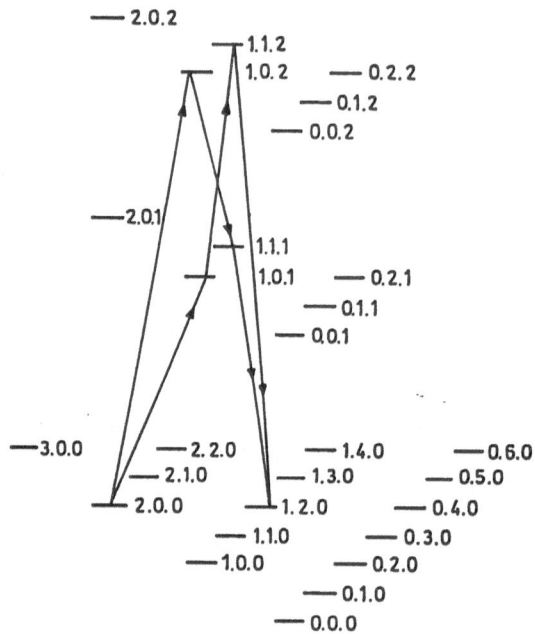

Fig. 13.4. The compound level scheme of two field oscillators ω_1 and ω_2 (the first two indices) and a three-level medium (third index) for the case $\omega_{01}, \omega_{02} \gg \omega_1, \omega_2$. Two possible paths of transition $|2.0.0\rangle \rightarrow |1.2.0\rangle$ corresponding to the generation of an undertone are shown

In the case of the field quanta energy comparable to the typical inter-level distances of the medium one should take into consideration the field frequencies along with the medium frequencies when determining the transition probability. In this case the maximum contribution to the transition probability is rendered by the trajectory which suits the conditions of intermediate resonances best, which is, as we know, the common property of multiphoton transitions. This implies, as one would anticipate, the coincidence of at least one field frequency with one of the medium frequencies. Note that it clearly shows the spectroscopic applications of the method of the generation of harmonics (the second, third ones, etc.). Measuring the frequencies where the nonlinear polarizabilities have a maximum, one may determine the frequencies of the transitions in the system under investigation. The compound scheme of levels for the presently considered case is given in Fig. 13.5. In the example shown, transition $2.0.0 \rightarrow 1.2.0$ occurs along the trajectory $2.0.0 \rightarrow 1.0.1 \rightarrow 1.1.2 \rightarrow 1.2.0$, where the double field frequency coincides with that of transition $0 \rightarrow 2$ in the scheme of the medium energy levels.

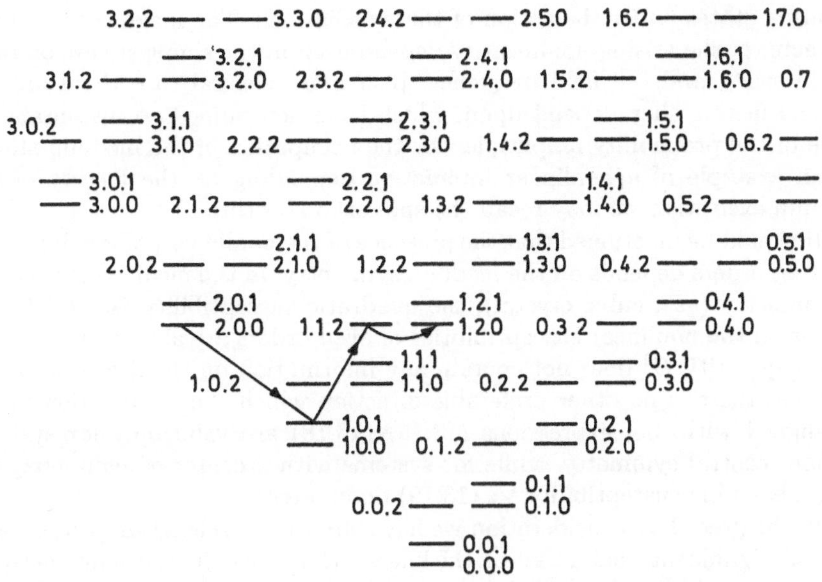

3.2.2 —— ——3.3.0   2.4.2 —— ——2.5.0   1.6.2 —— ——1.7.0

——'3.2.1   ——2.4.1   ——1.6.1
3.1.2 —— ——3.2.0   2.3.2 —— ——2.4.0   1.5.2 —— ——1.6.0   0.7

3.0.2   ——3.1.1   ——2.3.1   ——1.5.1
        ——3.1.0   2.2.2 —— ——2.3.0   1.4.2 —— ——1.5.0   0.6.2 ——

——3.0.1   ——2.2.1   ——1.4.1
——3.0.0   2.1.2 —— ——2.2.0   1.3.2 —— ——1.4.0   0.5.2 ——   ——

        ——2.1.1   ——1.3.1   ——0.5.1
2.0.2 —— ——2.1.0   1.2.2 —— ——1.3.0   0.4.2 —— ——0.5.0

——2.0.1   ——1.2.1   ——0.4.1
——2.0.0   1.1.2 ⟋⟍ ⟋⟍1.2.0   0.3.2 —— ——0.4.0

                ——1.1.1   ——0.3.1
1.0.2 ⟍   ——1.1.0   0.2.2 —— ——0.3.0

            ——1.0.1   ——0.2.1
            ——1.0.0   0.1.2 —— ——0.2.0

                ——0.1.1
        0.0.2 —— ——0.1.0

                ——0.0.1
                ——0.0.0

Fig. 13.5. The compound level scheme of two field oscillators and three-level medium for the case $2\omega = \omega_{02}$. Transition $|2.0.0\rangle \rightarrow |1.2.0\rangle$ (the generation of undertone $\omega/2$) is shown. Intermediate level $|1.1.2\rangle$ is resonant with respect to level 1.2.0⟩

Let us mention that if the field frequency is not small against the transition frequencies of the system the expressions for nonlinear susceptibilities become more complicated compared to (13.17). In this case it is necessary to perform a summation over the intermediate energy levels with allowance made for the energy of the field quanta. In the general case of a multilevel material medium this yields the expression for quantity χ_2 in the form

$$\chi_2 = \sum_{m,k,s,p} \frac{1}{\hbar^2} \frac{\mu_{nm}\mu_{mk}\mu_{kn}}{\Delta_{ms}\Delta_{kp}}, \tag{13.18}$$

where n is the starting level, while Δ_{ms} and Δ_{kp} are the detunings of the intermediate states in the compound system of medium levels (indices m and k) and those of the field (indices s and p).

In practice one often encounters nonlinear susceptibility of the third order χ_3, which is the coefficient before the next term of the polarization expansion over the field powers (13.9). We may easily generalize (13.18) to include the case of cubic polarizability:

$$\chi_3 = \sum_{m,k,l,s,p,q} \frac{1}{\hbar^3} \frac{\mu_{nm}\mu_{mk}\mu_{kl}\mu_{ln}}{\Delta_{ms}\Delta_{kp}\Delta_{lq}}. \tag{13.19}$$

The above consideration illustrates the fact that the nonlinear interaction of radiations occurs due to the changes in the populations of the medium quantum states under the action of these radiations. The magnitude and the character of the arising nonlinearity depend even in the simplest case on both the intensity and the field frequency. In a more general case there are additional factors they depend upon, which have determined an instantaneous value of the probability amplitudes for the occupation of the medium states. As an example of a nonlinear interaction depending on the history of the medium excitation we may recall the spin echo (Lecture 3).

It should be mentioned that the presence of the nonlinear polarizabilities of different orders depends on the medium symmetry. In the media with a center of symmetry (gas, cubic crystal), the quadratic susceptibility (see (13.9)) as well as all the nonlinear susceptibilities of even orders are absent. The reason is that quantity E^2 does not contain any information on the direction of the field and there is no other preferable direction which the polarization might be aligned with. So, expressions (13.15)–(13.18) are valid only for systems without central symmetry, while for systems with a center of symmetry, the next, i.e., cubic susceptibility χ_3 (13.19) dominates.

In the preceding consideration we have ignored all relaxation processes. If they are significant, their effect on the line width are usually taken into account by inserting additional imaginary terms into the resonant denominators of the formulae for the nonlinear polarizabilities of the type (13.17)–(13.19).

So far we have considered parametric processes which affect the state of the radiation field while the medium returns to its initial state. However, the processes, when states of both the radiation field and the system change, exist as well. If the system returns to its initial state, it occurs beyond the framework of the coherent process considered. An example of this kind is the induced Raman scattering (IRS).

In the IRS process the frequency of the emitted light (ω_s) differs from that of the exciting radiation (ω_1). Their difference corresponds to the resonant frequency of the internal medium motions (ω_m). The excitation of these motions ensures the energy conservation law, $\omega_1 = \omega_s + \omega_m$. The IRS interaction is linear over each field. Thus, in the energy expansion over the field, the quadratic term corresponds to this interaction. Nevertheless, it is parametric in the sense that the field-oscillator coupling, despite its linearity, is implemented at the expense of the changes in the medium state (changes in the medium parameters) because of the nonresonance of the field oscillators. In other words, the radiation quantum of high energy is changed into the lower-energy quantum and the quantum of medium vibrations. In the opposite case, where the quantum of the medium eigen vibrations and that of the field are added, the emission of the anti-Stokes radiation takes place.

Everything said above clarifies that the IRS process takes place only when accompanied by a change in the state of the medium. In the Stokes process the system gains energy and in the anti-Stokes one, loses it. Hence, somewhat prolonged pumping of the excitation energy in the Stokes (anti-Stokes) radiation may occur only in the system containing many identical particles. Let the energy spectrum of a single particle be described by the scheme shown in Fig. 13.6 a. Then the spectrum of a system of N particles may be represented

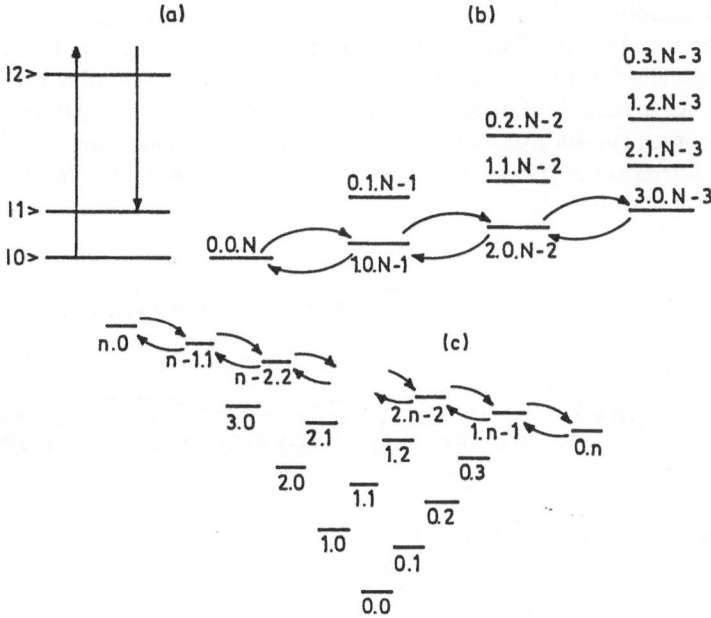

Fig. 13.6. IRS in the system of N identical particles: (a) the scheme of single-particle levels; (b) the case of N particles; (c) the compound system of the levels of the Stokes and anti-Stokes radiations. The representation of occupation numbers (b): the first number represents the number of particles in state 1; second, the number of particles in excited state 2; third, the number of particles in the ground state

by the scheme given in Fig. 13.6 b, where the first index denotes the number of particles promoted to the excited state 1 of the scheme shown in Fig. 13.6 a; the second index, to state 2; while the third one denotes the number of particles in the ground state. It is natural that each level of the scheme shown in Fig. 13.6 b is multiply degenerate since it contains many quantum states corresponding to the different ways the excitation can be distributed among the particles. The scheme of the energy levels of field oscillators corresponding to the exciting and Stokes (anti-Stokes) radiations is shown in Fig. 13.6 c. The distinction from the scheme shown in Fig. 13.6 b consists in the nonresonance of the field states with an equal total number of photons. In the IRS process the transitions between both the energy levels of the system of particles and between the field states take place, the latter shown in Fig. 13.6 b, c by arrows. As previously, the arrows going from left to right correspond to a decrease of the field frequency. Note, that the smaller frequency with respect to the higher one is the Stokes frequency, while the higher frequency is anti-Stokes with respect to the lower one. The compound scheme of the levels of the field oscillators and the system of many particles is given in Fig. 13.7 a. The first two indices specify the number of quanta in the field oscillators (Fig. 13.6 c), the last three denote the numbers of particles (Fig. 13.6 b). Two-quantum transitions of the type $k, p, 0, 0, N \rightarrow k-1, p+1, 1, 0, N-1$ passing through the intermediate level with minimum detuning $k-1, p, 0, 1, N-1$ correspond to the IRS process. The probabilities of such multiquantum processes may be determined in accord with the formulae of Lecture 9. Naturally, this probability is maximum when the intermediate detuning is minimum. To avoid the particulars of the multi-quantum mechanism of IRS excitation one may sometimes use a simplified representation of the level scheme of the resonant medium and transitions given in Fig. 13.7 b.

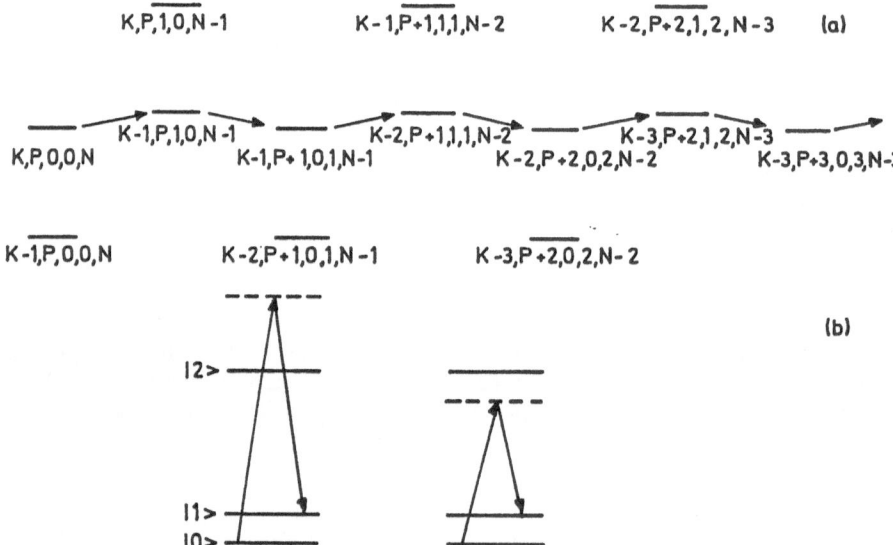

Fig. 13.7. (a) the IRS process in the compound level scheme of the field oscillators and the system of many particles. The representation of occupation numbers: first index, the number of pumping photons; second index, the number of the Stokes photons; third index, the number of particles in state $|2\rangle$; fourth index, the number of particles in state $|1\rangle$; fifth index, the number of particles in state $|0\rangle$. The energy levels are multiply degenerate since, at a fixed total number of excited particles the excitation may be differently distributed among the concrete particles. (b) the IRS process in the simplified level scheme of the resonant medium

As we have already mentioned, in the IRS both the Stokes and anti-Stokes emission components are possible. Nevertheless, it should be mentioned that the anti-Stokes radiation appears if the resonant levels of the medium are initially populated or become populated in the process of the Stokes emission. In addition, it is evident that sufficiently intense Stokes and anti-Stokes components may, in turn, cause the appearance of proper scattering components, i.e., be the reason for the emission of the Stokes and anti-Stokes lines of higher orders with respect to the initial radiation. The consideration of such processes, which are also multiquantum processes, is more involved since it requires a larger number of field oscillators to be taken into account. Note, however, one of them – the process of coherent anti-Stokes Raman scattering (CARS) – which has found many spectroscopic applications. Figure 13.8 a depicts a simplified scheme of the CARS generation. The population of state 1 – the starting state of the anti-Stokes radiation process – is implemented not by the stimulated emission of the Stokes component but as a result of the intense external field action at frequency ω_2. Since a single elementary act of CARS returns the system to its initial state, it is possible to carry out a single-particle consideration of the process. The replacement of the Stokes radiation by an intense external field, if it is resonant, $\omega_2 = \omega_s$, makes the process of the occupation of state 1 more effective, and while the tuning of the frequency of this field is available, permits spectroscopic studies to be performed. Figure 13.8 b shows the compound level scheme of the medium and the field oscillators where the arrows represent a four-photon process of the CARS generation. The probability of this process is maximum when the intermediate resonance is observed, i.e., when the difference of the energies of states $n, m, k, 0$ and $n - 1, n + 1, k, 1$ is small, where n, m and k are the numbers of the field quanta of frequencies ω_1, ω_2 and ω_s, respectively, while the fourth index refers to the system state.

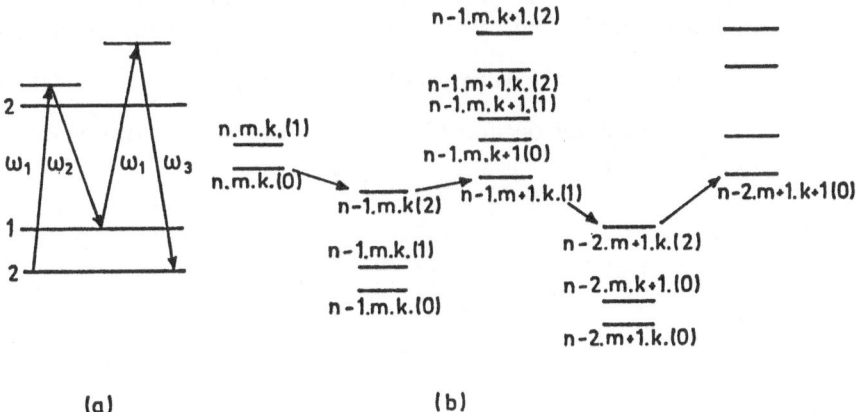

Fig. 13.8. The schemes of CARS generation: (a) simplified scheme; (b) compound level scheme of the field oscillators and the medium

The previous consideration was based on the assumption that the field oscillators participating in the interaction considered are given, which is, generally speaking, not always valid. In the analysis of the processes of small intensity or the initial stages of the intense processes one should not disregard the possibility of spontaneous emission of the photons with arbitrary non-preconditioned values of frequency and wave vector. The question of spontaneous emission is briefly discussed at the end of Lecture 4 (see also note 1 of Appendix). Now it is worthwhile investigating it in some more detail.

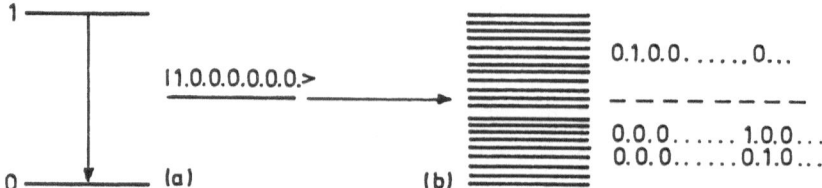

Fig. 13.9. Spontaneous emission: (**a**) transition in the scheme of the particle levels; (**b**) transition in the level scheme of the particle and the field. First index specifies the atomic states, all the others denoting the occupation numbers of different field oscillators

Figure 13.9 a depicts the traditional representation of spontaneous decay. The compound spectrum of the system and the field oscillators is more complex. This is the spectrum of the "single level-band of levels" type (Fig. 13.9 b). In the case of free space the spectrum of the field oscillators is the continuum. Then the transition dynamics is described by formula (10.8) for a dense band. According to (10.8) the probability of spontaneous transition $1 \rightarrow 0$ is

$$A_{10} = (\pi/\hbar^2)\mu^2\langle\mathcal{E}^2\rangle g(\omega)\,. \tag{13.20}$$

If we take into account that the mean-square matrix element of operator $\mathcal{E}$ for each field oscillator may be expressed in terms of the volume energy density,

$$\langle\mathcal{E}^2\rangle/7\pi = \hbar\omega(n+1/2)/V = \hbar\omega/2V\,, \tag{13.21}$$

where V is the space volume occupied by the field and the spectral density of the oscillators number is

$$g(\omega) = V\omega^2/\pi^2 c^3\,, \tag{13.22}$$

then

$$A_{10} = 4\frac{\omega^3}{c^3}\frac{\mu^2}{\hbar}\,. \tag{13.23}$$

This expression contains the mean-square value of the matrix element of the transition dipole moment operator $0 \rightarrow 1$, which is, in general, distinct from the usually introduced total matrix element of a spherically symmetric particle. In the latter case the factor $1/3$ should be inserted in formula (13.23). Then this expression will coincide with the well-known formula for Einstein's coefficient A.

Let us return to the combination scattering and consider the spontaneous scattering, i.e., the emission into many field oscillators. Here we may consider only a single particle, since the scattering events are rare and the medium has enough time to return to its initial state during the time span between them. Figure 13.10 gives simplified schemes showing single-particle energy levels and different possible radiative transitions. Figure 13.10 a refers to IRS; Fig. 13.10 b and 13.10 d pertain to the spontaneous combination scattering (RS); Fig. 13.10 c shows the resonant Rayleigh scattering; and Fig. 13.10 e corresponds to the process of spontaneous fluorescence which arises at RS occupation of an upper level. IRS (Fig. 13.10 a) and spontaneous decay (Fig. 13.10 e)

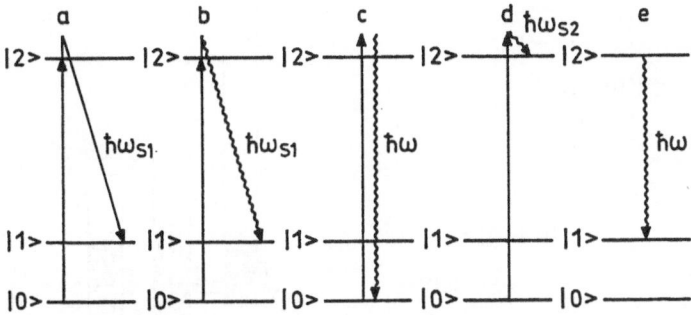

Fig. 13.10. Radiative transitions in the system of three energy levels: (**a**) IRS; (**b**) spontaneous RS; (**c**) resonant Rayleigh scattering; (**d**) spontaneous RS with large frequency change; (**e**) spontaneous fluorescence

have been considered above. To analyze the remaining three processes, let us turn to the compound spectrum of the field oscillators and the scattering-medium particle shown in Fig. 13.11. In the level notations in this figure the first index shows the number of photons of the exciting field oscillators; the second one indicates the energy level of a particle, the third denotes the number of spontaneous photons in all possible field oscillators. The spontaneous process represented in Fig. 13.10 b is seen to correspond to the two-photon transition of the "level-continuum" type from state $(n, 0, [0])$ to state $(n - 1, [1])$ via the intermediate level $(n - 1, 2, [0])$. The probability of this process equals the product of the mean-square composite matrix element by the density of the final state and differs from the probability of the spontaneous emission (13.23) by the factor $(\mu_{02}\mathcal{E}/2\hbar)^2(\omega_{02} - \omega)^{-2}$. This factor originates from the nature of the composite matrix element of a two-photon transition (see (9.23)). As a result, the probability of the spontaneous scattering is

$$W_s = \frac{\omega_s^3 \mu_{12}^2 \mu_{02}^2}{c^3 \hbar^3 (\omega_{02} - \omega)^2} \mathcal{E}^2, \tag{13.24}$$

where $\mathcal{E}$ is the strength of the radiation field to be scattered and ω is its frequency.

The two-photon transition $(n, 0, [0]) \to (n - 1, 2, [0]) \to (n - 1, 0, [1])$ corresponds to spontaneous Rayleigh scattering; transition $(n, 0, [0]) \to (n - 1, 1, [0]) \to (n - 1, 2, [1])$ corresponds to long-wave combination scattering. The probabilities of these two-quantum processes are given by the formulae similar to (13.24) with the appropriate replacement of the dipole moment values, the values of intermediate detunings and the frequencies of spontaneous emission. It is evident that the process shown in Fig. 13.10 d is unlikely since it occurs via a distantly detuned intermediate level and is accompanied by the emission of long-wave photons corresponding to the small density of field oscillators and the small strength of field of the zero-point oscillations. It should be mentioned here that, in reality, the population of level 2 required for spontaneous fluorescence to occur (Fig. 13.10 e) may take place due to many other

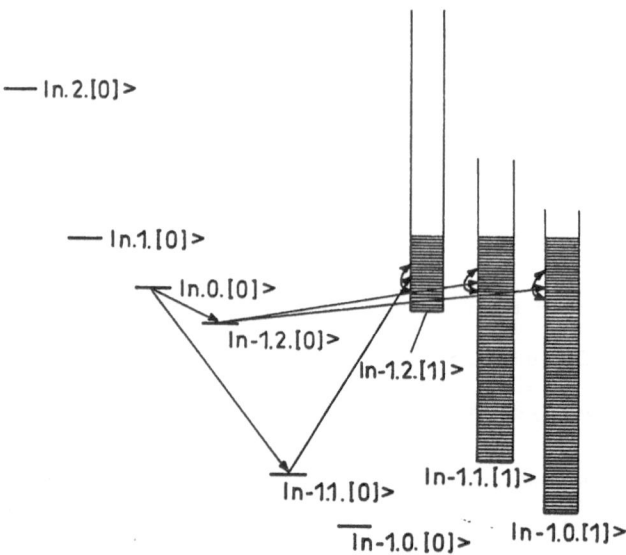

— $|n.2.[0]>$

— $|n.1.[0]>$

$|n.0.[0]>$

$|n-1.2.[0]>$

$|n-1.2.[1]>$

$|n-1.1.[0]>$ $|n-1.1.[1]>$

$\overline{|n}-1.0.[0]>$ $|n-1.0.[1]>$

Fig. 13.11. Compound spectrum of field oscillators and scattering medium. Transitions shown in Fig. 13.10, b, c, d are shown. Representation of occupation numbers: first index, number of photons of the radiation being scattered; second index, quantum state of a particle; third index, state of the field of all the other (spontaneous) photons

processes, and due to the radiative transition into the collisionally broadened line wing, in particular.

Let us turn again to the probability (13.24). Its magnitude sharply increases with a decrease in the detuning $\omega_{02} - \omega$. Naturally, this increase cannot be unlimited and in many cases it is confined by the width of level 2 accounted for by the collisions or some other processes not considered here. In the cases in which the relaxation may be ignored, the increase in the emission probability takes place until quantities $\mu\mathcal{E}$ and $\hbar(\omega_{02} - \omega)$ become of the same order of magnitude, and the coherent processes become essential. Then one speaks of spontaneous emission in the presence of a strong field.

For the description of spontaneous processes in a strong field, when coherent processes are taking place in the system, the quasi-energy representation is to be employed. In other words, the strong field containing many photons is treated classically, while the field oscillators occupied spontaneously are described in a quantum way. The consistent quantum description of these phenomena supposes the possibility of a change in the energy of spontaneous photons at the expense of a corresponding change in the statistics of the strong-field photons. Let us consider a three-level system inserted into a classical field with the frequency equal to that of transition $0 \to 2$ (Fig. 13.12 a). The quasi-energy spectrum of such a system is given in Fig. 13.12 b. In the field of zero strength the quasi-energies of states $(k, 0)$ and $(k - 1, 2)$ coincide. This degeneracy is removed with the field strength, and two stationary

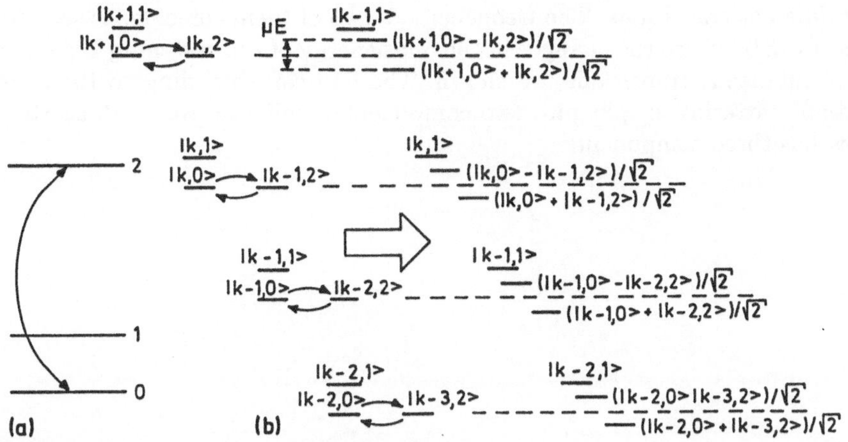

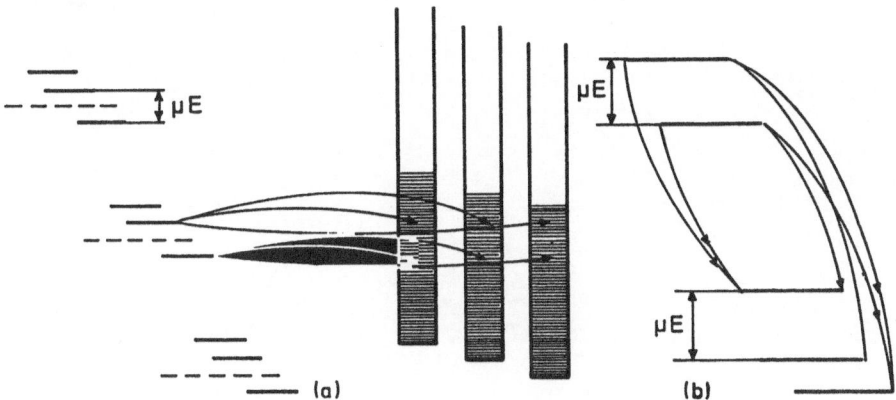

Fig. 13.12. (a) three-level system in a strong field. (b) the splitting of the quasi-energy levels

Fig. 13.13. The splitting of the combination and Rayleigh scattering lines in a strong field: (a) the compound spectrum of the quasi-energy levels of the system and the field oscillators; (b) spontaneous transitions between the quasi-energy levels of the system

quasi-energetic states, which are symmetric and antisymmetric combinations of the initial ones, are formed. The distance between these states comprises $\mu\mathcal{E}$. The compound spectrum of the quasi-energy levels of the system and the field oscillators is given in Fig. 13.13 a. This spectrum is similar to that given in Fig. 13.11, the only difference being that it refers to the quasi-energy states of the system in a strong field but not to the numbers of the exciting field photons and the system states. If prior to switching on the external field the lower level is populated, then after the field has been turned on the quasi-energy doublets turn out to be populated. These doublets may serve as the starting point for transitions into states corresponding to the quasi-energy levels lying below, accompanied by spontaneous photon emissions. There are six different transitions. The frequencies of two of them coincide, as seen from Fig. 13.13 b, where the changes in the quasienergy of the system in the process of spontaneous transitions are shown. The line corresponding to the combination scattering is split into two components, while the Rayleigh scattering now has three components.

Lecture Fourteen

Resonant Excitation of Atoms

Selective two-step ionization of atoms. Spectroscopic condition. Probability of excitation and ionization. Resonant transfer of excitation energy, resonant recharge. Laser isotope separation in atomic vapor, conditions of its effectiveness. Autoionization, methods of artificial autoionization. Multiphoton ionization of atoms. Example of three-photon ionization of metastable helium atoms. Parametric generation.

Now let us turn to the applications of the intense resonant interaction of laser light with matter. It is natural to begin our consideration with atoms. The relative simplicity and detailed specification of spectra are attractive and invite a description of the interaction of atoms with radiation. The point is that in the case of atoms the processes considered in the previous lectures may be easily taken into account with the least possible number of variables; besides, the results obtained are often used in experiments. This makes the conclusions of such a consideration most easily understandable and their physical meaning most clear.

Let us consider, for example, laser isotope separation in a monoatomic gas. The idea of such a separation is as follows. In the atomic spectra isotopic shifts are observed. The point is that the nuclei of isotopes differ somewhat in mass, size and shape. This distinction slightly, though noticeably, changes both the electrostatic potential of a nucleus and the position of the center of mass of the entire atom. This becomes apparent in the atomic energy spectrum as the isotopic shift. The presence of an isotopic shift in the absorption spectrum allows atoms of the only one specific type to be excited with monochromatic laser light. But this excitation, generally speaking, relaxes due to both the spontaneous emission and the collisional X–T-relaxation. Besides, a collisional resonant exchange takes place, causing an excitation transfer to an unwanted isotopic component. So, the selectivity of the excitation has to be fixed. A suitable method for such a fixation is the ionization of excited atoms by radiation with the energy of the quantum insufficient to ionize the unexcited atom. The ions may be easily extracted from the mixture, say, by a constant electric field. This idea is illustrated by the scheme in Fig. 14.1. The evident condition for the frequencies of these two stages ω_1 and ω_2 required for the selective ionization follows from this figure:

$$\hbar\omega_1 + \hbar\omega_2 > U, \quad \hbar\omega_1, \hbar\omega_2 < U, \tag{14.1}$$

where U is the ionization potential of an atom.

One may imagine the following scheme of then experimental implementation of such a process. In a chamber with two electrodes required for the extraction of ions a monoatomic gas (atomic vapor) is created. This gas is illuminated by two lasers working at frequencies satisfying condition (14.1). The frequency of the first stage ω_1 is resonant with respect to the corresponding transition of the desired isotope.

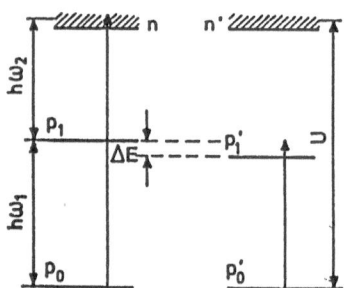

Fig. 14.1. The scheme of two-step selective ionization. The presence of an isotopic shift ΔE is shown

Two parameters – the separation coefficient and the mass productivity of the method – are most important in the problem of the laser isotope separation. It is evident that the mass productivity should increase with the gas pressure while the separation coefficient should decrease. A compromise between these tendencies is achieved at a quite definite pressure value. The corresponding density is usually found in the range 10^{13}–10^{16} cm^{-3}. At such densities of the atomic vapor an essential role is played by the transverse relaxation. Time T_2 belongs to the range 10^{-8}–10^{-10} s. Under such conditions the process of first-level excitation is of an incoherent nature. The excitation probability (see (6.3), (5.31) and (5.32)) is determined by the transition cross section σ_1 and the field intensity I_1:

$$W_1 = I_1\sigma_1/\hbar\omega_1 . \tag{14.2}$$

In some cases, especially at small pressures, the situation is aggravated by the presence of Doppler broadening which, in the visible region of the spectrum at the atomic vapor temperature of 1000–2000 K, usually has a value in the range of $(1-2)\cdot 10^9$ s^{-1}.

The transition from the excited state into the ionized one is that of the "level-continuum" type. If special care is not taken, this continuum is uniform and the spectral characteristic of the ionization transition is wide and, hence, the cross section is small:

$$\sigma_2 = 8\pi^2 \langle\mu^2\rangle g\hbar\omega/c\hbar^2 , \tag{14.3}$$

(see (10.8)). A sharp distinction between the cross-sections of the first-stage resonant transition (usually $\sigma_1 = 10^{-12}$–10^{-14} cm^2) and the ionization nonresonant transition (usually $\sigma_2 = 10^{-17}$–10^{-18} cm^2) presents one of the experimental complications of an efficient separation process.

Another difficulty is the resonant processes of collisional exchange. There are two of them (Fig. 14.2).

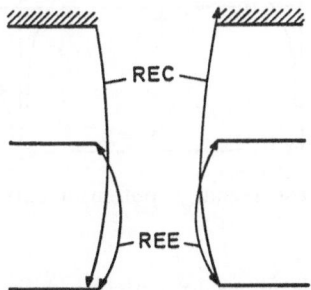

Fig. 14.2. Resonant processes of the collisional exchange: REC (resonant exchange by charge) – recharge; REE (resonant exchange by excitation) – excitation transfer

The first is the process of resonant excitation energy transfer (REE) from the atom of one isotope to that of another. We may call it the E–E-process. Since the difference between the energies of the excited levels (the isotopic shift) is much less than both the typical thermal motion energy kT and the inverse time of flight of one atom past another (in terms of $\hbar$), the E–E-process may be considered to be strictly resonant. Its cross section satisfies the relation

$$\sigma_{\text{REE}} = \pi r_W^2 , \tag{14.4}$$

where r_W is the Weisskopf radius (12.14). For the allowed dipole transitions the value of σ_{REE} is greater by two or three orders of magnitude than the gas-kinetic cross section. So, large cross sections of the resonant excitation transfer impose strict limitations on the intensity of the second-stage field. The point is that the parasitic influence of the resonant exchange may be eliminated only if its probability is less than that of the ionization by the field

$$W_2 = \frac{I_2 \sigma_2}{\hbar \omega_2} > \sigma_{\text{REE}} v_T N , \tag{14.5}$$

where v_T is the relative thermal velocity and N is the concentration of the colliding atoms.

The second, more important, effect leading to the loss of laser action selectivity in a relatively dense atomic vapor is the resonant recharge (REC). The ions of the selectively excited isotopes produced in the ionization process collide with the neutral atoms of other isotopes of the same elements, interchanging electrons with them. This process develops as follows. Approaching the ion, the outer electron of a neutral atom tunnels into the near-Coulomb potential well of the latter. Figure 14.3 shows the way in which the potential barrier decreases due to the long-range character of the Coulomb interaction.

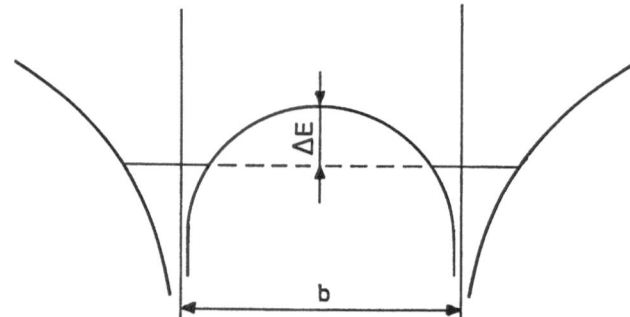

Fig. 14.3. The lowering of the recharge potential barrier at rapprochement of a neutral atom and an ion

As a result, the typical distances at which recharge processes are effective become several times larger than the atomic size. The barrier effectively disappears when the energy of the Coulomb interaction becomes equal to the ionization potential. This yields the lower limit for the recharge cross section in the form

$$\sigma_{\text{REC}} = \pi \left(\frac{e^2}{U} \right)^2 \approx \frac{4.5 \cdot 10^{-14}}{U^2 [\text{eV}]} \quad \text{cm}^2. \tag{14.6}$$

The allowance for under-barrier tunneling which takes place at distances somewhat exceeding e^2/U increases, this cross section yielding an additional term logarithmically depending on the relative velocity.

The recharge involves already-obtained products of the radiative action, so it cannot be controlled by laser light. Its influence may only be eliminated by the action upon the ions which should be extracted from the cloud of neutral atoms in less than the recharge time. This imposes limitations on the atomic vapor density and the geometry of the experiment:

$$N\sigma_{\text{REC}}l < 1, \tag{14.7}$$

where l is the size of the cloud of neutral atoms in the direction of ion extraction.

There is a great deal of experimental material on the laser separation of isotopes by the selective ionization of atoms. Take, for example, the separation of the isotopes of the rare earths carried out for all of them despite the differences in the isotopic and hyperfine structures of the spectra, in the melting temperatures and the chemical activity of the melt, and in the ionization potentials of the metals of this group.

Suitable for isotope separation by the method of selective ionization of atoms are the pulsed and the continuous-wave dye lasers. In principle, laser radiation may be completely utilized and, since for the extraction of one mole of a substance it is necessary to spend at least $6 \cdot 10^{23}$ quanta, then a laser energy no less than $N_A \hbar \omega / m_0$ is required for the extraction of one gram of an isotope, where N_A is the Avogadro number, and m_0 is the atomic mass.

This means that 1 gram of a heavy-element isotope requires no less than 1 kJ of laser energy.

A convenient experiment geometry may be implemented with the use of a slit-like source of atomic vapor, as shown in Fig. 14.4. This geometry does not only most fully utilize laser radiation in spite of the differences in the cross sections of excitation and ionization but permits use of the maximum density of atomic vapor limited only by the recharge processes.

The smallness of the ionization cross section and the large cross section of the resonant excitation transfer even at a vapor density of 10^{15} cm^{-3} require a second-stage radiation intensity of about 10^4–10^5 W/cm^2. Such intensities of laser radiation are natural for the pulsed mode of laser functioning. In this case a moderate repetition rate may ensure an acceptable mean power and a sufficient peak intensity. The repetition rate should be chosen so that in the interval between pulses, the gas in the irradiation region has time enough to change. Then the gas may be considered immobile during the pulse, and the extraction of ions may be described phenomenologically.

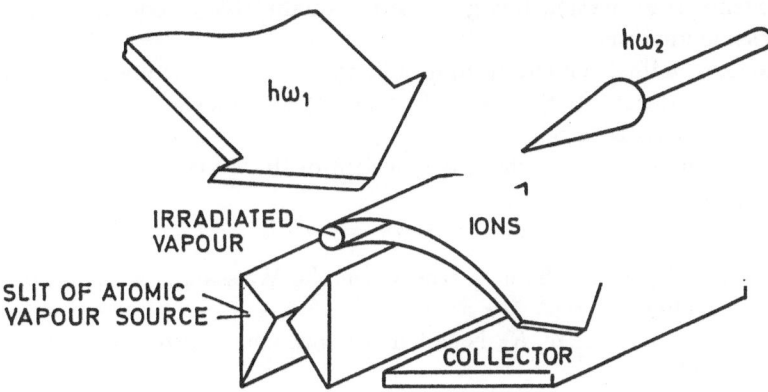

Fig. 14.4. Atomic beam ionization in crossed geometry

Then the process of the separation of the binary isotopic mixture by the method of two-step excitation is described by the following set of equations:

$$
\begin{aligned}
\dot{\rho}_0 &= -W_1(\rho_0 - \rho_1) + T_1^{-1}\rho_1 + K_1\rho_0'\rho_1 \,, \\
\dot{\rho}_1 &= W_1(\rho_0 - \rho_1) - (T_1^{-1} + W_2)\rho_1 - K_1(\rho_0\rho_1' + \rho_1\rho_0') \,, \\
\dot{\rho}_0' &= -W_1'(\rho_0' - \rho_1') + T_1^{-1}\rho_1' + K_1\rho_0\rho_1' \,, \\
\dot{\rho}_1' &= W_1'(\rho_0' - \rho_1') - (T_1^{-1} + W_2)\rho_1' - K_1(\rho_0\rho_1' + \rho_1'\rho_0') \,, \\
\dot{n} &= W_2\rho_1 - n(K_2^{(0)}\rho_0' + K_2^{(1)}\rho_1') + n'(K_2^{(0)}\rho_0 + K_2^{(1)}\rho_1) - n/\tau \,, \\
\dot{n}' &= W_2\rho_1' - n'(K_2^{(0)}\rho_0 + K_2^{(1)}\rho_1) + n(K_2^{(0)}\rho_0' + K_2^{(1)}\rho_1') - n'/\tau \,.
\end{aligned}
\tag{14.8}
$$

Here ρ_0 and ρ_0' are the concentrations of each isotopic modification in the ground state; ρ_1 and ρ_1' are the concentrations in the excited state; n and n'

are the concentrations of the corresponding ions. The probabilities of radiative transitions W_1 and W_1' corresponding to the proper isotopic components are determined by formula (5.7). In the notations used here we obtain

$$W_1(\omega) = I_1 \frac{\sigma_1(\omega)}{\hbar \omega_1} = I_1 \frac{8\pi}{c} \left(\frac{\mu_{01}}{2\hbar} \right)^2 \frac{2}{T_2} \left[\left(\frac{1}{T_2} \right)^2 + (\omega - \omega_{01})^2 \right]^{-1} , \quad (14.9)$$

$$W_1'(\omega) = I_1 \frac{\sigma_1'(\omega)}{\hbar \omega_1} = I_1 \frac{8\pi}{c} \left(\frac{\mu_{01}}{2\hbar} \right)^2 \frac{2}{T_2} \left[\left(\frac{1}{T_2} \right)^2 + (\omega - \omega_{01}')^2 \right]^{-1} . \quad (14.10)$$

It is easily seen that the probabilities and, hence, the transition cross sections may significantly differ. When the first-stage field is tuned in resonance with the frequency of one isotope (let it be frequency ω_{01}), the probability of its excitation considerably exceeds the excitation probability of the nonresonant isotope if the isotopic frequency shift $\omega_{01}' - \omega_{01}$ is greater than the transition line width. Hereafter we consider the case in which the homogeneous broadening exceeds the inhomogeneous one. The opposite case of an essentially inhomogeneous excitation being the same qualitatively requires much more cumbersome analysis.

The longitudinal relaxation in atomic systems is determined, as a rule, by a spontaneous decay. So, time T_1 in Eqs. (14.8) coincides with the spontaneous lifetime of an excited atom.

Coefficient K_1 is called the rate constant of the resonant excitation energy transfer and equals

$$K_1 = \sigma_{\mathrm{REE}} v_T = \pi \mu^2 / \hbar , \quad (14.11)$$

(see (14.4) and (12.14)). This is true when the Weisskopf radius exceeds the gas-kinetic radius.

By analogy, coefficient K_2 is called the rate constant of the recharge process and equals

$$K_2 = \langle \sigma_{\mathrm{REC}}(v)v \rangle , \quad (14.12)$$

where the average is taken over the distribution of extracted ions, while v is their velocity with respect to neutral atoms. From (14.6), quantity K_2 is seen to be inversely proportional to the ionization potential squared. Since in the scale of energies the distance from the excited level to the ionization is less than that from the ground state, the rate constants of recharge for excited and unexcited neutral atoms may differ significantly. That is taken into account in Eqs. (14.8) by the corresponding indices. Symbol τ denotes the effective time of ion extraction. This time is estimated by the ratio of size l to some typical extraction velocity v_i.

Set (14.8) may be easily seen to be nonlinear. However, if laser radiation acts on the rarer of the two isotopes and its concentration is small, set (14.8) may be linearized. In doing so we neglect the population of the excited level of the more abundant isotope against the population of its ground level, while all the parasitic effects like the recharge and the excitation transfer are taken to be accounted for only by the concentration of the more abundant isotope.

The equations thus obtained from (14.8) may be easily solved by the standard method. These solutions entail the conclusions that may be drawn from the qualitative analysis of the initial equations as well.

So, in addition to conditions (14.4) and (14.7), which ensure high selectivity of the process we should also mention a number of conditions which refer to the requirement of maximum possible utilization of laser radiation.

First, for the ionization to be faster than the spontaneous decay of the excited level, the following condition should be observed:

$$W_2 T_1 > 1. \tag{14.13}$$

Second, for the resonant atoms to be excited during the time of the first-stage laser pulse τ_{l1}, the following condition should be satisfied:

$$W_1 \tau_{l1} > 1. \tag{14.14}$$

Third, by analogy, for all the excited atoms to be ionized during the second-stage laser pulse τ_{l2}, the following condition should be satisfied:

$$W_2 \tau_{l2} > 1. \tag{14.15}$$

Further, for the spontaneous decay not to lead to the useless expenditure of exciting laser energy, time τ_{l1} should be less than time T_1:

$$\tau_{l1} < T_1. \tag{14.16}$$

It is also evident that it is of no use to illuminate a vapor with first-stage laser light in the absence of second-stage radiation. Hence, it is necessary that

$$\tau_{l1} \leq \tau_{l2}. \tag{14.17}$$

Under these conditions (we mean inequalities (14.4), (14.7), (14.13)–(14.17)) only resonant atoms are ionized by the radiation field, all of them being ionized during a single radiation pulse. The process selectivity is destroyed only at the stage of ion extraction when $nNK_2\tau = nN\sigma_{REC}l$ of non-resonant ion gather on the collector per every n ions of resonant modification. Here N means the concentration of the atoms of the undesired, more abundant isotope in their initial mixture. Hence, satisfying the above conditions, the final ratio of the concentrations of the desired and undesired isotopes is

$$\alpha = 1/N\sigma_{REC}l. \tag{14.18}$$

The selective two-step ionization of atoms is the most general method of laser separation of isotopes. It allows practically any degree of substance enrichment as long as the spectral differences of the absorption lines ensure the spectral selectivity.

A weak point of the selective two-step ionization is the second stage – the radiative ionization of the excited atoms. The reasons are as follows. The wavefunctions of an electron in bound states and those above the atom ionization potential are considerably different. While the former vary in space relatively smoothly, the latter rapidly oscillate. So, the matrix elements of the dipole moment of the transition to ionization are small and depend only slightly on the energy of the quantum. In other words, as we have already mentioned, the radiative ionization cross section is small and its spectral dependence is quite broad (see (14.3)). Therefore, for the selective ionization to be effective ((14.13) and (14.15)), high intensities of laser light are required. At $T_1 = 10^{-6}$ s and $\sigma_2 = 10^{-17}$–10^{-18} cm^2 intensity, I_2 should exceed $3 \cdot 10^4$–$3 \cdot 10^5$ W/cm^2.

There are many possible methods to lower so high a value of the ionizing field intensity. One of them is related to the phenomenon of autoionization.

Autoionization is the process of the spontaneous transition of an atom from the bound excited state to the state of ionization. The absorption spectra of atoms in the visible and near-UV regions usually correspond to the transitions between the states of an outer valence electron or, which is the same, between the states of an electron in the incomplete atomic shell. However, the transitions of inner-shell electrons take place in the optical spectra as well. As a rule, the photon energy necessary to excite an inner electron is greater than the photon energy needed to tear away an optical electron of an atom. So, the energy of the states corresponding to the transitions of an inner-shell electron may lie higher than the boundary of the ionization of an optical electron, i.e., they may get into the continuum of the energy levels corresponding to atom ionization. The internal atomic forces corresponding to the electrostatic interaction of the electrons*) bind the excited states of the inner-shell electrons with the continuum of the ionization states of the optical electron. In other words, we are dealing with a situation similar to the one considered at the beginning of Lecture 10 where we analyzed the decay of a single level into a continuous spectrum. In our case the decay is caused not by the radiative (μE) but by the internal atomic (V) interaction. The typical decay time τ (10.9) and the energy distribution of the emitted electrons (10.15) are determined by the value of V.

Evidently, one can speak of the autoionization states only when the product of the frequency of the transition to such a state multiplied by its decay time exceeds unity, $\omega\tau > 1$. The greater the value of $\omega\tau$, the narrower the line of the autoionization transition and the higher the cross section of the transition in the middle of the line. Such a situation takes place in the case of the so-called forbidden transitions between the autoionization states of the inner electrons of an atom and the continuum of the states of its optical electron.

For rare-earth atoms, autoionization states, which at $\omega\tau = 10^5$–10^6 possess cross sections up to 10^{-15} cm^2, have been observed experimentally. At such cross sections of ionization the required intensity of the second-stage field decreases to 300 W/cm^2.

Unfortunately, autoionization states with sufficiently large $\omega\tau$ exist neither for all atoms nor always in the convenient spectral region. In such a situation the artificial creation of autoionization states may be justified.

---

*) These forces have not been taken into account when distinguishing between inner and optical electrons.

One of the methods to create the autoionization state is as follows. A discrete level positioned sufficiently high, i.e., in the vicinity but not above the ionization potential, may be coupled with the ionization continuum, i.e., it may be ionized by a low-frequency field. Evidently, such a field acts as the internal interaction considered above (compare with Lecture 10). Therefore, we arrive at the three-step ionization scheme. In this scheme the third-step laser may be, say, the CO_2-laser (high power and high efficiency). The presence of the powerful third-stage field creates a peculiarity in the spectrum of ionization by the second-stage field similar to that observed at the excitation of the usual high-quality autoionization state. Such a similarity justifies the generalization of the autoionization concept for all cases of strong coupling of a discrete level with the ionization continuum. In other words, in the case of artificial autoionization the discrete states become the autoionization state; in our case, for the atoms placed in a strong third-stage field. Figure 14.5 illustrates the two-step photoionization without any autoionization (a), with the usual autoionization (b) and with the artificial autoionization by the third-step field (c). The width of the peculiarity in the ionization spectrum is determined in case (b) by the mean square of internal atomic interaction V^2 and by the density of states in continuum g^2; in case (c) it is determined by the probability of the radiative transition in the third-stage field.

Figure 14.5d shows still another possibility to create the artificial autoionization state. The third-stage field couples the highly positioned discrete level with the state of the continuum. This creates the spectral peculiarity in the density of states and in the ionization cross section for the second-stage field. As distinct from the previous case (Fig. 14.5 a) the discrete level is not excited directly by the second-stage field. This gives more freedom in choosing the transitions and lasers to be used. Generally speaking, within limits, one may do without the first stage, organizing the ionization process directly from the ground state.

The analogy between cases 14.5c and 14.5d becomes still more obvious if we mention that in the latter case, we have been dealing with the process of a two-photon excitation of discrete level 2 by the fields of the second and the third stages via the intermediate levels of the ionization continuum (Lecture 9). The composite matrix element of such a transition equals the sum of composite matrix elements taken over the continuous spectrum of the ionization continuum states similar to (9.29).

The autoionization state may also be created by means of the ionization continuum transformation, say, by placing an atom in an electrostatic field. The gist of the matter is as follows.

The optical electron of a highly excited atom moves from the ion core of an atom at a distance large compared to the Bohr radius. At such distances the electric field of the core is almost a Coulomb field. So the quantum states of such an atom are similar to the highly excited states of the hydrogen atom (the Rydberg states). An external electrostatic field distorts the Coulomb potential and makes tunnel ionization possible. The probability of this process rapidly grows with the principal quantum number n^* of the Rydberg state (Fig. 14.6).

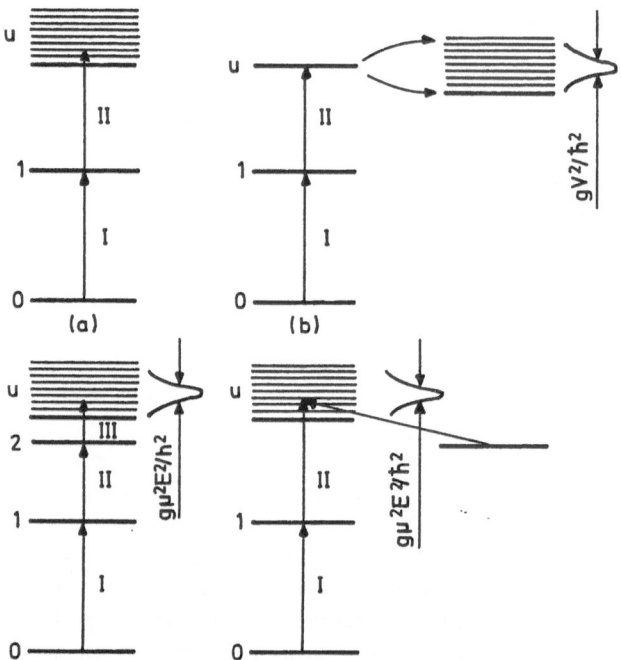

Fig. 14.5. Comparison of the two-step ionization schemes with (**a**) and without (**b**) autoionization

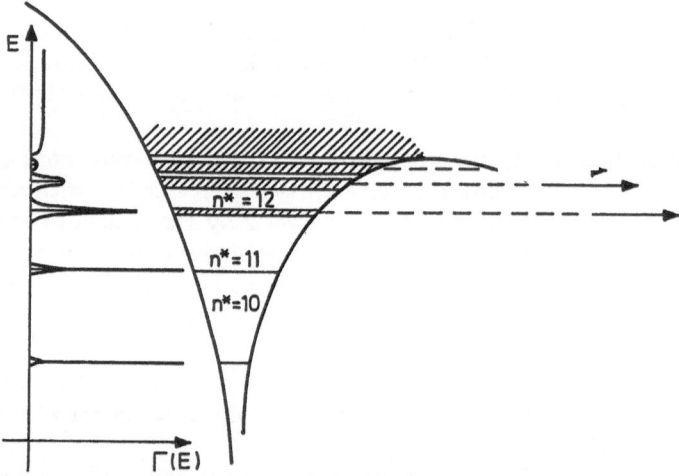

Fig. 14.6. Autoionization of the Rydberg atom in an external electric field. The energy dependence of the density of quantum states is shown on the left

Of special interest from the viewpoint of physics is the method of artificial ionization which combines both the distortion of the energy states of the atom and the quasi-energy state transformation as a result of the nonzero frequency of the action. This method may be realized using intense superhigh-frequency fields in which the energy of the interaction with the field μE for the Rydberg atoms is comparable with the distance between the quasi-energy levels $\hbar\omega$ for which the condition of quasi-continuum existence (10.12) is satisfied. Such a situation brings about strong mixing of the initial levels. On the one hand, the transitions to these levels from the discrete levels of the first and second steps positioned considerably lower are well allowed; on the another hand, they are strongly coupled with the ionization continuum.

The method of the artificial autoionization is significantly more complex than others mentioned above and requires special analysis. One of the possible approaches to such an analysis may be based on the results of the consideration of the multilevel-system-excitation dynamics. The point is that the quasi-energy spectrum of the Rydberg states in the SHF-field represents the multiband structure (Fig. 14.7). Without going into the particulars we may say that the ionization process in this case is similar to the diffusion of populations over the bands of the quasi-energy spectrum considered in Lecture 11. In other words the ionization continuum is reached from the starting Rydberg level excited by the second-stage laser field in the process of the stochastic absorption of the SHF-quanta.

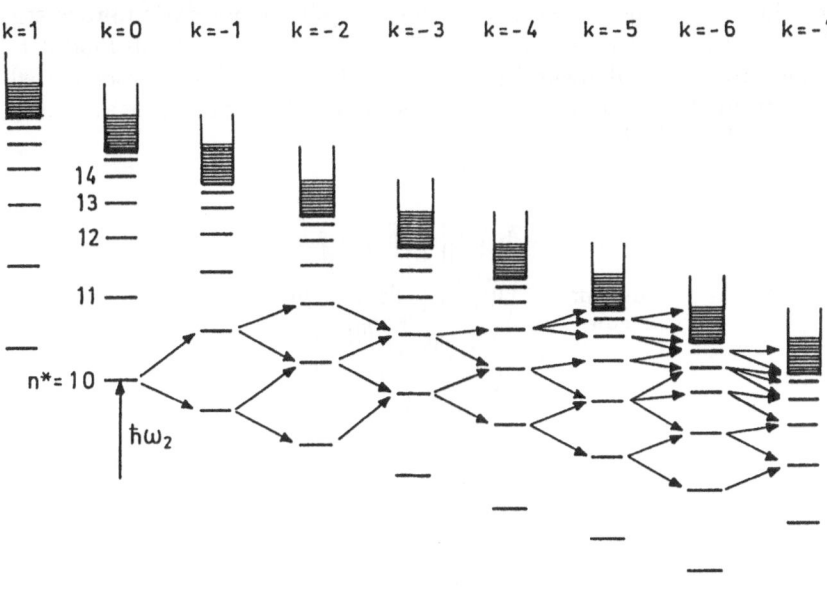

Fig. 14.7. Diffusion to the ionization over the quasi-energy spectrum states of the Rydberg levels of atoms in the SHF-field. Level $n^* = 10$ is excited by the second-stage laser field

Concluding the discussion of laser isotope separation in atomic vapor, we mention that there are many ways to ionize selectively excited atoms, including the collisional ionization of Rydberg's atoms and the ionization of such atoms by the electrostatic field turned on after the excitation process is over. In these cases ionization occurs owing to the distortion of the potential field in which the optical electron of an atom is placed, but after termination of the radiative action. In a number of cases the so-called Penning ionization may be essential. It occurs in the collisions between excited atoms. If the total energy of the excitation quanta of both atoms exceeds the ionization potential of one of them, their collision may result in the ionization of this atom and in the deactivation of the other participant. This process is similar to the E-E-exchange, the only difference being that one of the discrete levels is replaced by the continuum. So, the collision results not in the coherent interaction in a two-level compound-system (Lecture 12) but in the decay of the single level into the continuum (Lecture 10). The probability of this process as long as it retains its dipole-dipole character is given on the basis of Eqs. (12.7) and (10.8) by the formula

$$W = \int w d\Gamma \int_{-\infty}^{\infty} \frac{\mu^4}{R^6(t)} \frac{1}{\hbar^2} g dt. \tag{14.19}$$

Among the experimental studies of the intense resonant interactions of laser light with atoms, the studies of the multiphoton ionization processes are of pronounced importance. The interest in investigations of this kind is to be explained by the fundamental importance of the ionization processes which do not reveal the cut-off frequency of the photoeffect. Such processes may be observed only with the use of lasers.

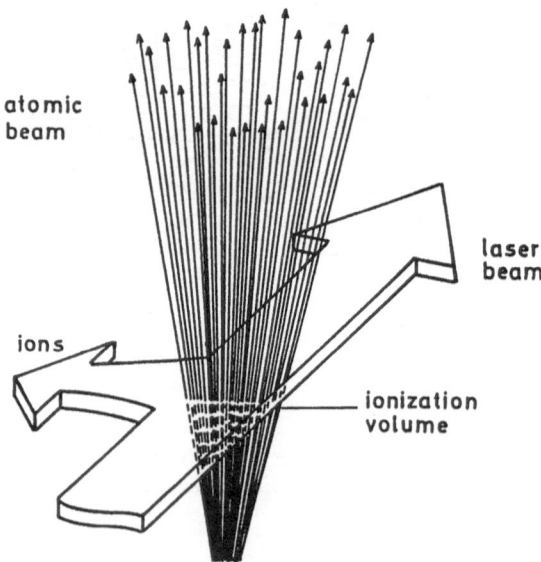

Fig. 14.8. Crossed geometry of an experiment on the multiphoton ionization of atoms

The experimental setup is shown in Fig. 14.8. This figure presents the most convenient crossed geometry in which the light beam, the atomic beam and the direction of ion extraction are mutually orthogonal. Such a geometry permits exclusion of the influence of the Doppler effect. At present a great deal of experimental information has been obtained, mainly, about the atoms of inert gases, alkali and alkaline-earth metals. The possibility of multiphoton ionization is safely established and a photonity degree sometimes exceeding the value of 10 has been measured for all of them.

It is convenient to show the typical features of the process described above using the example of a two-photon absorption of ruby-laser radiation by metastable helium atoms on the transition $2\,^1S \rightarrow 6\,^1S$. Since the same laser field has the effect of ionizing the helium atoms on the transition $6\,^1S$-continuum, this process may be called the three-photon ionization.

A specific feature of a helium atom is the fact that the $2\,^1S$-state is metastable. There is no P-state between this and the ^1S one, so its spontaneous decay turns out to be forbidden in the dipole approximation. This easily permits the accumulation of helium atoms in the metastable state, creating them, say, with the help of an electric discharge. (Note that this property is used in He-Ne lasers.) Therefore, the beam of metastable helium atoms in the $2\,^1S$-state is a suitable object to study two-photon absorption with.

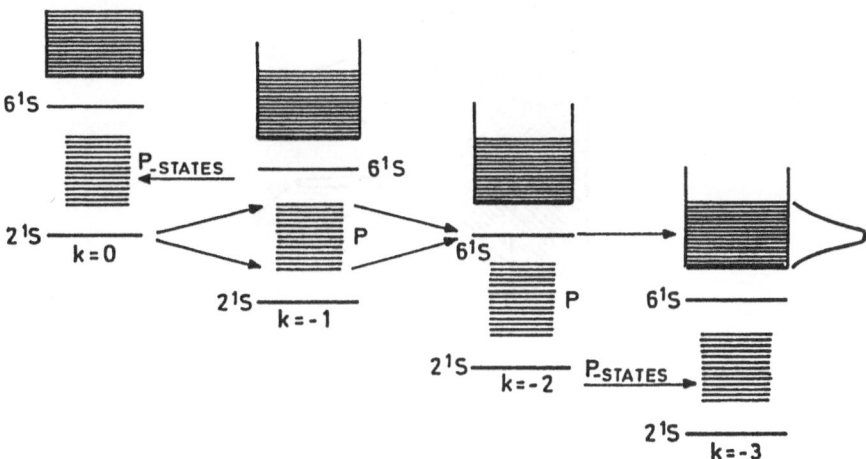

Fig. 14.9. Three-photon ionization of a metastable helium atom shown in the scheme of the quasi-energy levels

In accord with Lecture 9 let us consider the quasi-energy levels scheme given in Fig. 14.9. This figure, in fact, presents the details of the schemes given in Figs. 9.2–9.4 for the helium atoms. The two-photon process in this case occurs via the detuned intermediate levels belonging to the P-states. Due to the fast decay of level $6\,^1S$, which occurs both due to the ionization by the field and spontaneously, the process of two-photon excitation is incoherent. So, in accord with Lecture 9, population N of level $6\,^1S$ is proportional to the square of the composite matrix element V divided by the square of the decay rate W_p, $N \propto V^2/W_p^2$. An ionization yield, i.e., the number of ionization events per unit time, equals the product of the population multiplied by the ionization probability W_i, $N_i = NW_i$. Since the level decay rate is the sum of the probabilities of spontaneous decay Γ and W_i, then

$$N_i \propto V^2 W_i/(W_i + \Gamma)^2 . \tag{14.20}$$

In our case V and W_i are proportional to E^2, where E is the laser field strength. When E is relatively small the ionization yield turns out to be proportional to E^6, i.e., to the cube of the intensity. Testing this fact experimentally, one should not forget about the dependence of the positions of the resonant levels on the field intensity analogous to (9.24), which lead to a shift of the resonant frequency of the two-photon transition in the helium atom similar to (9.21). All this, including the dependence of the ion current on the intensity and the resonant frequency shifts proportional to the intensity, has been observed experimentally. Besides, the change in the power index of the dependence of the ion current on intensity in the region where the probability of the ionization decay coincides with that of the spontaneous decay has also been observed.

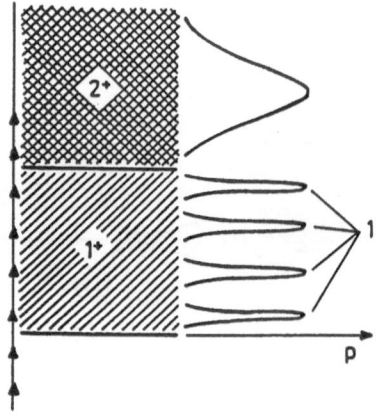

Fig. 14.10. Multiphoton two-electron ionization: 1 – long-living single-electron autoionization states, 1^+ and 2^+ – continuum of single- and twofold ionized atoms, ρ – density of states

Multiphoton ionization with the tearing off of several electrons simultaneously is also possible. It may also take place when in the ionization continuum many narrow long-living autoionization states exist, i.e., when the different electrons in a multielectron atom interact with each other, being excited to different orbits and having the total energy sufficient for the ionization. Figure 14.10 illustrates the possibility of the two-electron multiphoton ionization.

Narrow autoionization states of the first ionization continuum are the detuned intermediate levels of the multiphoton process of atom excitation to the region of the second ionization continuum. Processes of this kind have been observed in the alkaline-earth metal atoms and in the atoms of some inert gases.

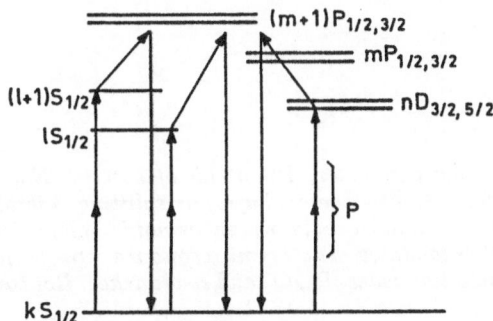

Fig. 14.11. Possible schemes of the parametric generation of the radiation with summed frequency in alkali metals vapors

Atoms or, more precisely, atomic vapors, present a convenient medium for the parametric generation of radiation on the higher harmonics and on the compound (summed, subtracted) frequencies. As an example we mention the generation of UV radiation in the vapors of alkali metals. A simplified scheme of the levels is given in Fig. 14.11. The matrix element of the generation transition by analogy with (13.16) is written in the form

$$V_4 = \frac{\mathcal{E}_1^2 \mathcal{E}_2 \mathcal{E}_3}{\hbar^3} \sum \frac{\mu_{0n}\mu_{nm}\mu_{mk}\mu_{k0}}{\Delta_{ns}\Delta_{mp}\Delta_{kq}} , \qquad (14.21)$$

where the summation is taken over all possible intermediate detunings for different quasi-energy levels s, p and q. Evidently, the largest contribution is rendered by the transitions closest to the strict resonances. Analogous schemes are successfully used for the visualization of IR radiation by means of the transformation of its frequency, and also for the generation of UV radiation with the help of CO-, CO_2-lasers and dye lasers. The extensive number of thoroughly studied resonant levels of alkali-metal atoms allows application of diverse schemes of parametric interaction.

Lecture Fifteen
Spectra of Molecules

The Born-Oppenheimer parameter. Hierarchy of spectra. Normal vibrations. Anharmonicity. Rotations. K-degeneracy, hyperfine splitting. Vibrational-rotational interactions. Three-dimensional oscillator, anharmonic removal of degeneracy. The Coriolis splitting. Hybrid states. The Fermi resonance. Dipole transitions. IR-active modes. Rotational selection rules. P-, Q- and R-branches. Hot bands. Stochastization of vibrations.

A molecule as compared to an atom is a much more complex quantum object. So the interactions of laser light with molecules are much more diverse. The reason is not only a greater number of electrons and the absence of central symmetry typical of atoms. The point is that in molecules there are additional – rotational and vibrational – degrees of freedom principally different from the electronic ones. They considerably complicate the energetic spectrum of molecules.

The motion of electrons in a molecule and its rotations and vibrations are coupled with each other. So, any stationary energy eigenstate of a molecule should, strictly speaking, be considered as an electronic-vibrational-rotational one. The spectrum of these states is extremely involved. Its interpretation would be practically impossible were there no hierarchy of internal interactions accounted for by the difference in the masses of electrons and nuclei. In a number of problems this hierarchy allows one to separate electronic, vibrational and rotational motions. To what extent this separation is possible and at which point it should be performed depend on the complexity of the molecule, the region of the energetic spectrum under examination, the presence of internal resonances in this region, and the problem at hand, etc.

The separation of molecular internal motions into electronic, vibrational, and rotational kinds is based on the assumption of a hierarchy of these motions. The vibrational motion of nuclei is assumed to be slow in comparison with the electronic motions, and the rigid rotation of the molecule is assumed to be slow as compared with the nuclear vibrations. The parameter of this so-called adiabatic approximation is the fourth route of the mass ratio of electron and nucleus $\Lambda = (m/M)^{1/4}$. It is referred to as the Born-Oppenheimer parameter.

Indeed, for the energy of electrons E_e, of vibrations E_v, and of rotations E_r, we have by order of magnitude

$$E_e \approx mv^2 \sim U_i \,,$$
$$E_V \approx MV^2 \sim a^2 \partial^2 U/\partial R^2 \sim a^2 U_i/R^2 \,, \tag{15.1}$$
$$E_r \approx MR^2 \Omega^2 \,.$$

Here v and V are the velocities of the electrons and nuclei, respectively; ω is the rotation frequency; U, the potential energy; U_i, the ionization potential; a, the amplitude of vibrations; and R, the typical molecular dimension.

The quantization conditions for the momentum yield

$$mvR \approx \hbar \,, \quad MVa \approx \hbar \,, \quad MR^2 \Omega \approx \hbar \,. \tag{15.2}$$

Relations (15.1) and (15.2) allow us to evaluate the ratio of energies with the formula

$$E_e/E_V \approx E_V/E_r \approx (M/m)^{1/2} = 1/\Lambda^2 \,; \tag{15.3}$$

the amplitude of vibrations with

$$a/R \approx (m/M)^{1/4} = \Lambda \,; \tag{15.4}$$

and the velocities ratio, with

$$v/V \approx (M/m)^{3/4} = 1/\Lambda^3 \,. \tag{15.5}$$

The usual procedure of the separation of electronic, vibrational and rotational motions can be described as follows. To begin with, we introduce the electron terms, assuming for this purpose that the nuclei of the atoms comprising the molecule are immobile in space. In this case we obtain the energy levels of a molecule determined by the interactions in a multielectron system. Every energy level thus obtained has an appropriate wavefunction. The change in the relative arrangement of nuclei modifies both the value of the level energy and the corresponding wavefunction. The dependence of the energy level of the electrons on the internuclear distance is called the electron term (Fig. 15.1).

Knowledge of the electron terms allows us to describe the vibrations of the molecule. If we assume that the vibrational motion does not result in a change of electronic state of the molecule, i.e., if the vibrations of the nuclei are interpreted by the electrons as adiabatic motions (Lecture 7), each of the electron terms is the potential curve of the vibrational motion of the nuclei. In other words, the electron term represents, generally speaking, the multidimensional potential well of the multidimensional vibrational motion. Thus, one specific vibrational motion corresponds to each electron term.

Let us consider the region near the bottom of the potential well, i.e., the region of small deviations of internuclear distances from the equilibrium values. Then the potential well can be considered parabolic. In the parabolic potential well, by means of rotations and scaling transformations, we can always choose the coordinate system in which the potential energy is represented by the sum of the squares of the coordinates multiplied by certain coefficients, while the kinetic energy is represented by the sum of the squares of their time

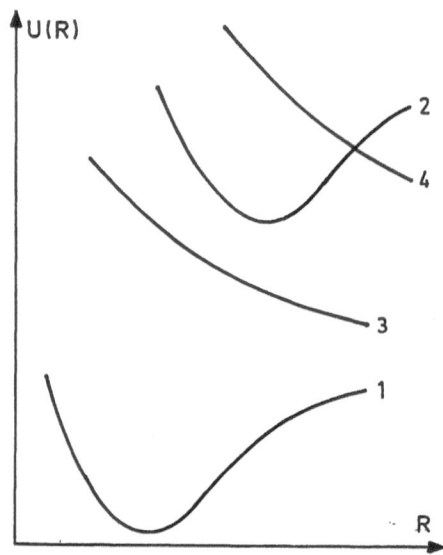

Fig. 15.1. Electron terms of the molecule: 1 and 2 – bonding terms; 3 and 4 – antibonding terms; $U(R)$, electronic energy; R, internuclear distance

derivatives multiplied by some other coefficients. These coefficients, called the restoring force coefficients and the reduced masses, respectively, determine the spectrum of normal vibrations. The coordinates thus chosen are referred to as the normal coordinates. The totality of the normal molecular vibrations of the same frequency is called the molecular vibrational mode. The number of vibrations in the mode is the degeneracy of the mode.

For the nondegenerate vibrational mode the molecular rotation is the adiabatically slow motion. Each vibrational level has its own nuclear wavefunction; hence, its own mass distribution and its own proper inertia tensor. The adiabaticity implies that in the process of rotation the vibrational state remains unchanged. Hence, the inertia tensor is also conserved, so the molecule rotates as the rigid top does. In this situation each vibrational level has its own rotational spectrum.

Let us emphasize once more that only the simultaneous validity of all the above conditions allows the complete separation of electronic, vibrational and rotational motions. Let us enumerate them once more. The distance between the electronic energy levels should be much greater everywhere than the distance between the vibrational levels. In turn, the latter distance should exceed the distance between the rotational levels. The violation of either requirement is sufficient for the separation procedure to be illegal. Such violations may be caused by the intersection of electron terms, by the degeneracy of the vibrational modes and states, by the rotational degeneracy accounted for by the spherical symmetry of the molecular inertia tensor, etc. Under these conditions the nonadiabaticity of the electronic motion and the nonlinearity of the vibrational motion and of the vibrational-rotational interactions become essential.

For example, the anharmonic interactions in the case of the degenerate modes
of polyatomic molecules lead to the fact that slow motions corresponding to
a slow energy transfer from one vibrational degree of freedom to another are
superimposed on the fast vibrations. The typical times of these slow motions
may be comparable with the rotation period, thus leading to a strong interaction and the coupling of vibrations and rotations. So, they turn out to be
inseparable.

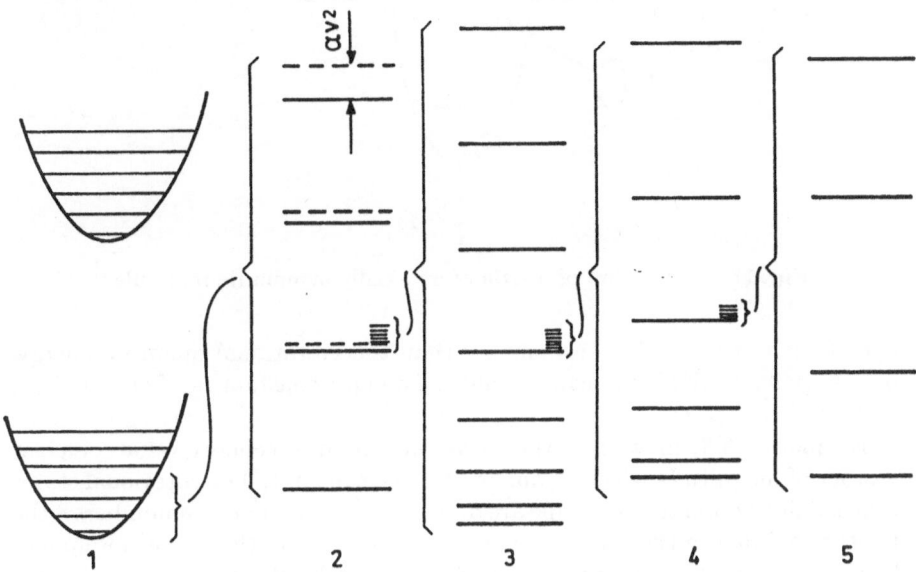

Fig. 15.2. Hierarchy of molecular spectra: 1 – electron terms; 2 – vibrational structure; 3 and 4 – rotational structure (3 – J-structure, 4 – K-structure); 5 – hyperfine
structure

An example of the spectrum hierarchy in the absence of degeneracies is
given in Fig. 15.2. The typical distance between the electron terms is 3–10
eV ($3 \cdot 10^4$–10^5 cm^{-1}). The energy levels of the vibrational motion close to
the harmonic vibrations are positioned near the bottom of these terms. The
typical distance between these levels in accordance with (15.3) is Λ^{-2} times
less than the distances between the electron terms and comprises 300–3000
cm^{-1}. The anharmonicity of vibrations becomes apparent in the displacement
of the real level from its harmonic positions. This displacement is quadratic
over the level number and is characterized by the anharmonicity constant α:

$$E_v - v\hbar\omega = \alpha v^2, \qquad (15.6)$$

where E_v is the energy of the vibrational level with the number v.

It is known that the anharmonic additions to the harmonic oscillator energy are accounted for by the cubic nonlinearity in second order perturbation
theory and for the nonlinearity of the fourth degree, in first order. Since, in
accord with (15.4) $\alpha/r \approx \Lambda$, the cubic nonlinearity is Λ^{-1} times- and that

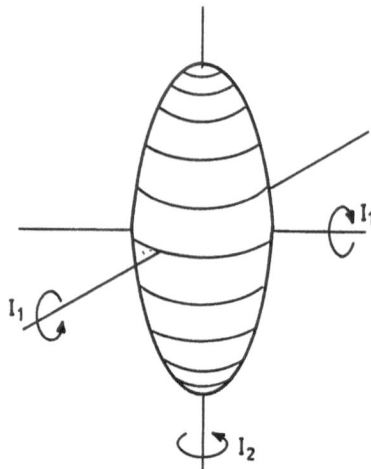

Fig. 15.3. Moments of inertia of an axially symmetric molecule

of the fourth order Λ^{-2} times- smaller than the vibrational quantum energy. So, $\alpha \approx \Lambda^2 \hbar\omega \approx \Lambda^4 U_i$. Hence, the anharmonicity constant is of the order of 0.1–10 cm^{-1}.

Estimate (15.3) shows that the vibrational quantum energy value coincides by order of magnitude with the anharmonicity constant. The rotational states of the axially symmetric molecule are described by two quantum numbers – the quantum number of the total angular momentum J and the quantum number of the angular momentum projection K on the molecular axis of symmetry[*]. The energy of rotation equals

$$E_{J,K} = BJ(J+1) + AK^2, \tag{15.7}$$

where A and B are the so-called rotational constants determined by the values of two moments of inertia of the axially symmetric molecule I_1 and I_2 (Fig. 15.3):

$$A = (\hbar^2/2)(1/I_2 - 1/I_1), \quad B = \hbar^2/2I_1. \tag{15.8}$$

The resulting energy level is the electronic-rotational-vibrational one. This level may also reveal its own structure accounted for by the interaction of the nucleus and electron spins, present in a molecule, with rigid rotation of the molecule. This splitting is not connected with the Born-Oppenheimer parameter and is of the order of one reciprocal second.

The presence of degeneracy essentially complicates matters and often even makes it impossible to separate the vibrational and rotational motions. This may be illustrated by the spectrum of the three-fold degenerate modes of the

---

[*] Projection J on the chosen axis of the laboratory coordinate system, i.e., the quantum number M, in the absence of any external field is of no significance and henceforth we shall ignore it.

highly symmetric molecules (tetrahedral symmetry, T_d; octahedral symmetry, O_h) such as the molecules of the type XY_4 (T_d), XY_6 (O_h), XY_8 (O_h). The vibrational motion corresponding to the three-fold degenerate mode of the XY_6-molecule is schematically represented in Fig. 15.4. In the harmonic approximation the energy levels of such three-dimensional vibrational motion are multiply degenerate. The degeneracy equals

$$N_v = (v+1)(v+2)/2\,. \tag{15.9}$$

The molecular symmetry does not require such high degeneracy, so even the anharmonic interaction is enough to remove it. The multiplets thus formed contain no more than the three-fold degenerate levels.

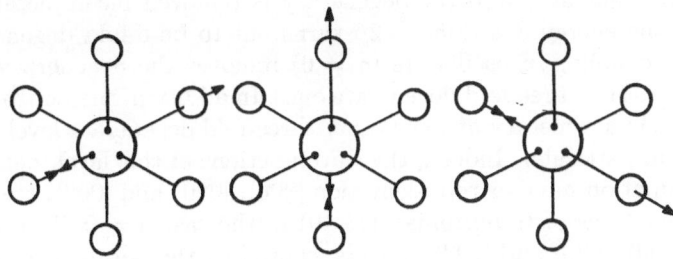

Fig. 15.4. Three spatial degrees of freedom of three-fold degenerate vibrational mode of XY_6-molecule (the so-called nonsymmetric valence vibration ν_3)

Let us dwell on this. The point is that in contrast to the one-dimensional oscillator the anharmonic interaction in the three-dimensional case has a more complex tensor structure. Let us take advantage of the creation and annihilation operators for the description of this interaction. In the case of the three-fold degenerate mode of the XY_6-molecule with O_h symmetry, the first nonvanishing anharmonic corrections are of the fourth order, i.e.,

$$\begin{aligned}
\widehat{H}_{\text{anh}} = {}& \alpha(a_1^+ a_1 a_1^+ a_1 + a_2^+ a_2 a_2^+ a_2 + a_3^+ a_3 a_3^+ a_3) \\
& + \beta(a_1^+ a_1 a_2^+ a_2 + a_1^+ a_1 a_3^+ a_3 + a_2^+ a_2 a_3^+ a_3) \\
& + \gamma(a_1^+ a_1^+ a_2 a_2 + a_1^+ a_1^+ a_3 a_3 + a_2^+ a_2^+ a_3 a_3 \\
& \quad + a_1 a_1 a_2^+ a_2^+ + a_1 a_1 a_3^+ a_3^+ + a_2 a_2 a_3^+ a_3^+)\,, \tag{15.10}
\end{aligned}$$

where indices 1, 2 and 3 numerate three spatial degrees of freedom of the three-fold degenerate mode (Fig. 15.4). Constant α corresponds to the anharmonicity of the vibrations of each degree of freedom by analogy with the one-dimensional oscillator anharmonicity constant. Constant β describes the change in the frequency of vibrations of one degree of freedom caused by the excitation of vibrations of another degree of freedom. Constant γ is responsible for the energy exchange between them, which is typical of the parametric processes to which the anharmonic interaction of the three-fold degenerate-mode degrees of freedom belongs. It is this tensor structure of anharmonicity characterized by three constants α, β, γ which leads to the splitting of the vibrational spectrum of the multiply degenerate modes. Most simply this may be seen in the case of $\gamma = 0$. Let us introduce three quantum numbers v_1, v_2, v_3 corresponding to the numbers of the vibrational quanta $\hbar\omega$ in each degree of freedom of the mode. Then the anharmonic correction to the energy of the level with the number $v = v_1 + v_2 + v_3$ equals

$$E_{\text{anh}} = \alpha(v_1^2 + v_2^2 + v_3^2) + \beta(v_1 v_2 + v_1 v_3 + v_2 v_3).\qquad(15.11)$$

The spectrum resulting from this correction is shown in Fig. 15.5 for the first four values of v: $v = 0$, $v = 1$, $v = 2$, $v = 3$. Each of the states of this spectrum is characterized by the occupation numbers of three mode oscillators $|v_1 v_2 v_2\rangle$. One can see that at $\gamma = 0$ the degeneracy is removed incompletely. So, the level with the energy $3\hbar\omega + 5\alpha + 2\beta$ turns out to be 6-fold degenerate. The parametric coupling of oscillators ($\gamma \neq 0$) removes the degeneracy, splitting this level into two three-fold degenerate ones. In addition, the nonzero value of γ removes the degeneracy of some other three-fold degenerate levels, the level $2\hbar\omega + 4\alpha$, in particular. Indeed, the wavefunctions of this level, being written in the occupation-number representation $|200\rangle$, $|020\rangle$ and $|002\rangle$, are no longer the eigenfunctions of Hamiltonian (15.10) in the case $\gamma \neq 0$. The interaction of states $|200\rangle$, $|020\rangle$ and $|002\rangle$ appears, caused by the parametric coupling of different degrees of freedom of the mode. This results in new eigenfunctions and in new eigenvalues. At $v = 2$, Hamiltonian (15.10) is written in the matrix form as

$$\widehat{H}_{\text{anh}} = \begin{pmatrix} 4\alpha & 2\gamma & 2\gamma & 0 & 0 & 0 \\ 2\gamma & 4\alpha & 2\gamma & 0 & 0 & 0 \\ 2\gamma & 2\gamma & 4\alpha & 0 & 0 & 0 \\ 0 & 0 & 0 & 2\alpha + \beta & 0 & 0 \\ 0 & 0 & 0 & 0 & 2\alpha + \beta & 0 \\ 0 & 0 & 0 & 0 & 0 & 2\alpha + \beta \end{pmatrix}\qquad(15.12)$$

The upper left-hand corner of this matrix contains the Hamiltonian corresponding to states $|200\rangle$, $|020\rangle$ and $|002\rangle$. This Hamiltonian has two eigenvalues, $E_E = 4\alpha - 2\gamma$ and $E_A = 4\alpha + 4\gamma$. Level E_E is doubly degenerate. Level E_A is nondegenerate. The eigenstates corresponding to these levels cannot be described by definite occupation numbers. At $\gamma \neq 0$ such a situation is typical of many other states lying higher. In the presence of the parametric interactions responsible for the energy exchange between the modes, the eigenenergy states cannot be considered as states with definite numbers of quanta in different modes. In turn, the states corresponding to the definite distribution of the energy among the degrees of freedom are no longer eigenenergy states. So, describing the vibrational spectra of symmetric polyatomic molecules, only the type of level symmetry is specified along with the level energy. The symmetry type is denoted by the letters A, E, F, which correspond to the one-, two- and threefold degeneracy shown in Fig. 15.5.

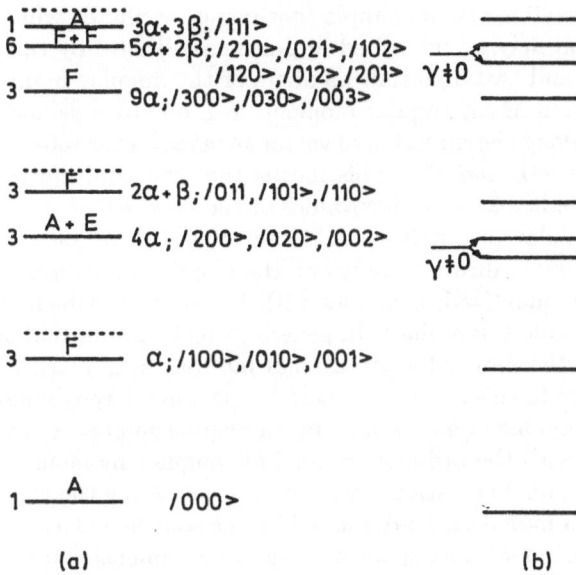

Fig. 15.5. The energy levels of the three-fold degenerate vibrational mode of the XY_6-molecule. Harmonic positions of energy levels are shown in dotted lines. The multiplet components are denoted using the occupation numbers $|v_1 v_2 v_3\rangle$. On the right of each level its shift with respect to harmonic position is shown, while on the left, the level degeneracy ((a) incomplete removal of degeneracy; (b) complete removal)

In general, every vibrational level has its own rotational structure. However, if the levels are degenerate or the distances between the multiplet components are small, the rotations may cause changes in the vibrational spectrum on account of the so-called vibrational-rotational interaction. Then the vibrational and rotational motions cannot be separated. The point is that in a polyatomic molecule there is a dynamic coupling between vibrations and rotations. Such a coupling originates from the presence of the angular momentum of a vibrational nature and brings about the nonconservation of the angular momentum corresponding to the rigid rotation of the molecule. Only the sum of these angular momenta is a conserved quantity. This implies that these two motions cannot be considered to be independent or, in other words, the angular momenta of vibrational and rotational origin interact with each other. Such an interaction leads to an additional term in the molecular Hamiltonian.

Indeed, let us pass on to the coordinate system rigidly attached to a molecule. All the vibrations are considered in this particular system. This is the noninertial coordinate system, rotating with the angular velocity Ω of the molecular rigid rotation. Then we should include in the Hamiltonian the Coriolis term ΩL, where L is the angular momentum of vibrational motion.

The nature of the angular momentum of vibrational motion can be illustrated by the three-dimensional vibrational motion of the central atom in the molecule shown in Fig. 15.4. (The positions of other atoms are assumed fixed.) The three-dimensional simple (harmonic) oscillator which corresponds to such a motion may be described in polar coordinates by the orbital quantum number n and two quantum numbers of the angular momentum, L and M. The projection $\widehat{M}$ on angular momentum $\widehat{L}$ on any axis and, particularly, its component along the direction of vector Ω takes on the integer values in the interval between $-L$ and $+L$. This means that the addition to Hamiltonian mentioned above has $2L + 1$ eigenvalues of the form $\hbar\Omega M$.

In the particular case being considered of the vibrations of central atom, the space of normal vibration, in which the angular momentum of the vibrational motion is quantized, coincides with the space in which the rigid rotation of the molecule takes place. In general, this is an unusual situation. The quantization of the molecular vibrational motion takes place in the normal-coordinate space. In this space of, generally speaking, large dimensionality, the vibrational motion may be described by an angular momentum which is quantized in accord with the ordinary rules. This angular momentum differs from the three-dimensional vibrational momentum $\widehat{L}$ entering the Coriolis addition. The two kinds of momentum are related by the so-called Coriolis tensor which projects the eigencoordinate space into the three-dimensional Euclidean space of the molecular axes. In the case of the three-fold degenerate mode the Coriolis tensor components for all three normal coordinates are the same. Then the allowance for this tensor when passing on to the noninertial coordinate system is reduced to the mere multiplication of vector $\widehat{L}$ by the quantity ζ, usually called the Coriolis constant.

Now let us consider how the vibrational-rotational interaction $\zeta\omega L$ affects the spectrum of the three-fold degenerate mode. Should the anharmonic removal of the degeneracy shown in Fig. 15.5 be absent, the Coriolis interaction will remove the degeneracy of the harmonic states and form multiplets with a distance ζBJ (Fig. 15.6) between its components in the harmonic energy level system. The eigenenergy states shown in Figs. 15.5 and 15.6 are different since the eigenfunctions of the anharmonic and the Coriolis additional terms to the molecular Hamiltonian are distinct, while the additions do not commute. So if both additions are of importance, the spectrum of all the states except the first excited one becomes considerably more complicated. This results in a more complete removal of the degeneracy and in a violation of the dipole transition selection rules in a three-dimensional harmonic oscillator.

Everything which has been said above testifies to the fact that the hierarchy of the spectra corresponding to different types of molecular motions is essentially more complex than that presented in Fig. 15.2. Each electron term has its own multidimensional potential well with many vibrational modes. The spectrum of degenerate modes is a set of multiplets formed after the removal of the degeneracy of harmonic oscillations by the anharmonicity. The rotational structure is superposed on this spectrum, each level with rotational quantum number J possessing its own vibrational structure caused by the vibrational-rotational (Coriolis) interaction.

The interaction of vibrations and rotations leads to a more serious complication of the molecular spectrum caused by the loss of rigidity when there is

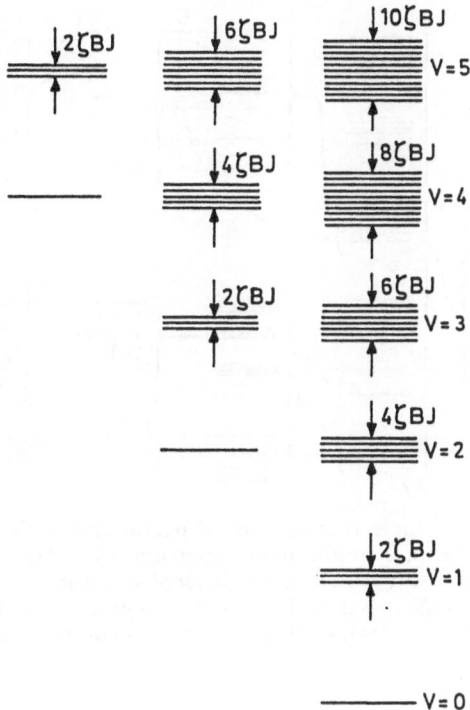

Fig. 15.6. Removal of degeneracy of vibrational motion of three-fold degenerate mode by Coriolis interaction

no definite quantum number J which can be assigned to the rotational motion of the molecule. The point is as follows. In complex molecules the difference in the energy states with different J may be compensated for by anharmonic splitting. The vibrational-rotational interaction links these resonant states and mixes their wavefunctions, thus bringing about the loss of rotational individuality. The position of this effect in the general hierarchy of the spectra of polyatomic molecules is shown in Fig. 15.7.

One should not forget that slightly different notations are usually used when describing nonrigid molecules in the case in which the vibrational-rotational interactions are essential. Symbol J denotes the conservative quantity – the total angular momentum of the molecule, which besides the angular momentum of the molecule motion as a whole, includes the angular momentum of the vibrational motion. Symbol R denotes the nonconservative angular momentum of the motion of the molecule as a whole.

We consider the molecules of symmetry T_d and Q_h which are, in fact, the spherical tops. The rotational constant $A = 0$ for these molecules (see (11.8)), so the vibrational levels are $2J + 1$-fold degenerate over quantum number K (K-degeneracy). The loss of molecular rigidity breaks the spherical symmetry of its inertia tensor. Thus the K-degeneracy is removed. The spectrum

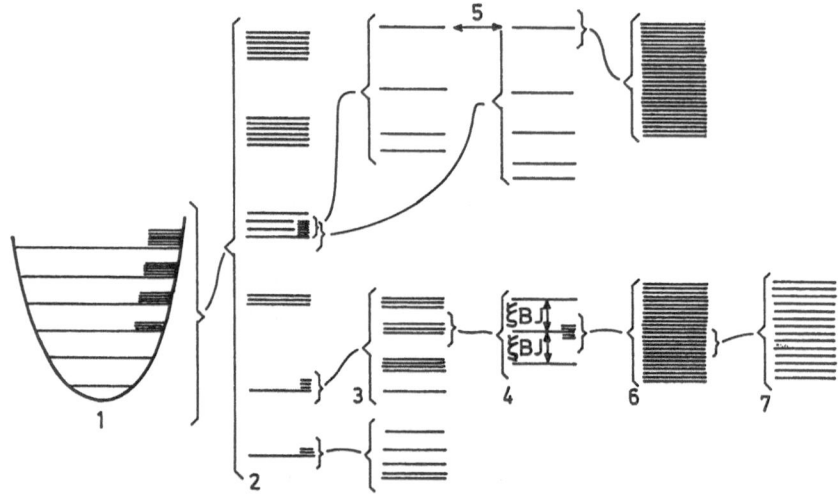

Fig. 15.7. Hierarchy of molecular spectra of polyatomic molecule: 1 – vibrational levels in the electron term; 2 – anharmonic splitting, $E_v = \hbar\omega(v_1 + v_2 + v_4) + \alpha(v_1^2 + v_2^2 + v_3^2) + \beta(v_1v_2 + v_1v_3 + v_2v_3) + \ldots$; 3 – rotational structure, $E_r = BJ(J+1) + E_{cor}$; 4 – Coriolis splitting, $\Delta E_{cor} = \pm\zeta BJ$; 5 – loss of rigidity caused by the vibrational-rotational interaction; 6 – octahedral splitting; 7 – hyperfine splitting

thus obtained is complicated not only vibrationally but rotationally, too. The eigenfunctions of the spectrum states have no definite values of the quantum number K.

The vibrational-rotational interaction removes the K-degeneracy of both the upper and the lower vibrational levels. This becomes apparent in the so-called fine tetrahedral or octahedral (depending on the molecular type) splitting. Indeed, in the ground vibrational state the rotational energy level $BJ(J + 1)$ is $(2J + 1)$-fold degenerate, while in the first vibrationally excited state, three $(2J + 1)$-fold degenerate over K levels correspond to each vibrational level with a definite value of J. The energies of these three levels are $BJ(J+1)$, $(BJ(J+1)\pm\zeta BJ$. The Coriolis interaction does not yet remove the K-degeneracy. To progress still further in the framework of first-order perturbation theory is impossible. In second order we should take into account the small mixing of the states with different total numbers of quanta in the mode caused by the Coriolis addition. This mixing brings about the interaction of different K-components of the same vibrational-rotational state. This interaction is weak, its mechanism being analogous to the two-quantum transition between the levels with a highly detuned intermediate state (Lecture 9). One should not, of course, forget that the intermediate detunings may be much less than the frequency of the quantum of vibrations being excited, since in a polyatomic molecule there are always other vibrational modes or hybrid vibrations whose levels are relatively closer to the state considered, as shown in Fig. 15.8.

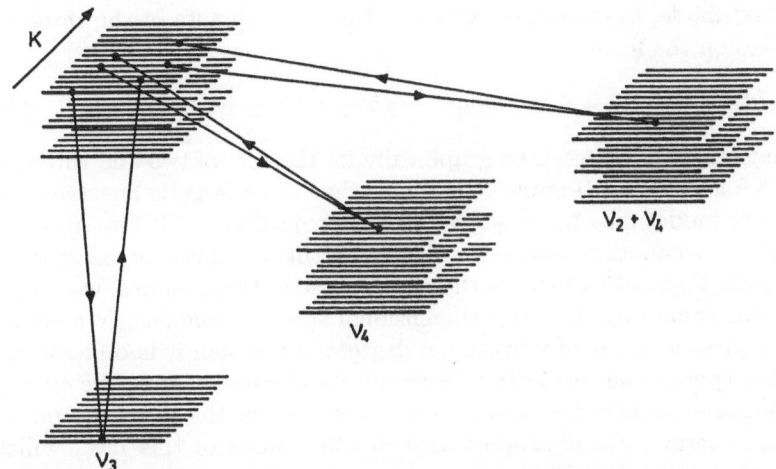

Fig. 15.8. Removal of K-degeneracy due to interaction of vibrational levels

The fine spectrum of the first vibrational level determined by the removal of the K-degeneracy is of a rather specific structure that reflects the molecular symmetry. The simplest way to understand this is to realize that the very existence of this fine spectrum is a demonstration of the anisotropy of the elastic properties of the molecule. The extent of tension and the energy of rotational motion associated with it are significantly dependent on the orientation of the rotational axis with respect to the axes of symmetry of the molecule. The minimum energy of rotation corresponds to the orientation of the axis along the direction of the easiest tensility. In molecules of the XY_6-type there are six such directions. They coincide with the vectors drawn from the central atom X towards the six peripheral atoms Y[*]. So the levels corresponding to these directions should be 6-fold degenerate. The removal of this degeneracy due to the small tunnel interaction leads to the formation of the so-called clusters of lines in fine vibrational-rotational spectra. The presence of clusters is typical of the octahedral and tetrahedral spectra. The distance between the components of a fine spectrum usually equals several thousandths of reciprocal centimeter. Note that each of the fine-spectrum components has a hyperfine spin-origin splitting (compare Figs. 15.7 and 15.2).

Let us now turn to the well-known fact that the presence of many modes is typical of polyatomic molecules. This complicates the molecular spectra to a considerable degree. The energy levels corresponding to the simultaneous excitation of the states of several vibrational modes are called hybrid. Consider the spectrum of hybrid energy levels, ignoring their rotational, Coriolis, fine and hyperfine structures. This is possible if the vibrational-rotational interaction does not mix different hybrid states together. If the modes are considered to be harmonic the energy of the hybrid state $|n, m, k, \ldots\rangle$, containing n quanta

---

[*] Sometimes the directions of easiest tensility coincide with the bisectrices of solid angles.

of the first mode, m quanta of the second mode, k quanta of the third mode, etc., takes on the form

$$E_{n,m,k,\ldots} = n\hbar\omega_1 + m\hbar\omega_2 + k\hbar\omega_3 + \ldots. \qquad (15.13)$$

This may be easily represented graphically for the case of two and three modes (Figs. 15.9 and 15.10). Figure 15.9 shows that the energetic spectrum in the case of two modes may be visualized as the projection of the two-dimensional space of the occupation numbers n and m on the one-dimensional space – the energy axis. Figure 15.10 shows that in the case of three modes, we project in exactly the same way the three-dimensional space on the energy axis. In the case of a great number of vibrational degrees of freedom it is difficult to represent the space being projected. However, we can easily draw the conclusion that the state density, i.e., the number of states on the layer of unit width along the energy axis, is proportional to the volume of this layer which, in turn, is proportional to E^{s-1}, where s is the number of vibrational degrees of freedom of the molecules (Fig. 15.10). Note that if $\omega_1 = \omega_2 = \omega_3$, i.e., all the degrees of freedom presented in Fig. 15.10 correspond to one three-fold degenerate vibrational mode, the states with the same total number of quanta have the same energy. Hence, the vibrational energy levels are degenerate in this case. The degeneracy multiplicity and the state density are directly proportional to each other and increase quadratically with the energy (compare with (15.9)).

We know that the anharmonic interaction shifts the energy levels and mixes the harmonic quantum states of the three-fold degenerate mode. In exactly the same way, anharmonicity is capable of shifting the energy levels of the hybrid harmonic states and of mixing those among them whose energies are sufficiently close.

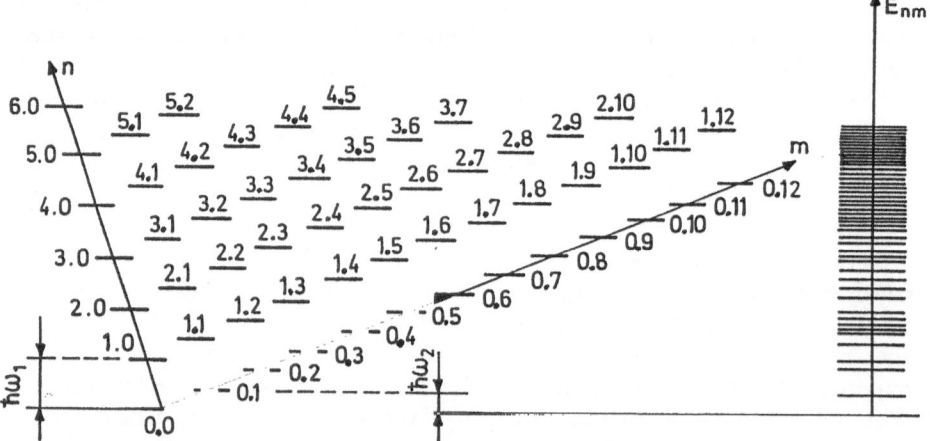

Fig. 15.9. Levels of hybrid (combination) vibrations for the case of two harmonic vibrational modes. The position of these levels on the energy axis is shown on the right

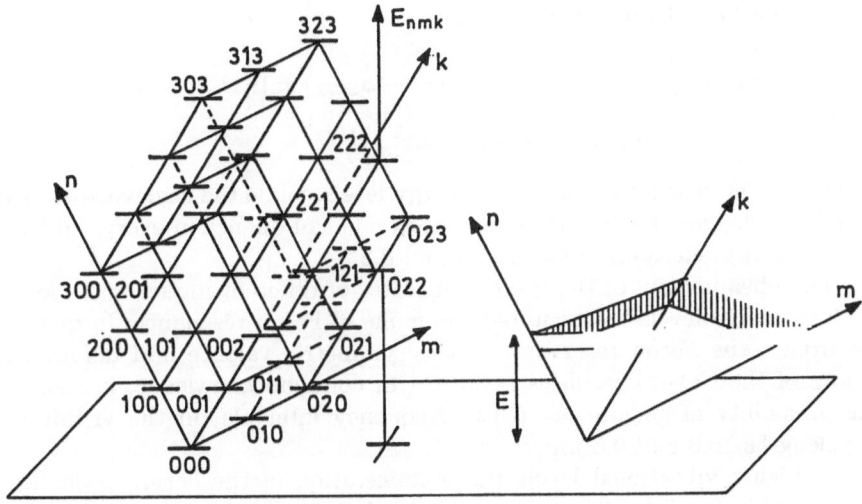

Fig. 15.10. Hybrid (combination) vibrations for the case of three harmonic vibrational modes. The layer of unit thickness corresponding to level E is shown on the right

Let us consider these changes exemplified by the hybrid levels shown in Fig. 15.9, when the frequency of one of the modes is about twice as large as the energy of the other. The Hamiltonian for these modes has the form

$$\widehat{H}_{\text{anh}} = \hbar\omega_1 a_1^+ a_1 + \hbar\omega_2 a_2^+ a_2 + \alpha_{122}(a_1^+ a_2 a_2 + a_1 a_2^+ a_2^+)$$
$$+ \alpha_{1111} a_1^+ a_1 a_1^+ a_1 + \alpha_{2222} a_2^+ a_2 a_2^+ a_2 + \alpha_{1122} a_1^+ a_1 a_2^+ a_2, \quad (15.14)$$

where $\omega_1 \approx 2\omega_2$, α_{122} is the constant of the cubic anharmonicity responsible for the parametric coupling of modes 1 and 2 leading to the exchange of one quantum of frequency ω_1 for two quanta of frequency ω_2, α_{1111}; α_{2222} are the constants of internal anharmonicity of modes 1 and 2, respectively; and α_{1122} is the constant of the intermode anharmonicity of fourth order, responsible for the interchange of frequencies of the interacting modes.

Hamiltonian (15.14) is much like (15.10) for the three-fold degenerate mode. Constants α_{1111} and α_{2222} are the analogues of constant α; α_{1122} corresponds to constant β, while constant α_{122} stands for constant γ responsible for the parametric interaction. Note, that, strictly speaking, in the anharmonic Hamiltonian (15.14) other terms such as, for example, $\alpha_{122}(a_1 a_1 a_2^+ + a_1^+ a_1 a_2)$ should be taken into account. However, the levels coupled via these additional terms strongly differ in energies, so, their effect is unessential.

Now let us turn to Fig. 15.9 and consider the changes induced by Hamiltonian (15.14) in the levels close in energy. Most simply we may trace these changes for the levels 1.0 and 0.2. Hamiltonian (15.14) in the basis of states 1.0 and 0.2 takes on the form

$$\widehat{H}_{\text{anh}} = \begin{vmatrix} \hbar\omega_1 + \alpha_{1111} & 2^{1/2}\alpha_{122} \\ 2^{1/2}\alpha_{122} & 2\hbar\omega_2 + 4\alpha_{2222} \end{vmatrix}. \quad (15.15)$$

The eigenvalues of this Hamiltonian,

$$E_{1,2} = \frac{1}{2}(\hbar\omega_1 + \alpha_{1111} + 2\hbar\omega_2 + 4\alpha_{2222}) \pm \left[\frac{1}{4}(\hbar\omega_1 + \alpha_{1111}\right.$$
$$\left. - 2\hbar\omega_2 - 4\alpha_{2222})^2 + 2\alpha_{122}^2\right]^{1/2},$$

determine the new positions of the energy levels, while the eigenvectors of the matrix, no longer states with a definite number of quanta in every mode, are the linear combinations of states 1.0 and 0.2.

The phenomenon of the mixing of states of different modes and the shift of their energy levels has acquired the name of Fermi resonance. In quantum electronics the Fermi resonance of the symmetric valence and deformation modes of the CO_2-molecule is a subject of common knowledge; it results in the possibility of generation in two frequency intervals (in the vicinities of wavelengths 10.6 and 9.6 μm).

In higher vibrational levels the manifestation of the Fermi resonance is more complex and results in an energy level shift and the mixing of different hybrid states. So, for example, for levels 3.0, 2.2, 1.4 and 0.6 Hamiltonian (15.14) in the basis of these states takes on the form

$$\widehat{H}_{\text{anh}} = \begin{vmatrix} a_{11} & a_{12} & a_{13} & a_{14} \\ a_{21} & a_{22} & a_{23} & a_{24} \\ a_{31} & a_{32} & a_{33} & a_{34} \\ a_{41} & a_{42} & a_{43} & a_{44} \end{vmatrix}, \tag{15.16}$$

$$a_{11} = 3\hbar\omega_1 + 9\alpha_{1111}, \quad a_{12} = 6^{1/2}\alpha_{122}, \quad a_{13} = 0, \quad a_{14} = 0,$$

$$a_{21} = 6^{1/2}\alpha_{122}, \quad a_{22} = 2\hbar\omega_1 + 2\hbar\omega_2 + 4\alpha_{1111} + 4\alpha_{2222} + 4\alpha_{11}\alpha_{22},$$

$$a_{23} = 24^{1/2}\alpha_{122}, \quad a_{24} = 0, \quad a_{31} = 0, \quad a_{32} = 24^{1/2}\alpha_{122},$$

$$\alpha_{33} = \hbar\omega_1 + 4\hbar\omega_2\alpha_{1111} + 16\alpha_{2222} + 4\alpha_{11}\alpha_{22}, \quad a_{34} = 30^{1/2}\alpha_{122},$$

$$a_{41} = 0, \quad a_{42} = 0, \quad a_{43} = 30^{1/2}\alpha_{122}, \quad a_{444}6\hbar\omega_2 + 36\alpha_{2222}.$$

This list reflects the fact that in the initial Hamiltonian (15.14), all the terms except the third are diagonal in occupation-number representation. The effect of the third term (the parametric one) is an increase of the number of quanta in the first mode by one, accompanied by a decrease of the number of quanta in the second mode by two. In other words,

$$\alpha_{122}a_1^+a_2a_2|n_1n_2\rangle = \alpha_{122}[(n_1+1)(n_2-1)n_2]^{1/2}|n_1+1, n_2-2\rangle,$$

i.e., the corresponding matrix element of the Hamiltonian is

$$H_{n_1,n_2}^{n_1+1,n_2-2} = \alpha_{122}[(n_1+1)(n_2-1)n_2]^{1/2}.$$

Hamiltonian (15.16) has four eigenvalues distinct from the harmonic energies. The appropriate four eigenfunctions at $\alpha_{122} \neq 0$ are the linear combinations of the states with a definite numbers of quanta in the modes. The very mixing of the states is the most important property of the Fermi resonance. It brings about a considerable increase in the number of possible lines in the spectra of the molecular dipole transitions. We shall conclude this lecture by considering them.

The probability of any radiative transition is determined by the value of the matrix element of the dipole moment operator μ_{ik}. The spectrum of the transitions depends on the specific pairs of states for which μ_{ik} is nonzero. The molecular dipole moment operator $\widehat{\mu}$ can be represented in the form

$$\widehat{\mu} = \sum_i q_i e_i + \sum_{ik} c_{ik} q_i q_k + \sum_{ijk} c_{ijk} q_i q_j q_k + \dots , \qquad (15.17)$$

where q_i is the normal coordinate; e_i, the so-called effective charge in the mode i; c_{ik}, the constant of quadratic correction in dependence of the molecule dipole moment on the internuclear distances; and c_{ijk}, the cubic ones. These higher-order corrections are caused by changes in the configuration of the electronic cloud induced by the displacements of the nuclei. Henceforth, these corrections (of the 4th, 6th and higher orders in the Born-Oppenheimer parameter Λ) will be disregarded. The modes with $e_i \neq 0$ are called dipole-active.

In the harmonic approximation there are only a few nonzero matrix elements of the dipole moment operator. These are the matrix elements of the transitions between the states which differ in the number of quanta in just a single mode and just by unity. So, for example, state 2.3 (Fig. 15.9) admits the transitions between states 3.3, 1.3, 2.4 and 2.2. Closely spaced to them in energy are levels 0.5, 0.7, 4.1, 3.0, etc. However, transitions to them from state 2.3 are impossible. Sometimes such transitions are said to be forbidden, since the corresponding states are not linked with state 2.3 in a dipole way. It is known that the reason for this is that for a one-dimensional harmonic oscillator, the only nonzero matrix elements of the coordinate operator q are those closest to the diagonal, i.e., $q_{n,n\pm1}$, and in the harmonic approximation the hybrid states are direct products of harmonic oscillators. The frequency spectrum of the absorption (emission) lines of the molecule in the harmonic approximation coincides therefore with the spectrum of the frequencies of normal vibrations of the dipole active modes (sometimes the term *IR active* is used).

The rotational structure of the vibrational states also becomes apparent in the dipole transition spectra. For free rotation not coupled with vibrations, i.e., for a rigid top, there are proper selection rules as for the harmonic oscillator. These rules state that the transitions with $\Delta J = 0 \pm 1$ and $\Delta K = 0^{*)}$ are

---

[*)] Among the vibrational selection rules there is also the limitation imposed on the magnetic quantum number M, i.e., on the eigenvalue of the molecule angular momentum projection on a fixed axis in the laboratory coordinate system. If, as assumed, external static fields are absent, M-degeneracy exists and the rules corresponding to this quantum number do not become apparent spectrally. Note, however, that in the case of the circular polarization, when the quantization axis is oriented along the direction of propagation, requirement $\Delta M = \pm 1$ should be satisfied, and in the case of the linear polarization, when the quantization axis is oriented along the electric field of the wave, $\Delta M = 0$.

allowed. If we ignore the vibrational-rotational interaction, the energy eigen-states of a molecule appear to be the direct products of the rotational and vibrational wavefunctions, the eigenvalues of energy corresponding to them being the sums of the rotational and vibrational energies. Therefore the vibrational transition has its own rotational structure which becomes apparent in the spectra. The totality of the vibrational-rotational transition lines corresponding to the same vibration is called the vibrational band. In Fig. 15.11 we present the transitions comprising the vibrational band:

$$|0,0,0,\ldots,0\rangle \rightarrow |1,0,0,\ldots,0\rangle.$$

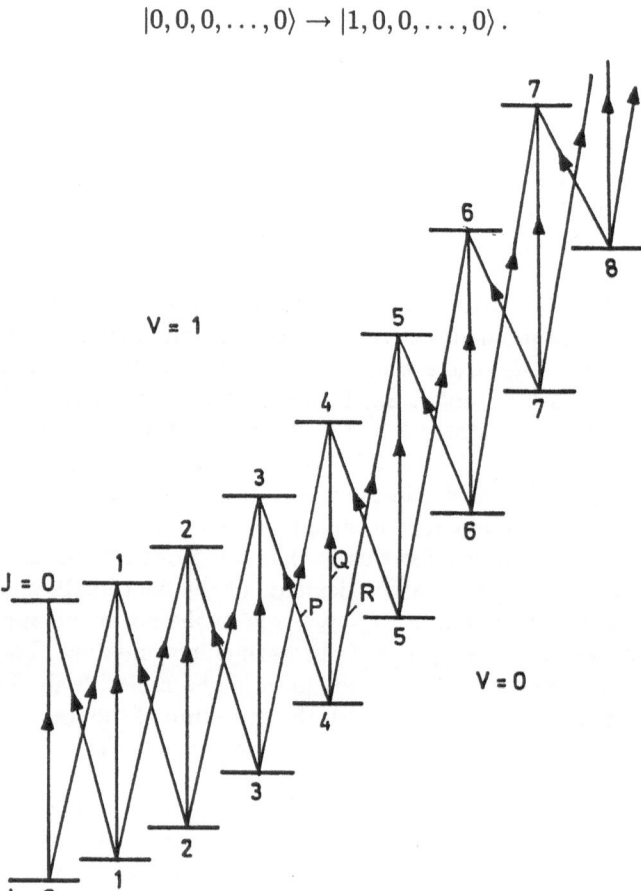

Fig. 15.11. $P(J \rightarrow J - 1)$-, $Q(J \rightarrow J)$, $R(J \rightarrow J + 1)$-transitions of vibrational band $V = 0 \rightarrow V = 1$

The transition with $\Delta J = 0$ which is referred to as the Q-transition does not change the rotational state; its frequency is independent of the number J, being determined by the vibrational quantum energy. The transition with $\Delta J = -1$, which is referred to as the P-transition, is detuned from the Q-transition toward lower frequencies by the value of $2BJ/\hbar$, while the transition with $\Delta J = +1$, which is referred to as the R-transition, is detuned toward higher frequencies by the value of $2B(J+1)/\hbar$ (see (15.7)). These transitions in the ensemble of molecules form the P-, Q- and R-branches of the vibrational band. The typical form of these branches is given in Fig. 15.2. The present consideration does not take into account the vibrational-rotational interaction. Then the vibrational constant B is the same for all the vibrational states and the Q-branch is the δ-function. The form of the P- and R-branches is determined by the distribution of molecules among the vibrational states. In the case of thermal equilibrium this distribution is the Boltzmannian

$$\rho_{JJ} = \frac{1}{Z}(2J+1)\exp\left[-\frac{BJ(j+1)+AK^2}{kT}\right], \qquad (15.18)$$

where Z is the statistical sum. Factor $2J+1$ in this formula is the consequence of the degeneracy over the quantum number M. Since the transitions considered conserve the quantum number K, this number is spectrally unobservable. So, we ought to sum (15.18) over K from $-J$ to $+J$ to obtain the shape of the P- and R-branches. At small A, $AB \ll (kT)^2$, the summation results in an additional factor $2J+1$:

$$\rho_{JJ} = \frac{1}{Z}(2J+1)^2\exp\left[-\frac{BJ(J+1)}{kT}\right]. \qquad (15.19)$$

Since the transitions of the P- and R-branches are detuned with regard to the Q-branch by $2BJ/\hbar$ and $2B(J+1)/\hbar$, respectively, (15.19) represents, in fact, the usually observed shape of these branches. When the states coupled via the transition are degenerate in K, this shape is the envelope of the equidistant lines separated by $2B/\hbar$, of the isolated P- and R-transitions, which are infinitesimally narrow if we neglect the vibrational-rotational interaction. The removal of the K-degeneracy affects the spectra although the shape of the envelopes of the P- and R-branches remains unchanged in the sense that the number of transitions within the small but finite spectral interval is the same as in the case of the unremoved degeneracy. However, the fine structure of the line of the vibrational-rotational transition is modified essentially. The most striking demonstration of this effect is the Q-branch, which becomes broadened on account of the superposition of the K-structures corresponding to the levels with different J (Figs. 15.7 and 15.2). For large J the removal of the K-degeneracy leads to the formation of the level bands of a width exceeding the rotational constant B. As a result the P-, Q- and R-transitions in molecules with such J take on the form of bands of a width greater than $2B$. The centers of these bands correspond to the positions of the lines of P-, Q- and R-transitions in the case of unremoved degeneracy. Under these conditions the bands appropriate to the molecules with the neighboring values of J may overlap.

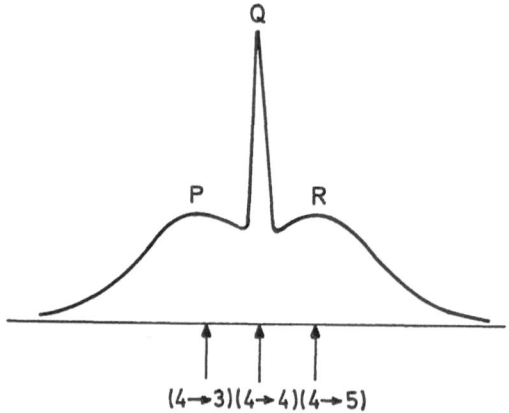

Fig. 15.12. P-, Q-, R-branches

So far we only touched upon the transitions in the bonds of the vibrational band $|0,0,0,\ldots,0\rangle \to |1,0,0,\ldots,0\rangle$. However, upward dipole transitions may start from the excited vibrational states, for example, the transitions from the first excited state of mode ν_2 to the hybrid state $\nu_1 + \nu_2$: $|0,1,0,\ldots,0\rangle \to |1,1,0,\ldots,0\rangle$. Due to the intermode anharmonicity this results in the displacement of the position of the Q-branch transition usually directed toward the lower frequency. The bands thus obtained, i.e., the bands of transitions starting from the excited and not from the ground vibrational state, have acquired the name of hot bands. This name originates from the fact that corresponding absorption bands have been observed when heating the gas of polyatomic molecules. an Actually, the spectra of hot bands are close to the absorption spectra of transitions from the highly excited vibrational states. In this case the excitation from the mode directly excited by radiation is transferred to other modes due to the parametric coupling of the modes. An extreme manifestation of intermode couplings is very large molecules with dozens of atoms. The resonant excitation of a single mode causes the heating of the molecule as a whole, which, in turn, increases the number of hot bands. If the number of atoms in a molecule is not so large, the process of the excitation transfer from one mode to another may be neither uniform in modes nor equilibrium thermally and may be even quite different from the thermal. The corresponding distribution function f may noticeably differ from the Boltzmann distribution. In such cases, when describing the absorption spectra of the excited molecules, one should bear in mind the specific structure of the molecular anharmonicity and the specific intermode parametric couplings which, in fact, form the energy distribution among the molecular degrees of freedom.

In the classical limit, i.e., in case the total number of the vibrational quanta in the molecule is large, one may write down the following expression for the cross section of the radiation absorption at frequency ω, the latter being close to the frequency of the main transition $0 \to 1$ of the mode with number i:

$$\sigma_i(\omega) = \frac{8\pi\hbar\omega}{c}(\mu_{01}^{(i)})^2 \int \delta(\omega - \omega_i)(\{v\}, \{\theta\}, J, K))$$
$$\times f(\{v\}, \{\theta\}, J, K)\{dv\}\{d\theta\}dJdK \,, \tag{15.20}$$

where $\{v\}$ denotes the aggregate of all the numbers of quanta in each mode, v being continuous in the classical limit; $\{\theta\}$ is the aggregate of the phases of all the vibrations of all the modes; and f is the distribution function of molecules in the phase volume.

The expression thus written contains the dipole moment of mode $\mu_{01}^{(i)}$. On the high levels the states are significantly mixed due to the parametric interaction, so, attributing the transition to a single definite mode becomes rather conditional. The choice of the value of $\mu_{01}^{(i)}$ is dictated by the closeness of frequency ω to the frequency of transition $0 \to 1$ of some mode. At a relatively lower excitation extent where the levels are essentially nonresonant the parametric interaction is incapable of mixing the states of different modes completely. In this situation one may observe the phenomenon of induced-dipole activity which can be described as follows.

Let the hybrid vibration consisting, say, of the sum of the motions of modes ν_1 and ν_2, have a frequency close to the frequency of some dipole active mode ν_3 but distinct from it by a value greater than the energy of the parametric interaction between them (in terms of $\hbar$). This weak but finite interaction will give a small but nonzero addition to the wavefunction of the hybrid state. This implies that the vibrations of mode $\nu_1 + \nu_2$ are accompanied by the synchronous small-amplitude vibrations of mode ν_3.

Hence, the hybrid vibration $\nu_1 + \nu_2$, being weakly parametrically coupled with vibration ν_3, provides the charge motions in the molecule corresponding to the nonzero dipole moment of transition.

Concluding this lecture, let us dwell upon the important question of the stochastization of the molecular vibrational motion. This phenomenon takes place when the parametric (or Coriolis, etc.) interaction strongly mixes the states corresponding to the definite mode occupation numbers (or the definite rotational quantum numbers), the latter ceasing to be the energy eigenstates. New energy eigenstates appear, which are linear combinations of the initial states. In case the old states corresponding to the molecular vibrational modes are represented in these linear combinations more or less uniformly, one speaks of the stochastization of the vibrations. Stochastization differs from thermalization. Generally speaking, not all the modes participate in it. Moreover, the mere presence of some mode in the stochastized motion does not automatically mean that all the hybrid states of these modes are completely mixed.

In principle, the hybrids with definite occupation numbers slightly coupled with other states may exist, thus being the energy eigenstates. For these states the coupling with any other state is less than the corresponding detuning. The part of the phase volume, in which the hybrid states are mixed, is called the stochasticity region.

Completing our discussion of the vast and important question of the spectra of molecules, we note that the usual IR spectroscopic apparatus with a resolution of up to 0.1 cm^{-1} does not, as a rule, permit observation of the fine structure of the vibrational-rotational spectra caused by the vibrational-rotational interaction, the removal of K-degeneracy, etc. In order to observe them, the more sophisticated methods such as Fourier-spectroscopy, diode laser spectroscopy, CARS, etc., are required.

Lecture Sixteen
Excitation of Molecules

Simple-oscillator model. Anharmonicity and overcoming it by means of a strong field. Polyatomic molecules. Model of the excited mode damped by the interaction with a thermostat. Classical and quantum approaches. Large polyatomic molecules. Longitudinal and transverse relaxations. Fermi resonances, soft modes. Small polyatomic molecules. Number of atoms. Three stages of excitation. Excitation of quasicontinuum. Kinetic coefficients. Phase-volume average. Single degree of freedom of three-fold degenerate mode excitation. Spectral dependencies.

Resonant laser radiation exites molecules. It is evident that in accordance with the laser-wavelength range we may excite electronic, vibrational or rotational transitions. Let us first consider the vibrational transitions since they are of great interest.

The simplest possible model describing the excitation of molecular vibrations is the harmonic oscillator. The manner in which excitation of a simple oscillator occurs under the action of an external harmonic force is well known. The amplitude of forced oscillations is proportional to the amplitude of the exciting force divided by the frequency detuning. In the case of strict resonance, the amplitude of the forced oscillations increases linearly with time. However, as a molecule is essentially a quantum object, the adequate treatment of its excitation process involvs, naturally, the quantum approach. Then the excitation of a simple oscillator means exciting the equidistant ladder of the energy levels by a periodic external field. The quasi-energy method considered in Lecture 9 may be easily applied to this problem. The use of this method in the situation considered leads to the set of Eqs. (9.25) in which $E_k = k\hbar\omega_0$, where ω_0 is the eigenfrequency of the oscillator and, besides, $\mu_{k,k+1} = \mu_0\sqrt{k+1}$, where μ_0 is the dipole moment of the main transition $(0 \to 1)$. The description of the vibrational excitation dynamics of a molecule by Eqs. (9.25) is much more complex than in the classical limit and forms a special branch of theoretical physics called "the dynamics of coherent states."

Both the classical and the quantum models of a simple oscillator are well developed. However, due to anharmonicity, their applicability to the description of the excitation of molecular vibrations is restricted. The presence of anharmonicity implies that the radiation resonant with respect to one pair of levels becomes out-of-resonance for another transition. Hence, such a molecule may be efficiently excited either through a multiphoton process (but then with a small cross-section, i.e., slowly), or in fields strong enough for the Stark broadenings of levels to cover their detunings from resonance. Let us evaluate the field strength required for the latter process. Evidently, the dynamic Stark broadening $\mu\mathcal{E}/\hbar$ should exceed the anharmonic detuning Δ. For the vibrational level v of a diatomic molecule, $\Delta = \alpha v^2$. Thus we obtain the minimum value of intensity I required for the effective population of the vibrational level with number v in the form $I = 10^{-7} c^3 \hbar^2 \alpha^2 v^4 / 8\pi\mu^2$, where I is expressed in watts per cm^2, provided α is given in wavenumbers, and all the other quantities, in CGSE units. This implies that the effective excitation of, say, the thirtieth vibrational level which usually corresponds to the dissociation limit, occurs at $I = 10^{11}$ W/cm^2, given $\alpha = 1$ cm^{-1} and $\mu = 1$ D.

The same estimate remains valid in the case of the excitation of molecules with a small number of atoms, in which the Fermi-resonances, if any, belong to rather high levels close to the dissociation energy. The situation may be exemplified with deuteroformaldehyde, D_2CO, which admits the excitation of levels with $v = 10$–20 in the IR laser fields of intensity 10^9 W/cm^2. For level numbers $v > 10$ the Fermi-resonances become essential.

Polyatomic molecules may be effectively excited at much lower intensities. The reasons are a higher density of their vibrational spectrum, the presence of the intermediate resonances, intermode couplings, the Fermi resonances, etc. (see Lecture 15). The simplest description of the vibrational spectrum may be given in the case of rather big polyatomic molecules where the density of the vibrational spectrum is high enough to allow the typical energy difference of quanta of neighboring modes (or combinations of the quanta) to be less than the typical constant of the corresponding intermode anharmonicity, i.e., when practically all the modes are coupled by the Fermi resonances of different orders, even at the lowest levels of the vibrational spectrum. Then the model of the excited mode interacting with a thermostat may be used. In the classical treatment this model corresponds to the oscillator with friction, the only complication being the dependence of the friction coefficient on the thermostat temperature, the latter being determined by all the details of the history of the oscillator excitation process. Let us write down the corresponding equations:

$$\ddot{x} + \gamma(T)\dot{x} + \omega_0^2 x + \alpha x^3 = \mu\mathcal{E}(t)\cos\omega t, \quad c\dot{T} = -\gamma(T)x\dot{x}, \qquad (16.1)$$

where c is the heat capacity; T, the temperature; γ, the friction coefficient, and all other notations common. This equation allows only for anharmonicity of fourth order, which yields the cubic term in the equation for the coordinate. The oscillator motion is much faster than the heating of the thermostat. So, we may look for a quasi-stationary solution of the oscillator equation, assuming $\gamma(T)$ to be constant. Let us try the solution in the form $x = x_0 \cos(\omega t + \phi)$. The straightforward substitution in the original equation produces the following equation for x_0:

$$x_0^2 = \frac{\mu^2 \mathcal{E}^2}{\left[\omega^2 - \omega_0^2 + 3\alpha x_0^2/4\right]^2 + 4\gamma^2\omega^2}. \qquad (16.2)$$

Then

$$c\dot{T} = \gamma(T)\frac{\omega^2\mu^2\mathcal{E}^2}{\left[\omega^2 - \omega_0^2 + 3\alpha x_0^2/4\right]^2 + 4\gamma^2\omega^2}. \tag{16.3}$$

In fact, we have obtained an equation describing how the molecule accumulates energy. It may be rewritten explicitly, provided the bicubic Eq. (16.2) for x_0 is solved. Without dwelling on a general analysis we consider only some particular cases under the assumption of the linear dependence $\gamma(T)$, $\gamma(T) = \gamma_0 T/T_0$. This assumption is made just for simplicity, so the following analysis is only illustrative.

If the system is in resonance and the anharmonicity is small, only the last term in the denominator of (16.3) is significant, leading under the above assumption to the equation,

$$2T\dot{T} = \frac{\mu^2\mathcal{E}^2}{2\gamma_0 c}T_0. \tag{16.4}$$

Its solution is

$$T = \mu\mathcal{E}\left(\frac{T_0}{2\gamma_0 c}t\right)^{1/2} \propto t^{1/2}. \tag{16.5}$$

In other words, this is the often encountered case of the diffusive buildup of molecular energy. Another possible case is when the field frequency is detuned from the resonant one by a value significantly exceeding both the anharmonicity and the line width,

$$\dot{T} = \frac{\omega^2\gamma_0\mu^2\mathcal{E}^2}{2(\omega^2 - \omega_0^2)T_0 c}T. \tag{16.6}$$

This implies the exponential heating of the molecule with the typical growth rate value of $\gamma_0\omega_2\mu^2\mathcal{E}^2/(\omega^2 - \omega_0^2)cT_0$. The third limit is where the oscillation anharmonicity is the largest term. Then, as it follows from (16.2), $x_0^2 = (4\mu\mathcal{E}/3\alpha)^{2/3}$ and

$$\dot{T} = \frac{\gamma_0\omega^2}{cT_0}\left[\frac{4\mu\mathcal{E}}{3\alpha}\right]^{2/3}T. \tag{16.7}$$

This also means an exponential increase of the molecular temperature, but with a different value of the growth rate.

Therefore, in the model of a single anharmonic vibrational mode coupled with the aggregate of other vibrational modes serving as a thermostat, we observe both the thermostat heating at the expense of the energy transferred to it from the mode being excited, and the backward action of the thermostat on the mode. This action brings about changes in the damping, the oscillation amplitude, and the position of the resonance.

A molecule, generally speaking, is essentially a quantum object. The classical oscillator is not directly related to it. However, the qualitative features of the excitation dynamics of the classical oscillator simulate rather well the motion of large molecules and so are worth our attention. The quantum nature of a large molecule is to be taken into account when describing the interaction of the mode being excited with the radiation. If all the other modes may be treated as a thermostat (see (4.36)), then (4.7) for the density matrix with the relaxation matrix elements (4.41) should be employed.

In relatively weak fields where the corresponding Rabi frequencies are much less than the partial relaxation rates for the off-diagonal matrix elements closest to the main matrix diagonal, the stationary equations are valid and the more distant off-diagonal elements may be neglected. Then for the diagonal elements we can write:

$$
\begin{aligned}
\dot{\rho}_{n,n} = {} & D_{n,n+1}(\rho_{n+1,n+1} - \rho_{n,n}) + D_{n,n-1}(\rho_{n-1,n-1} - \rho_{n,n}) \\
& + R_{n,n,n+1,n+1}\rho_{n+1,n+1} + R_{n,n,n-1,n-1}\rho_{n-1,n-1} \\
& - (R_{n-1,n-1,n,n} + R_{n+1,n+1,n,n})\rho_{n,n} \,,
\end{aligned}
\tag{16.8}
$$

where R_{ijkl} are the relaxation matrix (4.41) elements. The kinetic coefficients for the upward $D_{n,n-1}$ and downward $D_{n,n+1}$ transitions are given by

$$
D_{n,n-1} = (n-1)\mu^2 \mathcal{E}^2 \frac{2R_{n,n-1,n,n-1}}{(E_n - E_{n-1})^2 + (\hbar R_{n,n-1,n,n-1})^2} \,,
\tag{16.9}
$$

$$
D_{n,n+1} = n\mu^2 \mathcal{E}^2 \frac{2R_{n,n+1,n,n+1}}{(E_{n+1} - E_n)^2 + (\hbar R_{n,n+1,n,n+1})^2} \,.
\tag{16.10}
$$

Expressions (16.8) are the kinetic equations of the balance type which have already been discussed many times. The kinetic coefficients $D_{n,n-1}$ and $D_{n,n+1}$ are proportional to the cross sections of radiative transitions possessing, as a rule, resonant denominators. In these denominators the corresponding elements of the relaxation matrix determine the values of the partial times of transverse relaxations. The kinetic coefficients are proportional to the field intensity and to the matrix element squared of the transition $0 \to 1$ operator.

The longitudinal relaxation is described by the four last terms of (16.8). A rather big polyatomic molecule with the number of degrees of freedom $s \geq 30$ cannot be stable at temperatures exceeding the dissociation energy divided by the number of degrees of freedom, since otherwise the molecule decays due to the fluctuations of the energy distribution among the degrees of freedom. Hence, $kT \leq U_d/s$, where U_d is the dissociation energy. If the radiation quantum energy noticeably exceeds the value U_d/s, then we have $kT \ll \hbar\omega$. In this case only the downward relaxation is essential, and (16.8) takes on the form

$$
\begin{aligned}
\dot{\rho}_{nn} = {} & D_{n,n+1}(\rho_{n+1,n+1} - \rho_{nn}) + D_{n,n-1}(\rho_{n-1,n-1} - \rho_{nn}) \\
& + R_{n,n,n+1,n+1}\rho_{n+1,n+1} - R_{n-1,n-1,n,n}\rho_{nn} \,.
\end{aligned}
\tag{16.11}
$$

It is natural that (16.11) for the diagonal elements of the density matrix of the mode being excited should be supplemented with an equation for the thermostat temperature,

$$
c\dot{T} = \sum_{n,k}(E_n - E_k) R_{nnkk}\rho_{kk} \,,
\tag{16.12}
$$

where c is the thermostat heat capacity dependent, generally speaking, on the temperature.

Now let us determine the elements of the relaxation matrix responsible for the longitudinal relaxation. In the case considered the longitudinal relaxation arises from the interaction of the excited mode with some other modes of the same molecule jointly comprising the thermostat. This interaction is of the parametric type and originates from the anharmonicity. The cubic anharmonicity is the main term by order of magnitude and corresponds to the operator

$$\widehat{V} = \alpha \widehat{X}\widehat{Y}\widehat{Z},\tag{16.13}$$

where $\widehat{X}$ is the coordinate operator of the mode being excited, and $\widehat{Y}$ and $\widehat{Z}$ are the coordinate operators of two thermostat modes. Let us denote the mean-square values of the amplitudes of the zero-point oscillations corresponding to these modes by the symbols X_0, Y_0 and Z_0, respectively; i.e., we assume that

$$\langle n|X^2|n\rangle = nX_0^2, \quad \langle n_y|Y^2|n_y\rangle = n_yY_0^2, \quad \langle n_z|Z^2|n_z\rangle = n_zZ_0^2\tag{16.14}$$

and substitute (16.13) in (4.41). Then

$$\begin{aligned}
R_{n-1,n-1,n,n} &= 2J_{n-1,n,n,n-1}\\
&= 2\pi\alpha^2\int\left[\frac{d\Gamma}{dE}\right]^2 X_0^2Y_0^2Z_0^2\exp\left[-\frac{n_z\hbar\omega_z + n_y\hbar\omega_y}{kT}\right]\frac{1}{Z}n\\
&\quad \times(n_y-1)(n_z-1)\delta(\omega_y+\omega_z-\omega) = 2\pi\alpha^2 X_0^2Y_0^2Z_0^2n\\
&\quad \times\int\frac{\delta(\omega_y+\omega_z-\omega)d\omega_yd\omega_z}{[1-\exp(-\hbar\omega_y/kT)][1-\exp(-\hbar\omega_z/kT)]}\\
&= 2\pi n\gamma_3^2 g_F(\omega),
\end{aligned}\tag{16.15}$$

where $g_F(\omega)$ is the density of the Fermi resonances, while $\gamma_3^2 = \alpha^2 X_0^2Y_0^2Z_0^2$ is the square of the matrix element of the cubic anharmonic interaction.

Expression (16.15) is analogous to that obtained in Lecture 10 when considering the decay of a single level into the band. This coincidence is not just accidental. When we say that the resonant mode decays into the thermostat, we mean the level corresponding to a certain thermostat state and a fixed number n of quanta in the mode being excited, decaying into the band, each level of which corresponds, in turn, to the mode state with $n-1$ quanta and the thermostat state obtained as a result of the excitation transfer via the channel of one of the Fermi resonances.

Now let us turn to the transverse relaxation. Its matrix elements are given by the same expression (4.41) and besides, the contribution determined by the longitudinal relaxation rate contain an additional term caused by the chaotic modulation of the positions of the energy levels of the excited mode as a result of the internal motion of the thermostat. These positions are determined by the even powers of the mode coordinates. The lowest order anharmonicity still leading to the transverse relaxation is, therefore, of the type

$$\widehat{V} = \beta\widehat{X}^2\widehat{Y}\widehat{Z}.\tag{16.16}$$

The interaction of the type of (16.16) is significant for the transverse relaxation only if in the polyatomic molecule considered there is a sufficient number of soft modes with the energy of the quantum $\hbar\omega \ll KT$. To avoid any misunderstanding, note that we are dealing with such a large molecule that it has a average energy of the vibrational quantum exceeding the typical thermal energy, $\hbar\omega \gg kT$, and the mode being excited by the resonant laser radiation is also rather hard, $\hbar\omega \gg kT$. But among the variety of the molecular modes there are many soft ones for which $\hbar\omega \ll kT$. The transitions between the latter modes occur owing to their parametric interaction, which exists in the presence of the excitation of the resonant mode and the anharmonicity of the type of (16.16). It is these transitions that account for the transverse relaxation. We calculate the matrix element of such a transition with the help of (4.41). The result obtained may be illustrated by the following qualitative considerations. As in the case of the longitudinal relaxation, the whole process may be represented as the decay of the level into the band. The scheme of this process is

$$|n, n_y, \ldots, n_z, n_{z'}, n_{z''}, \ldots\rangle$$

$$\mapsto |\ldots, \quad \ldots \quad , \quad \ldots\ldots\ldots\ldots\ldots\ldots\ldots\ldots\ldots\rangle$$
$$\mapsto |n \quad , \quad n_y - 1, \quad \ldots\ldots\ldots\ldots\ldots\ldots\ldots\ldots\ldots\ldots\rangle$$
$$\mapsto |n \quad , \quad n_y - 1, \quad \ldots, \quad n_z + 1, \quad n_{z'} \quad , \quad n_{z''} \quad , \quad \ldots \quad \rangle$$
$$\Rightarrow \mapsto |n \quad , \quad n_y - 1, \quad \ldots, \quad n_z \quad , \quad n_{z'} + 1, \quad n_{z''} \quad , \quad \ldots \quad \rangle, \quad (16.17)$$
$$\mapsto |n \quad , \quad n_y - 1, \quad \ldots, \quad n_z \quad , \quad n_{z'} \quad , \quad n_{z''} + 1, \quad \ldots \quad \rangle$$
$$\mapsto |n \quad , \quad n_y - 1, \quad \ldots, \quad n_z \quad , \quad n_{z'} \quad , \quad n_{z''} \quad , \quad \ldots \quad \rangle$$
$$\mapsto |\ldots, \quad \ldots \quad , \quad \ldots\ldots\ldots\ldots\ldots\ldots\ldots\ldots\ldots\ldots\rangle$$

where $\omega_y \approx \ldots \approx \omega_z \approx \omega_{z'} \approx \omega_{z''} \approx \ldots$. It visualizes the fact that the number of quanta in the resonant mode remains unchanged, while the number of quanta in one of the soft modes increases at the expense of another soft mode. The probability of such a process is

$$W = 2\pi\beta^2 n(n+1) X_0^4 (n_z + 1) Z_0^2 n_y Y_0^2 g(\omega_z). \quad (16.18)$$

Since modes y and z may be chosen arbitrarily, this expression is to be averaged over all such pairs of modes. Then, taking into account that

$$n_y n_z = (kT/\hbar\omega_z)^2 \exp(-\hbar\omega_z/kT), \quad (16.19)$$

we obtain an expression for the additional term being searched for:

$$\Delta R_{n,n+1,n,n+1} = 2\pi\gamma_4^2 n(n+1) \int \left[\frac{kT}{\hbar\omega}\right]^2 \exp\left[-\frac{\hbar\omega}{kT}\right] d\omega, \quad (16.20)$$

where $\gamma_4^2 = \beta^2 X_0^4 Y_0^2 Z_0^2$ is the square of the matrix element of the fourth-order anharmonic interaction.

By comparing (16.10) and (16.15) we see that these elements of the relaxation matrix are accounted for by quite different spectral characteristics of a molecule. For the longitudinal relaxation only the spectral density of the Fermi resonances in the vicinity of the frequency of the mode being excited is essential, while the transverse relaxation depends on the spectral density of the soft modes frequencies as well. Besides, in large molecules the longitudinal relaxation is independent of the temperature, contrary to the transverse one, which contains an additional term essentially dependent on temperature. This is why the rate of the latter may considerably exceed the longitudinal relaxation rate determined by the cubic anharmonicity despite the fact that an addition to the transverse relaxation rate is determined by the fourth-order anharmonicity.

Let us estimate the rate of the longitudinal relaxation. At $s = 30$ and $\hbar\omega = 30$ cm^{-1} the density of the Fermi resonances is $2\pi g_F(\omega) \simeq s/\hbar\omega = 3 \cdot 10^{-2}$ cm. Further, the constant of the cubic anharmonicity $\gamma_3 \approx \Lambda\hbar\omega$, where Λ is the Born-Oppenheimer parameter, is about 20 cm^{-1} by order of magnitude. As a result, $R_{n-1,n-1,n,n} \approx 30$ cm^{-1}, which corresponds to a relaxation time of about 10^{-12} s.

If the Rabi oscillation period exceeds the longitudinal relaxation time the dynamics of molecular excitation reduces to a simple process, where an external field excites the first vibrational level of the resonant mode which immediately renders its energy to the thermostat. The higher levels of the mode remain unexcited. The amount of energy per thermostat mode is small, being less than the energy of the quantum of the excited mode. However, the nonlinear interactions between the thermostat modes bring about fluctuations of the energy stored in each of them. As a rule, these fluctuations are not large, but sometimes (quite rarely in the time scale of the oscillation period) and for a short time (of the same order) the energy may be concentrated in one of the modes, due to the large fluctuation. Such a concentration of a large amount of energy in a single degree of freedom may break the corresponding bond in the molecule. The probability of such an event is higher, the lower the dissociation threshold of the bond is. So, usually, it is sufficient to consider the fluctuational concentration of energy only in the weakest bond.

If the intensity is so large that the Rabi oscillation period is shorter than the longitudinal relaxation time, the equation for the density matrix of the system with the relaxation matrix of the type of (4.7) should be solved. Bearing in mind that the intensities required are quite large (they typically reach the value of 10^{11}–10^{13} W/cm^2) we shall not consider these cases.

Note still another interesting possibility. In the polyatomic molecules, we may come across situations in which the density of the Fermi resonances is small, while the density of the soft modes is large. This is the case for the coordinational compounds in which a simple and rigid group of atoms forms a kern surrounded by ligands slightly bound with it (coordinated with respect to it). Each ligand represents a rather complex molecular formation. An example is the complex compound 3-butil-phosphate of uranil which demonstrates an explicitly expressed resonance at a frequency of about 10^3 cm^{-1}, thus leading itself conveniently to investigation by CO_2-lasers.

In such cases the rate of the transverse relaxation considerably exceeds that of the longitudinal one which, in the collisionless case considered here, may be accounted for only by Fermi resonances of higher orders. The dynamics of this system is described by Eqs. (16.11) and (16.12), with the transverse relaxation rate given by (16.20), and reduces to the following. After the field has been switched on, the resonant levels are excited. This process is a diffusion in time with the diffusion coefficients being the kinetic coefficients (16.9) and (16.10). They are Lorentzian contours centered around the frequency determined by the anharmonic shift of the corresponding transition with the width dependent on the initial temperature of the soft-mode thermostat. The diffusion over the levels of the excited mode persists until a quasi-stationary distribution is established due to the longitudinal relaxation. This process also results in the slow heating of the thermostat, which, in turn, leads to the broadening of the spectral dependence of the kinetic coefficients. The temperature increase eventually causes the weakest molecular bonds to be broken; in our case, these are the coordinational ones.

Note, that in the above consideration of the excitation dynamics of big polyatomic molecules one can see an analogy with the dynamics of the simple-oscillator excitation in the presence of friction.

A qualitatively different (nonthermal) situation, irreducible to the case of the oscillator coupled with a thermostat, arises when considering comparatively small polyatomics with $s < 30$, for which the existence of the Fermi resonances is an exception rather than a rule. The model of a thermostat for molecules with such a low number of atoms ($N < 10$–12) is inapplicable, not only because the coupling between the excited mode and such a thermostat is weak, irregular or even absent at all (condition (4.36) or (4.37) is not satisfied). The point is that if the number of atoms in a molecule is small, the thermostat has a small heat capacity and the energy transmitted to it may be returned back to the mode being excited in a short time. This occurs due to the comparatively more important role of the fluctuations in the system with a small number of degrees of freedom. So the molecular modes distinct from the one being excited cannot be treated as a thermostat since the backward flows of energy are essential. In the framework of classical dynamics the process of the excitation of small polyatomics may be treated as follows. The resonant radiation interacts with one of the molecular degrees of freedom (modes) and swings it. This degree of freedom is coupled via a nonlinear parametric interaction with other modes, although not with all of them, and transfers energy to them in the course of the exchange process. Due to other modes the molecular motion ceases to be one-dimensional and as a result of nonlinearity becomes stochastic. This also makes the oscillation phase undergo rapid random fluctuations. Thus, the motion of the mode being excited is no longer periodic. The frequency spectrum becomes broad and continuous and the excitation may be performed by radiation at different frequencies. The irregularity of the phase beats makes the excitation process lose its coherence. It becomes analogous to the random walk, when during some time intervals the molecular energy rises while during another, it falls. To determine the spectral dependence of the excitation efficiency we should know the frequency spectrum of motion of the vibrational mode interacting with radiation. Its determination is a complicated procedure, even in the realm of classical dynamics.For this purpose it is necessary to determine the position of the domain in the phase space of molecular vibrational motion that becomes involved in the laser excitation process.

However, as molecules are essentially quantum objects, they should be treated quantum mechanically. But the spectral complexity of these molecules does not admit a description of their excitation dynamics on the basis of the Schrödinger equation.

The point is as follows. Considering the excitation of the molecule possessing no Fermi resonances, one should bear in mind that each molecular state corresponds to a definite number of quanta in every molecular mode (Fig. 16.1). The upward transition starting from any state is associated with an increase of the number of quanta in some mode or with simultaneous increments in several modes, if we are dealing with a hybrid vibration. In the simplest case the number of quanta increases only in the excited mode and each transition has a strictly determined frequency. In a more general case the number of frequencies at which an upward transition may take place is of course greater, being, however, still limited. This number coincides with the number of dipole-active normal and hybrid vibrations. Thus, the density of the states attainable via dipole transition is rather small. It is significantly less than the total density of states at a given excitation energy. The excitation of such a molecule is an example of the relay-type occupation of a system of isolated, almost equidistant, levels.

Considering the excitation of a molecule with modes subjected to Fermi resonances one should keep in mind that the states corresponding to definite numbers of quanta in the modes are no longer energy eigenstates. The energy eigenstates in this situation turn out to be the complex combinations of the number states of oscillators, $|k\rangle = \sum c_k^n |n, 1 - n\rangle$, and the selection rules over the dipole moment cease to forbid transitions among the majority of the energy levels present. This results in a great number of levels in the vicinity of the resonant energy $n\hbar\omega$ attainable via a dipole transition. The frequency spectrum of the transition probability becomes rather wide and contains many lines. In the classical limit it tends towards a continuous spectrum of vibrational frequencies. It is evident that such a spectrally involved system may be excited much more easily than a system of isolated levels. However, it hardly admits a description based on the Schrödinger equation.

In order to describe the excitation dynamics of small polyatomics the following approach may be used. First we solve the model problems concerned with the excitation of complicated multilevel systems that represent the typical complicated structure of the molecular spectrum and the operator of the interaction with radiation. This structure reveals the stochasticity of the molecular motion when the problem is treated quantum-mechanically. The solution of such model problems aims at finding the molecular characteristics responsible for the dynamics of excitation by the field. The next stage is to calculate these

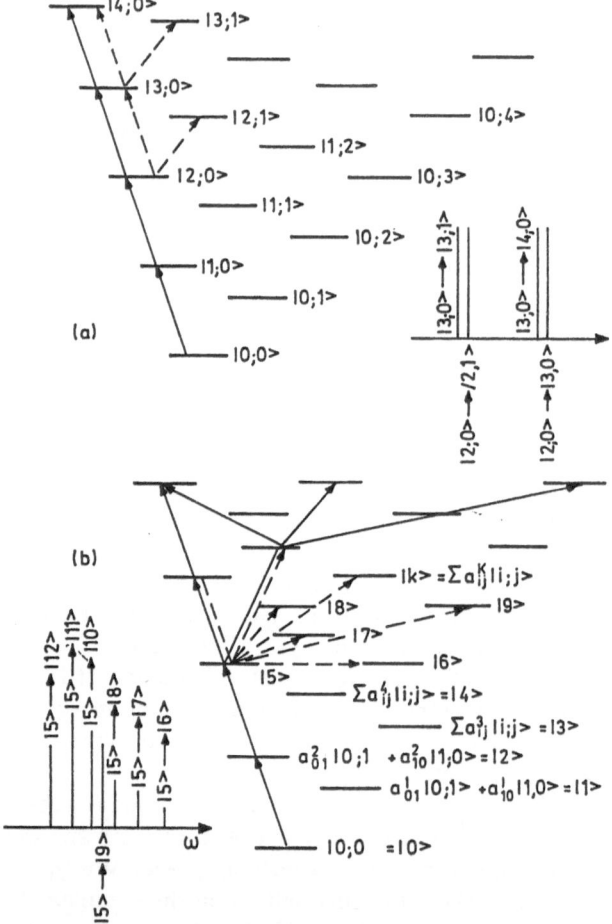

Fig. 16.1. The schemes of the molecular levels and the transitions between them. Continuous arrows represent the transitions in the course of the excitation process; the dotted ones, other transitions forming the spectral dependence of the absorption line. (a) The absence of the Fermi resonances. Only the states of a single molecule become populated. The frequency characteristic of the absorption consists of two lines corresponding to two types of vibrations (for states $|2,0\rangle$ and $|3,0\rangle$ on the right). (b) The Fermi resonance of two modes. Energy eigenstates denoted by their ordinal numbers $|k\rangle$ no longer correspond to fixed numbers of quanta in the modes. There are more excitation channels, while the spectral characteristic has many lines. Their number is significantly greater than the number of modes (on the left)

characteristics using the spectroscopic information on the molecule. This approach is helpful if the spectrum of the molecular states is rather dense and the internal molecular interactions forming it exceed the interaction with the radiation field. In fact, we already know from Lectures 10 and 11 that there are two such basic characteristics: the kinetic coefficient of transition D, and the density of states g. If the number of levels falling into the resonance region is large, $\hbar Dg \gg 1$, the process develops effectively; otherwise the system is hardly excitable.

Developing this approach we should keep in mind that the density of the states of the polyatomic molecule considered increases with the extent of vibration excitation at the rate of E^{s-1}, where E is the excitation energy (Fig. 15.10). So, at the lower and upper levels the character of the excitation process is different. At the lower levels, where the conditions of resonance of the successive transitions are not provided, the excitation occurs in accord with the multiphoton mechanism (Lecture 9). At the upper levels, where the density of the states is large the requirement of quasicontinuum existence (11.3) for the Stark broadening of the levels to overlap the typical interlevel distance is valid already at field intensities of 10^5–10^9 W/cm^2. In this region the excitation dynamics may be described by the modes of a multiband structure (see (11.19)). In addition, in case the energy of the molecule exceeds the dissociation threshold, (11.19) should be supplemented by the term (11.46) describing the molecule decay. All this leads to eqn (11.47).

To avoid a misunderstanding, we ought to emphasize the difference between Eqs. (16.11) and (11.19). In the former case, index n corresponds to the definite energy content in the mode being directly excited. In the latter case, index n describes the state in which the whole molecule, not a single mode, has the excitation energy of n quanta of the laser field.

Therefore, one can specify three stages of the excitation of a small polyatomic in an intense laser field. First, the lower levels get excited by means of the multiphoton mechanism. This stage is followed by the excitation of the molecule in the quasi-continuum region described by the balance equations. At the final stage, the molecule, overexcited above the dissociation threshold, decays.

Let us proceed with the stage of the quasi-continuum excitation. Its mechanism has many features in common with those previously described in the present lecture. However, before we go on calculating the kinetic coefficients, note that, as we have already mentioned, we are dealing with comparatively small polyatomic molecules which bind no more that 10–20 atoms. Nevertheless, they should still be essentially polyatomic, with the number of atoms exceeding three. The point is that the density of levels in the energy space is nonuniform. As we already know (Lecture 15), there are bands in the vibrational spectrum, where the levels are spaced closer than on the average; there, the density of levels is greater. The distinction is about equal to the ratio of the quantum energy to the energetic width of the band. Hence, the density of levels in the band is to be estimated as $g \approx (E/\omega)^{s-1}/\Delta_B(E)$, where Δ_B is the band width. In accordance with the summation rule, $\sum_i \mu_i^2 = \mu^2 N \propto E$, where N is the number of transitions in the spectral width of the resonance Δ_r, which equals the product of the level density by this width, $N = g\Delta_r(E)$. Then, $\mu^2 g \propto E/\Delta_r(E)$ and, hence, $\mu^2 g^2 \propto E^s/\Delta_B(E)\Delta_r(E)$. Of all the nonlinear interactions, the cubic ones grow most slowly with the excitation energy. In case the spectrum of states and the transition lines are formed by these cubic interactions, $\Delta_B(E) \propto E^{3/2}$ and $\Delta_r(E) \propto E^{3/2}$. Then $\mu^2 g^2 \propto E^{s-3}$.

If the criterion of the quasicontinuum existence (11.3) is satisfied for some
level of the molecular excitation, it will automatically hold for all the levels
lying higher if only $s > 3$. If the lowest anharmonic interaction forming the
spectra and the lines is the anharmonicity of fourth order, the requirement
$s > 4$ should be fulfilled. In both cases the molecule should contain more than
three atoms. This is why we call the molecules containing 4–12 atoms small
polyatomics.

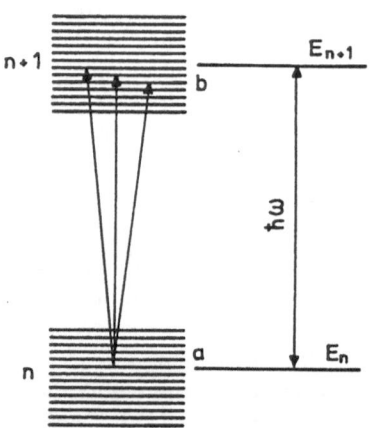

Fig. 16.2. Transition between bands n and $n + 1$

Let us now turn to calculating the kinetic coefficients. Consider the transition from band n to band $n + 1$ (Fig. 16.2). In order to obtain the probability
of such a transition in accord with (11.14a) we should select some level a
in the lower band lying in the vicinity of the resonance and determine the
mean-square matrix element corresponding to the transition from this level to
other levels b in the resonance vicinity in the upper band, $\langle |\mathcal{E}\mu_a^b|^2 \rangle_b$. Then this
mean-square matrix element should be multiplied by the density of levels b
in the resonance vicinity $g(E_{n+1})$, this operation being followed by averaging
over all the starting levels a of the lower band which are occupied in the course
of the excitation process. As a result, we obtain

$$D_{n,n+1} = \pi \left\langle \left\langle \left(\frac{\mathcal{E}\mu_a^b}{\hbar} \right)^2 \right\rangle_b g(E_{n+1}) \right\rangle_a. \tag{16.21}$$

Expression (16.21) assumes the positions of all the molecular levels in the
bands as well as all the matrix elements of the dipole moment operator to
be known, i.e., we should possess information on all the vibrational-rotational
wavefunctions of the molecule. Evidently, this information is not available.
So, we have to express the kinetic coefficients $D_{n,n+1}$ in terms of other more
convenient variables, say, in terms of the mode ones. Most simply this may
be accomplished in case the semiclassical approach applies and the average
over the levels may be replaced by the average over the phase volume of the

molecule. The latter is valid if the vibrational motion of the molecule is rather intricate and covers appreciable domains of the phase space containing many levels. Since a single quantum state occupies, on the average, the phase volume of $(2\pi\hbar)^s$, the volume covered by the classical (semiclassical) trajectory of vibrational motion in the phase space should essentially exceed this quantity, i.e. $V \gg (2\pi\hbar)^s$. The volume covered by the phase trajectory is the space inside the tube with the diameter of the order of the zero-point vibration amplitude enclosing the trajectory.

In calculating we shall utilize the fact that the probability of the upward transition from level a of the lower band W_a^{n+1} is invariant by virtue of the optical theorem, i.e., it is independent of details of the upper-band-spectrum structure and may be represented as the integral over the phase volume $d\Gamma$:

$$W_a^{n+1} = \hbar \int \left(\frac{\mathcal{E}\mu_a^b}{\hbar}\right)^2 \delta(E_b - E_{n+1})d\Gamma_b, \qquad (16.22)$$

where we have taken into account that $\hbar d\Gamma_b = g(E_b)dE_b$. This expression needs further averaging over those starting levels a of the lower band that become populated in the excitation process.

Let us introduce the distribution function f of the molecule in the phase space volume which determines the part of the phase space involved in the vibrational motion.

Then the required average over the lower-band levels lying in the vicinity of energy E_n is equivalent to the average over the phase space with the distribution function f, and we obtain

$$D_{n,n+1} = \pi \int f(\Gamma_a)\delta(E_a - E_n) \int \mathcal{E}^2(\mu_a^b)^2\delta(E_b - E_{n+1})d\Gamma_a d\Gamma_b \qquad (16.23)$$

under the normalization condition for the function f,

$$\hbar \int f(\Gamma_a)\delta(E_a - E_n)d\Gamma_a = 1. \qquad (16.24)$$

The physical meaning of these expressions is as follows. In the course of the excitation process the states with energy $E_a = E_n = n\hbar\omega + E_0$ turn out to be populated after the absorption of n quanta. However, these states differ in the probability of occupation, the latter being taken into account by the distribution function f. From each state a it is possible to reach the states b coupled with a in a dipole way, $\mu_a^b \neq 0$. However, we are not interested in all of them, but just in those states resonant with respect to the radiation field, $E_a = E_b - \hbar\omega$ or $E_b = E_{n+1} = (n+1)\hbar\omega + E_0$. The probability of the transition to these states is determined by their density and the mean-square matrix element of the transition $\mathcal{E}^2(\mu_a^b)^2 g \simeq \int (\mathcal{E}\mu_a^b)^2\delta(E_b - E_{n+1})d\Gamma_b$.

Equation (16.23) was derived considering the transitions between the energy levels, i.e., in energy-eigenstate representation. However, the integrals entering (16.23) have the remarkable, though evident, property of being independent of the variables they are taken over. Hence, they are invariant and do not depend on the particular representation (basis) being used. Therefore, we may take advantage of the oscillator-wavefunction basis instead of that of the energy eigenstates. In this case the computation of the kinetic coefficients is considerably simplified and may be performed as follows.

Expression (16.23) is an analogue of formula (15.26) from the previous lecture. Indeed, we use the following identity: $\delta(x)\delta(y) = \delta(x)\delta(y-x)$, subtract the argument of one δ-function from the argument of the other and bear in mind that $E_{n+1} - E_n = \hbar\omega_l$ and $E_b - E_a = \hbar\omega(\Gamma_a, \Gamma_b)$, where ω_l is the laser field frequency and ω, that of transition[*]. In the expression thus obtained,

$$D_{n,n+1} = \pi \int f(\Gamma_a)\delta(E_a - E_n)\mathcal{E}^2(\mu_a^b)^2\delta(\hbar\omega(\Gamma_a, \Gamma_b) - \hbar\omega_l)d\Gamma_a d\Gamma_b \quad (16.25)$$

we pass on to the "energy-phase" ("action-angle") oscillator variables. We obtain

$$D_{n,n+1} = (\pi/\hbar) \int f(\{I_i\}, \{\theta_i\})\mathcal{E}^2\mu_{01}^2 I_1\delta_E\delta_\omega \prod_{i=1}^{s}(d\theta_i dI_i),$$

$$\delta_E \equiv \delta(E(\{I_i\}, \{\theta_i\}) - E_n), \quad \delta_\omega \equiv \delta(\omega(\{I_i\}, \{\theta_i\}) - \omega_l). \quad (16.26)$$

Here θ_i is the phase of the oscillations in mode i; I_i, the number of quanta in mode i (in the classical limit – the continuous variable of action). The mode being excited is that with $i = 1$, variable I_1 corresponding to it. The dipole moment of the transition $\mu_{01}\sqrt{I_1}$, where μ_{01} is the dipole moment of the general transition of this mode. In this case the normalization condition takes on the form

$$\int f(\{I_i\}, \{\theta_i\})\delta(E(\{I_i\}, \{\theta_i\}) - E_n) \prod_{i=1}^{s}(d\theta_i dI_i) = 1. \quad (16.27)$$

It should be noted that in (16.26) in contrast to (16.25) the integration is performed only over the phase space of the initial states. The integration over the final states is not carried out. It is the choice of oscillator variables rather than eigenenergy variables as the basis the integration is performed over that essentially facilitates the calculations (especially in the multidimensional case). The well-known selection rules for dipole transitions which tightly link the final and the initial states are valid in this basis. When the anharmonic frequency shifts of the mode being excited are independent of the oscillation phases and the consideration of the phase space may, therefore, be reduced to considering its subspace, the occupation number space $I_1, I_2, I_3, \ldots$, the derivation performed may be illustrated as follows.

Let many different states, mostly hybrid ones (Fig. 16.3), lie on the molecular excitation energy level $E_a = $ const. Let us emphasize that the states with fixed numbers of quanta in the mode are energy eigenstates b virtue of the assumption concerning the vibrational energy being independent of the phases.

---

[*] This may be done, given that operator $\hat{\mu}$ and the Hamiltonian do not commute, which is unessential.

In fact, this implies the absence of parametric pumping of energy from one mode into another. The selection rules are strictly observed in this situation. At a frequency ω_l close to that of mode 1 the only possible upward transition from the initial state $|I_1, I_2, I_3, \ldots\rangle$ is the one to the state $|I_1+1, I_2, I_3, \ldots\rangle$. At a fixed frequency ω_l, the transition is only possible for states $|I_1, I_2, I_3, \ldots\rangle$ for which the transition with an increase in the number of quanta in mode I_1 is in strict resonance with radiation, but not for all the states. The regions containing the initial and final states of such transitions are denoted in Fig. 16.2 by the symbols B and B', respectively. The region of states $|I_1, I_2, I_3, \ldots\ldots\rangle$ that become populated in the process of the laser excitation of the molecule is denoted in Fig. 16.2 by letter A. Were the selection rules observed absolutely strictly (which implies the absence of any intermode parametric coupling), this region would consist of a single level unambiguously determined by the initial molecular state prior to the excitation. However, even very weak parametric coupling between the modes, which leaves the resonant conditions for the levels in regions B and B' practically unchanged, is sufficient for region A to embrace a large number of levels.

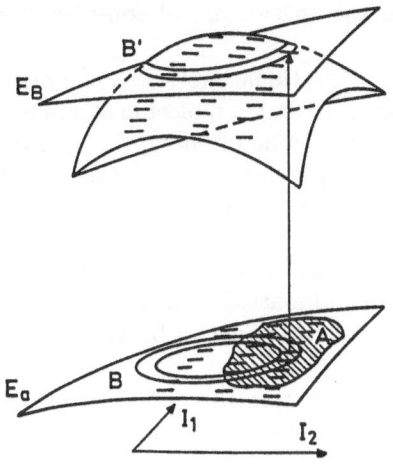

Fig. 16.3. Disposition of resonant levels in the space of occupation numbers. Vertical arrow shows transition $I_1, I_2, I_3, \ldots\rangle \to |I_1 + 1, I_2, I_3, \ldots\rangle$, $\hbar\omega = \hbar\omega_l$

In fact, region A is the region of stochasticity of vibrational motion which has been discussed at the beginning of this lecture and at the end of the previous one. One should remember that along with intrinsic molecular parametric interactions (which are presumably weak), the interactions caused by the presence of an external field may be significant among the factors affecting the formation of this region. There are two such interactions. The first is the multiphoton interaction between the levels of band n on account of the transitions via the intermediate levels lying in bands $n - 1$ and $n + 1$. The second is the parametric interaction between the modes arising from the nonlinear dependence of the molecular dipole moment operator on the normal coordinates (see (15.17)).

It seems evident that the distribution function f introduced above significantly differs from zero only in region A. In fact, region A itself is determined by the requirement that the distribution function f should be nonzero.

It is natural that region B changes with frequency ω_l. This change entails a change in the intersection of manifolds A and B and, hence, the change in the kinetic coefficient $D_{n,n+1}$. In case manifolds A and B do not intersect, $D_{n,n+1}$ turns to zero. This is how the spectral dependence of the kinetic coefficients is formed. Let us consider several examples. First we consider the three-fold degenerate mode and determine this dependence for the case when only one degree of freedom is excited. In this case the anharmonic interaction has the form of (15.11), and the anharmonic frequency shift entering (16.26) is given by the relation

$$\omega(I_i, \theta_i) - \omega_{\text{harm}} = 2\alpha I_1 + \beta(I_2 + I_3) \,. \tag{16.28}$$

Note that notations I_i and v in this and the previous lectures are identical. (Symbol I is preferable in the quasi-classical consideration and symbol v, in the quantum one.)

Let us assume that the vibrational motion is completely stochastic. This implies that the probability of finding a molecule at some point in phase space which corresponds to a certain value of energy is the same everywhere. The distribution function f entering expression (16.26) is in this case a constant determined by the normalization condition (16.27). Therefore, the average over levels a reduces to that over the microcanonic distribution. If the number of the degrees of freedom for the molecule considered is $S \gg 3$, it follows from the common statements of statistical physics, that the average over the microcanonic distribution may be replaced by the average over the canonic one.

In integral (16.26) the integration successively runs over all the variables of interest I_1, I_2, I_3 influencing the transition frequency (see(16.28)) and all the other variables corresponding to other molecular degrees of freedom. The replacement of the microcanonic distribution by the canonic one means that by integrating (16.26) we can replace the integral

$$\int f(\{I_i\}, \{\theta_i\}) \delta(E(\{I_i\}, \{\theta_i\}) - E_n) \prod_{i=4}^{s} (dI_i d\theta_i)$$

by

$$Z^{-1} \exp\left\{ -(s/E_n) \sum_{i=1}^{3} E_i \right\} \,,$$

where Z^{-1} is the normalization factor.

Then (16.26) takes on the form

$$D_{n,n+1} = \frac{\pi}{\hbar}\mathcal{E}^2\mu_{01}^2 \int I_1\delta_\omega \exp\left\{-\frac{\hbar\omega}{kT}(I_1 + i_2 + I_3)\right\}\left[\frac{\hbar\omega}{kT}\right]^3 dI_1 dI_2 dI_3,$$

$$\delta_\omega \equiv \delta(2\alpha I_1 + \beta I_2 + \beta I - 3 + \omega_{\text{harm}} - \omega_l). \tag{16.29}$$

The computation of this integral gives different results for different ratios of the anharmonicity constants α and β and the detuning of the laser frequency ω_l from the harmonic one, ω_{harm}, $\Delta \equiv \omega_l - \omega_{\text{harm}}$. Namely, we obtain at $\Delta/2\alpha < 0$, $\beta/2\alpha > 0$,

$$D_{n,n+1} = 0; \tag{16.30}$$

at $\Delta/2\alpha > 0$, $\beta/2\alpha > 0$,

$$D_{n,n+1} = \frac{\pi}{\hbar}(\mathcal{E}\mu_{01})^2 \frac{4\alpha\beta - \frac{\hbar\omega}{kT}\Delta(2\alpha - \beta)}{(2\alpha - \beta)^3}\left\{\exp\left[-\frac{\Delta}{\beta}\frac{\hbar\omega}{kT}\right] - \exp\left[-\frac{\Delta}{2\alpha}\frac{\hbar\omega}{kT}\right]\right\}; \tag{16.31}$$

at $\Delta/2\alpha < 0$, $\beta/2\alpha < 0$,

$$D_{n,n+1} = \frac{\pi}{\hbar}(\mathcal{E}\mu_{01})^2 \frac{4\alpha\beta - \frac{\hbar\omega}{kT}\Delta(2\alpha - \beta)}{(2\alpha - \beta)^3}\exp\left[-\frac{\Delta}{\beta}\frac{\hbar\omega}{kT}\right]; \tag{16.32}$$

at $\Delta/2\alpha > 0$, $\beta/2\alpha < 0$,

$$D_{n,n+1} = \frac{\pi}{\hbar}(\mathcal{E}\mu_{01})^2 \frac{4\alpha\beta - \frac{\hbar\omega}{kT}\Delta(2\alpha - \beta)}{(2\alpha - \beta)^3}\exp\left[-\frac{\Delta}{2\alpha}\frac{\hbar\omega}{kT}\right] \tag{16.33}$$

In these expressions we use the notation

$$kT = E_n/s, \tag{16.34}$$

which naturally arises at the transition from the microcanonic distribution to the canonic one.

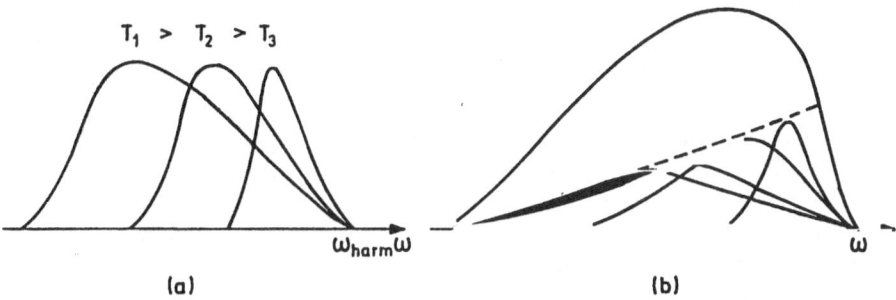

Fig. 16.4. (a) Spectral dependence (16.32) at $\alpha = 0$ for three temperatures $T_1 > T_2 > T_3$. (b) Superposition of the Lorentzian envelopes corresponding to transitions $0 \to 1$, $1 \to 2$, $2 \to 3$, etc.

The formulae obtained show that the spectral dependence of the upward transition probabilities in contrast to (16.10) exhibits essentially non-Lorentzian shape. As an example, Fig. 16.4 a illustrates the spectral dependence (16.32) for $\alpha = 0$. It is useful to explain the reason for such a distinction. The point is that indices n in (16.10) and (16.30)–(16.33) have different meanings. For large polyatomic molecules (i.e., in the case of (16.10)), it is assigned to a certain energy level of the molecular degree of freedom being excited, while for small polyatomics ((16.30)–(16.33)), index n corresponds to the number of the absorbed quanta, i.e., to the excitation energy of the whole molecule. If a large polyatomic was able to accumulate, without dissociation, the energy sufficient to provide each of its degrees of freedom with an average energy considerably exceeding the excited mode quantum ($kT \gg \hbar\omega_l$), the spectral dependence of the probability of the upward transition for such a molecule at a fixed total energy would be the superposition of the Lorentzian contours taken with statistical weights of the initial states 0, 1, 2, 3, ... corresponding to transitions $0 \to 1$, $1 \to 2$, $2 \to 3$, etc. of the mode under consideration (Fig. 16.4 b). The envelope of such contours would not, of course, be the Lorentzian one, but would have a form determined by the distribution of the populations among the initial starting states of hot bands somewhat resembling contours (16.31)–(16.33). But for large polyatomics (because of dissociation) only the lines of one, two or, at most, three transitions are significant. So, the hypothetic contour of the line (Fig. 16.4 b) is not formed. In the case of small polyatomics the envelope of a large number of line contours has the possibility of being formed. However, the contours constituting it have, in general, non-Lorentzian shapes formed as a result of the process for which the returns of the energy into the mode being excited are important. The latter processes are more complex than the exponential decay into the thermostat which is responsible for the Lorentzian line.

Let us note yet another circumstance. In (16.31) there is a singularity at $2\alpha = \beta$. It is connected with the fact that at $2\alpha = \beta$, as it follows from (15.11), the anharmonic additional term to the energy degenerates into

$$E_{\text{anh}} = \alpha(I_1 + I_2 + I_3)^2 \,. \tag{16.35}$$

Then in this order of anharmonicity there is no splitting of the three-fold degenerate mode producing the singularity in the density of states that may be removed just by taking into account the anharmonicity of a higher order.

Let us emphasize again the general rule one should bear in mind when calculating the kinetic coefficients. Their spectral dependence is determined by the anharmonic shift of frequencies and the distribution of the molecule over the phase space. Here one should remember that though it is sufficient to consider only the strongest anharmonic interactions in calculating the anharmonic shifts, all the nonlinear interactions present in the molecule which may become essential in the time-span of the laser radiation pulse are to be taken into account when determining the distribution function f. This function f may principally differ from the the microcanonic function at thermal equilibrium.

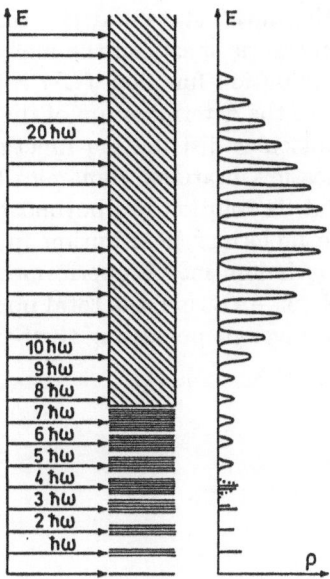

Fig. 16.5. Distribution of populations over the molecular vibrational spectrum formed in the laser field

In conclusion, let us turn to the excitation of high vibrational states in small polyatomics on the whole. Laser radiation makes the states belonging to the narrow vicinity of resonant energy values populated. These states, as a rule, are stochastized, i.e., they cannot be assigned any definite numbers of quanta in the modes of the molecule. One should keep in mind that the stochastization process itself develops rapidly. Its duration is much less than the time of the radiative transition. The involved internal vibrational dynamics determines the detailed form of the eigenenergy functions. However, some common "coarse-grain" characteristics correspond to these functions, such as, for example, the mean numbers of quanta in the modes. In other words, all of them have approximately the same quasi-classical probability of being found in different parts of the phase space $f(\Gamma)$. The probability of the radiation-induced transition from the vicinity of resonance into the vicinity of the succeeding resonances is determined by the local density of the levels attainable by a dipole transition in the vicinity of those resonances, and the total cross section of the transition to them. This quantity no longer depends on the vibrational dynamics details but is determined just by the "coarse-grain" characteristics of the molecule – function $f(\Gamma)$ and the most intense anharmonic interaction. In order to calculate this quantity one may use the quasi-classical approach. Indeed, the knowledge of the distribution function $f(\Gamma)$ allows the determination of the energy distribution among the oscillator states of the IR-active modes and the broadening of the transition lines that arises from their anharmonic coupling with other modes via the hot-band mechanism.

Thus, the choice of the initial state and the radiation frequency determines the regions of the energetic spectrum the molecules are promoted into. The knowledge of the distribution function $f(\Gamma)$ corresponding to these regions of the spectrum allows the determination of the transition probabilities. The calculation of the molecular distribution function over the phase space which is formed by stochastic vibrational dynamics belongs to the problems of stochasticity theory and its solution lies beyond the scope of this course. The determination of the molecular distribution function over the energies in the vibrational quasicontinuum may be performed via the solution of the population balance equations. As a rule, the total population of the groups of levels has a rather narrow envelope possessing the bell-shaped Gaussian form (Fig. 16.5).

Lecture Seventeen
Excitation of Molecules (continued)

Lower levels. Multiphoton resonances. Fraction of excited molecules. Two-photon Q-branch of a spherical top. Saturation. Large molecules. RRKM model. Small polyatomic molecules, multidimensional motion in a potential well, stochasticity, probability of dissociation. Experimental studies. Examples of CF_3I- and SF_6-molecules. Fermi resonances. Spectral dependencies of kinetic coefficients. Red shift. Sharp resonant structure of the spectrum of lower levels.

Now let us turn to the processes taking place on the lower levels. As a rule, they are the multiphoton processes. In the simplest case of a one-dimensional anharmonic oscillator the situation is evident. If the radiation is in resonance with the main transition $0 \to 1$, transition $1 \to 2$ is out of resonance and the anharmonic detuning increases with the number of the excited level. On the lower levels of polyatomic molecules at a moderate extent of excitation there are many states in the vicinity of the resonant-radiation-quantum energy and their density is high. However, these states are grouped in such a way that they cover only a small part of the energy space (Lecture 15). Therefore, the successive transitions in the system of lower levels are not automatically resonant in the case of polyatomic molecules, too. This may be exemplified by the spectrum of a triply degenerate mode of a symmetric molecule (Fig. 15.5) given in Fig. 17.1, taking into account the vibrational structure and the fine splitting. If transition $0 \to 1$ in the P-branch is resonant with respect to radiation, transition $1 \to 2$ is out of resonance. If we adjust the radiation frequency to satisfy the condition of the two-photon resonance for transition $0 \to 2$, then transition $0 \to 1$ falls out of resonance and we obtain the scheme of a two-photon resonance with a detuned intermediate level (Lecture 9). The probability of transition $0 \to 2$ is described by the composite matrix element. The field broadening corresponding to this matrix element of a two-photon transition is less than the spacing between the levels of the vibrational multiplet, but it usually exceeds the distance between the fine-structure levels (a quasicontinuum exists). Then transition $0 \to 2$ is excited while transitions $2 \to 3$ and $2 \to 4$ fall, as a rule, out of resonance and, generally speaking, are not excited.

It is possible to adjust the radiation frequency to satisfy the resonant conditions for the three-photon resonance $0 \to 3$ (two intermediate detunings). The composite matrix element of a three-photon transition at the same intensity is noticeably less than that of a two-photon one. The condition for the quasicontinuum to exist in the fine structure of spectrum of the state $v = 3$ requires special testing, even though at high intensity it may, in principle, be satisfied.

Therefore, the situation in which the resonant conditions for successive transitions $0 \rightarrow 1 \rightarrow 2 \rightarrow 3 \rightarrow 4$, etc. are not satisfied on the lower levels of small polyatomics is the typical one. Transitions $0 \rightarrow 1$, $0 \rightarrow 2$, $0 \rightarrow 3$, etc. have different resonant frequencies. The values of these frequencies are determined by the initial quantum state of the molecule. In the given example the values of the transition frequencies depend on the vibrational quantum number J of the molecular ground state. However, one should also allow for the dependence on the extent of excitation of other molecular modes.

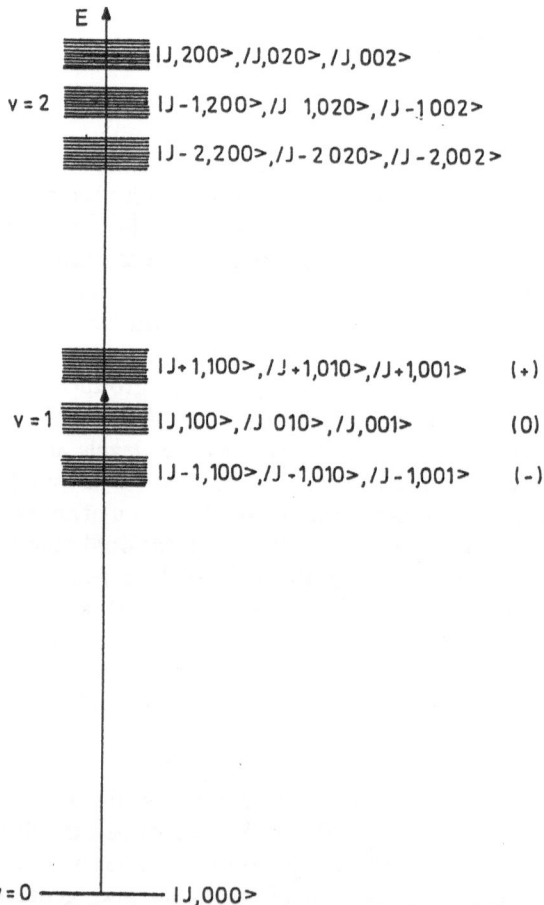

Fig. 17.1. Fine splitting and rotational structure at lower energy levels of triply degenerate vibrational mode of XY_6-molecule. Figure shows only those components of the rotational and Coriolis spectra to which transitions from the starting state are allowed by the selection rules. Symbols (+), (0), (−) denote the upper, middle and lower components of the Coriolis splitting

Quite different starting states are present in the Boltzmann ensemble of molecules. Therefore, for any value of the monochromatic radiation frequency corresponding to the absorption band of the main vibrational transition we can always find the molecules for which the resonant conditions for one or another successive transition are fulfilled ($0 \rightarrow 1$, $1 \rightarrow 2$, $2 \rightarrow 3$, ..., $0 \rightarrow 2$, $2 \rightarrow 4$, ...). However, the relative part of these molecules (fraction) is small. In other words, the majority of molecules remain unexcited, although there always are molecules which absorb a certain number of radiation quanta. This phenomenon is referred to as the bottleneck effect.

It is natural to expect the fraction of excited molecules q to vary with the radiation frequency. At high temperatures there are a great number of starting states. They appear due to the Boltzmann distribution of molecules among the vibrational states and the thermal population of the lower levels of soft modes. In this case, the smooth continuous dependence $q(\omega)$ is observed. At low temperatures there is a small number of initial states and a rather sharp structure of single resonances may be detected in the $q(\omega)$ dependence.

In Lecture 15 we introduced the concepts of P-, Q- and R-branches of the vibrational transition. We also mentioned that the Q-branch is infinitely narrow provided the vibrational-rotational interaction is neglected. Hence, a large fraction of molecules gets excited at the Q-branch frequency of transition $0 \rightarrow 1$. The two-photon transition $0 \rightarrow 2$ possesses five analogous branches. They correspond to the selection rules $\Delta J = \pm 2, \pm 1, 0$ and are called O-, P-, Q-, R- and S-branches. The Q-branch of the two-photon transition is also rather narrow. Hence, a large fraction of molecules may be excited when the radiation frequency is tuned at this Q-branch. Evidently, the Q-branches exist for all the transitions of higher orders as well. It is the Q-branches of the vibrational transitions of different orders which comprise the nonuniform resonant structure of the excitation spectrum of the lower levels of small polyatomics.

The excitation of multiphoton transitions in molecules differs from that in atoms. In the case of molecules the number of levels falling into the multiphoton resonance region increases with the radiation intensity. Indeed, the Stark width of resonance for a band system equals $\pi V^2 g$, where g is the energetic density of states, while V^2 is the square of the matrix element of the interaction operator ($E\mu$, $(E\mu)^2/\Delta$, etc.). This width increases with the radiation intensity and cause the retraction of more and more levels. So, the power dependencies typical of atoms are not obligatory for molecules in which the probability of a multiphoton transition may depend on the radiation intensity in a different way (not as the intensity raised to the power of the transition order). The powers are not necessarily integers and the dependencies themselves may not even be power dependencies.

Let us consider an example of the excitation of the two-photon Q-branch corresponding to the molecule of the spherical-top type which has fine (octahedric) splitting. The excitation is the superposition of the molecular transitions from the ground vibrational state with different rotational quantum numbers J to the second excited vibrational state with the same J. Let this vibrational state be of symmetry F. Let us fix the quantum number J. Then,

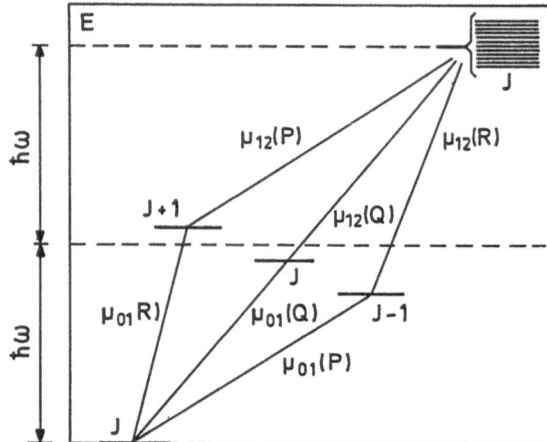

Fig. 17.2. R–P-, Q–Q- and P–R-channels of Q-branch of two-photon resonant transition

the two-photon transition is of the "degenerate level–band of levels" type. Indeed, the lower level $|J, 000\rangle$ is $2J+1$-fold degenerate with respect to the quantum number K (Lecture 15). In the excited state which is triply degenerate $|J, 110\rangle, |J, 011\rangle, |J, 101\rangle$, the K-degeneracy is removed due to the octahedric interaction. When such a system is being excited, saturation occurs when the number of the upper-band levels withdrawn by the Stark width resulting from the composite matrix element of the two-photon transition becomes about equal to the number of components in the $2J+1$-times degenerate manifold of the lower level. In fact, this statement follows from formula (11.17), with the only distinction that the interaction matrix element should be substituted by the composite matrix element of the two-photon transition. Its value varies with J. Indeed, this matrix element is the superposition of the matrix elements of the transitions through three different channels: R–Q, Q–Q, P–R (Fig. 17.2). Then it may be written as

$$V_{02} = \mathcal{E}^2 \left[\frac{\mu_{01}(Q)\mu_{21}(Q)}{\Delta_0} + \frac{\mu_{01}(P)\mu_{12}(R)}{\Delta_0 - 2BJ} + \frac{\mu_{01}(R)\mu_{12}(P)}{\Delta_0 + 2BJ} \right], \qquad (17.1)$$

where Δ_0 is the detuning of the intermediate level $|J, 100\rangle$, $|J, 010\rangle$, $|J, 001\rangle$ from resonance. This expression shows that the number of the upper-band levels of density g in the Stark withdrawal width equal to gV_{02} depends upon J in a non-monotonous way. This dependence, taking its maximum value at $J = \Delta_0/2B$, which corresponds to the rotational compensation of the intermediate detuning, is given in Fig. 17.3 for two values of the vibration intensity. The straight line in this figure shows the degeneracy of the ground state $2J + 1$. In the regions where $gV_{02}(J) > 2J + 1$ the transitions are saturated. The boundaries of these regions correspond to the intersection points of the straight line $2J + 1$ with the gV_{02} curve and are denoted in Fig. 17.3 by the symbols A, A', B and B'. The number of molecules for which transition $0 \to 2$ turns out to be saturated is determined by the rotational Boltzmann distribution function. At the maximum of this distribution the value of J exceeds that corresponding to the rotational compensation of the intermediate detuning, i.e., if

$$(kT/B)^{1/2} \gg \Delta_0/2B, \tag{17.2}$$

the number of molecules for which the two-photon transition is saturated grows up with intensity I as I^3.

Indeed, at $J \gg \Delta_0/2B$ only one term in (17.1) is essential, in particular, that corresponding to the Q–Q transition. Then it follows from the condition on the boundary of the saturated transition region $gV_{02}(J_B) \approx 2J_B$ $(J_B \gg 1)$ that $J_B = \mu_{01}(Q)\mu_{12}(Q)\mathcal{E}^2/2\Delta_0 g \propto I$. Then the total number of saturated transitions is

$$N = \int_{J_A}^{J_B} \rho B(J)dJ \approx \int_0^{J_B} \rho_B(J)dJ \propto J_B^3 \propto I^3 \quad \text{at} \quad J_B < (kT/B)^{1/2}.$$
$$\tag{17.3}$$

Here $\rho_B(J) = Z^{-1}(2J+1)^2 \exp[-B(J+1)/kT]$ is the Boltzmann function of the rotational distribution. It is natural that in case the right margin of the saturation region extends beyond the maximum of the Boltzmann distribution, i.e., at $J_B > (kT/B)^{1/2}$, the transition is completely saturated.

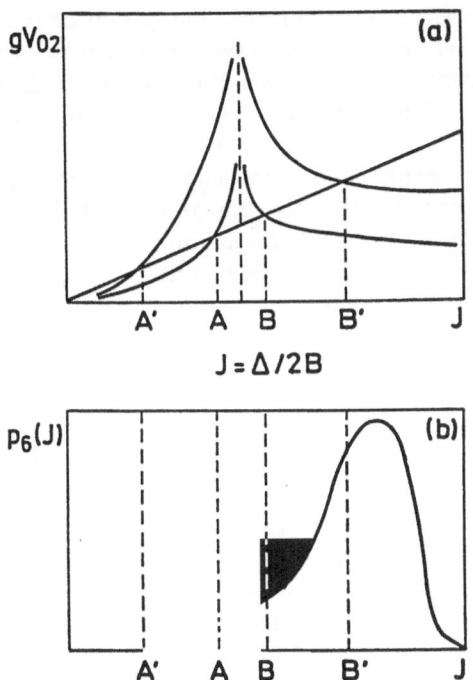

Fig. 17.3. (a) The number of levels falling into two-photon resonance as the function of rotational quantum number. (b) The distribution of molecules over J at fixed temperature

The example considered illustrates the above statement that in the system possessing a complex spectrum the dependence observed of the absorbed energy on the irradiation intensity may not correspond to the photonity of the transition. This fact considerably complicates the investigation of the processes taking place on the lower levels of small polyatomics when they are being excited. It also makes it difficult to predict and analyze the experimentally observed spectral dependencies of the lower level excitation at different temperatures and intensities.

It is the complexity of the processes developing at the lower levels of a molecule that distinguishes this range of the energetic spectrum. This range starts from the ground state and stretches up to a lower boundary of the vibrational quasicontinuum, where the Stark broadening becomes greater than the distance to the closest vibrational level, and the resonant conditions for successive single-quantum transitions are automatically fulfilled.

If before the laser irradiation only the low-lying vibrational states and their rotational sublevels are populated due to the thermal excitation, in the field of laser light only a fraction of molecules reaches the vibrational quasicontinuum region. An appreciable amount of them gets stuck (localized) on the lower levels due to the fact that the resonant conditions of successive single- or multiphoton transitions are not satisfied for them. The energetic distribution function of these molecules depends on the spectrum details and is sensitive to the radiation intensity and frequency. The group of molecules persisting in the lower-level system despite the action of laser light is sometimes called the cold ensemble, while the molecules which have reached the vibrational quasicontinuum are said to comprise the hot ensemble, their fraction being called the q-factor.

Everything said above is valid if collisions are of no importance for sure. The presence of collisions significantly changes the distribution function of the cold-ensemble molecules. If during the action of laser light these molecules undergo collisions and transit into other states, the process in which they absorb radiation quanta may recommence. This, in turn, may considerably increase the q-factor. The answer to the question whether the collisions are essential for a particular molecular type depends upon two factors – the gas pressure p and the laser pulse duration τ. For the collisions to be considered unessential the product $p\tau$ should be smaller than a certain value quite definite for the particular molecular type, which is referred to as the collision typical time. One of the fastest collisional processes is the V–V-exchange. In case this particular process turns out to be most probable, the collisions may be ignored at $p\tau < 10^{-9}$ tor·s (Lecture 19).

Now let us treat the third stage of the process under consideration: excitation of small polyatomics. Let us discuss the dissociation of small polyatomics, which is more complex than that of diatomic or large polyatomic molecules. The diatomic molecule, which has reached the dissociation limit, decays during a single oscillation period. To dissociate during a single oscillation the large polyatomic molecule should be essentially overexcited above the lowest dissociation threshold, so that its temperature, i.e., the total energy divided by the number of degrees of freedom, is of the order of the dissociation energy. In a more general case the probability of the dissociation process which is, in fact, the fluctuational concentration of the molecular energy in the bond being broken, is described by the well-known Boltzmann formula:

$$\Gamma_d \approx \omega \exp(-U_t/kT), \qquad (17.4)$$

where ω is the typical frequency of vibrations; $kT = U_t/s$; and U_t is the total molecule energy.

This approach cannot be applied for a small polyatomic molecule, $3 < s < 12$. The point is that not even (17.4) gives the noticeable value of the dissociation probability equal to we^{-s} for the molecules whose energy is close to, but less than, the dissociation energy, $U_t \leq U_d$. This circumstance, caused by the fact that the transition from the microcanonic distribution to the canonical one is not possible for small polyatomics, might be corrected by the appropriate choice of the factor multiplying the exponent in the (17.4)-type formula[*]. The point is that for small polyatomic molecules the applicability of the very microcanonic distribution is not evident *apriori*, i.e., their motions are nonergodic. The regions where the vibrational motion of these molecules is stochastic may not completely coincide with the whole surface $E(\{I_i\}, \{\theta_i\}) = U_t = \text{const}$. So, it is expedient to describe the dissociation of such molecules using other concepts not associated with the assumption of the equilibrium thermal distribution and fluctuations.

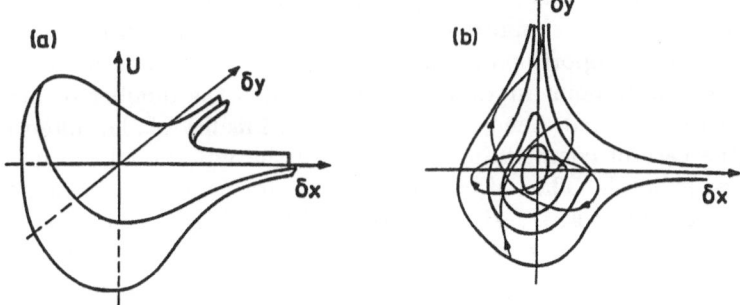

Fig. 17.4. (a) Two-dimensional potential well. (b) Top view of the representing-point trajectory getting into the dissociation valley along the *y*-axis

The physical image which is to be used, describing the dissociation of small polyatomic molecules, is the motion of the representing point in the space of normal-coordinates. This point represents the state of the molecule in the multidimensional potential well. If the motion of the molecule is stochastic, the classical trajectory of the point is a complex curve. Above the dissociation threshold the molecule may get to the region of normal-coordinate space

---

[*] Such a model is called the Reiss-Ramsperger-Kassel-Marcus model (or, briefly, the RRKM model).

which corresponds to the tearing away of an atom (or atoms). This implies that the point representing the molecular state has entered the so-called dissociation valley. In the two-dimensional case we may give a pictorial representation of this situation. Let us take, for example, the valence vibrations of the O–C–S-molecule and denote the distances O–C and C–S by the symbols x and y, respectively. Figure 17.4 gives the qualitative representation of the two-dimensional potential well $U(x,y)$ and shows the particle entering the dissociation valley along the y-axis.

The time of the molecular dissociation, i.e., the quantity reciprocal to the dissociation probability, is, in fact, the time the molecule excited above the dissociation threshold spends travelling in the potential well before it enters the dissociation valley. In case the molecular motion is stochastic this time depends only slightly upon the initial trajectory point in the sense that for the majority of the initial conditions the times of the arrival at the dissociation valley are of the same order. In this situation the probability of dissociation is determined by the typical relative transverse dimension, i.e., by the dissociation-valley width, and may be evaluated as a solid angle at which the exit to the valley is seen from the middle point of the potential well at the given level of the total molecular energy.

In the two-dimensional case when the excess of the total molecular energy U_t above the dissociation threshold U_d is small, the vertical cross section of the potential surface taken across the dissociation valley is parabolic. Hence, the valley width is proportional to $(U_t - U_d)^{1/2}$. The solid angle of the exit to the valley may be taken to be proportional to $[(U_t - U_d)/U_d]^{1/2}$. In the case of multidimensional motion in multidimensional potential well we obviously have the proportionality $[(U_t - U_d)/U_d]^{(s'-1)/2}$. The dimensionality of the representing-point motion s' coincides with the number of molecular degrees of freedom and, hence, with the dimensionality of the potential well only in the case of complete stochasticity of the vibrational motion of the overexcited molecule. Then $s' = s$. But, if the stochasticity is incomplete the situation aggravates. First, s' may not coincide with s. Second, the stochastized vibrational motion may embrace only those degrees of freedom which do not correspond to the dissociation with the lowest threshold. In this case the molecule will decay only when these degrees of freedom are sufficiently involved in the stochastized motion. The latter occurs when the total molecular energy exceeds some threshold value U_s typical of such a process (this value is sometimes called the activation energy). Energy U_s may be greater than the dissociation energy. Then, at $U_d < U_t < U_s$ the molecule may retain its overexcited state until collisions force it to dissociate. At $U_t > U_s$ the molecule will dissociate at a rate determined by the excess of U_t above U_s, the latter being the analogue for the dissociation threshold in this case.

Therefore, a steep power dependence of the dissociation probability upon the total molecular energy excess above the dissociation threshold is typical of small polyatomics. The proportionality coefficient of this power dependence may be estimated by the typical value of the vibration frequency. As a result, we have

$$\Gamma_d \simeq \omega[(U_t - U_d)/U_d]^{(s-1)/2}. \tag{17.5}$$

It is evident that this strong power dependence does not allow the laser light to overexcite the molecule even to the least noticeable degree above the energy level, where the exit to the dissociation begins. So, in the whole succession of molecular states with constantly increasing energy, which are occupied due to the laser irradiation, only a small fraction of the states are, in fact, decaying. Hence, the dissociation of small polyatomics may be described by (11.19)-type equations, in which the influence of the molecular decay is accounted for by an appropriate choice of the boundary conditions, $\rho_n = 0$ at $n\hbar\omega > U_d$. This is a distinction between the small and large polyatomics, the latter capable of being highly overexcited by laser light above the dissociation threshold due to the small probability of dissociation. This statement is equivalent to another: for small polyatomics the pre-exponential (entropy) factor is more significant than the Boltzmann exponent, as opposed to the case of the large molecules, where the exponential cofactor is dominant.

The above considerations allow us to gain a general understanding of a great number of the experimental results on intense laser action upon molecules. We restrict our further discussion to the experimental study of small polyatomics. There are very many molecules of this kind and many of them have already been investigated. As a rule, the experiments are carried out with CO_2-lasers. The radiation of these lasers falls into the region 900–1100 cm^{-1}, where the absorption lines of many molecules may be found. The radiation frequency may be adjusted both in discrete steps of 1–2 cm^{-1} and continuously. The radiation intensity in the pulses of duration of 1 ns–1 μs usually comprises 10^{10}–10^5 W/cm^3.

The pressure in the gas of molecules being investigated is usually 0.1–0.01 Tor, which at the most typical laser pulse duration of 100 ns corresponds to $\rho\tau = 10^{-8}$–10^{-9} tor·s. This implies that the experiments are carried out on the boundary of the region, where the collisions may be ignored, as is usually the case. The most complete elimination of the collision influence is achieved in molecular beams; however, in this case it becomes difficult to record the parameters of the process being investigated.

The classical objects for the experimental study of the strong laser excitation of small polyatomics are the CF_3I- and SF_6-molecules.

Let us start by considering a CF_3I-molecule. For this molecule to dissociate, an absorption of about 20 quanta of CO_2-laser radiation is sufficient. With this the molecule breaks into two neutral fragments – the CF_3-radical and the iodine atom. The molecule is excited by acting at the frequency 1074.5 cm^{-1}, that is, the resonant frequency of the Q-branch of the vibrational mode ν_1. The molecule dissociates at an energy flux of about 0.2 J/cm^2, the q-factor of this molecule at a typical duration of a laser pulse of 100 ns being close to unity. The energy flux density of several tenths of a joule per cm^2 is typical for molecular excitation up to the dissociation threshold. At these particular densities the quantity $\pi\mathcal{E}^2\mu_2 g\tau/\hbar$ at $g = 1$ cm, $\mu = 0.3$ D and $\tau = 10^{-7}$ s comprises several dozens of units (see (11.4a)). The proximity of the q-factor to unity at the excitation of the non-degenerate IR-active mode and the correspondence of the energy required for the effective dissociation to a fairly optimistic theoretical estimate are explained by the presence of the Fermi resonances in a CF_3I-molecule, which lead to the complete stochastization of the vibrational motion starting almost from the third vibrational level of the ν_1-mode. The main role in this process is played by resonance $\nu_1 = 2\nu_5$, which is accurate up to several reciprocal centimeters, i.e., up to the value of the order of the anharmonicity constant. Such a situation has been considered in Lecture 15 (see (15.15) and Fig. 15.9).

The Fermi resonances provide a high density of the dipole-coupled levels required for the effective excitation of the molecule. Modes ν_1 and ν_5 are coupled via the cubic interaction. This interaction being the strongest one determines the frequency dependence of the excitation. This characteristic has been studied experimentally using the method of the two-frequency laser action which amounts to the following. The gas of the molecules studied is excited by a laser pulse of moderate intensity which is sufficient for the molecules to reach relatively high levels but too low to dissociate them. Then the energy absorption efficiency is investigated as a function of the frequency for another laser field. The frequency characteristic of the absorption efficiency for the molecules of CF_3I thus excited appears to be smooth, wide, and falling symmetrically with the detuning of the second laser field from the main transition frequency. It is this dependence that should be observed for the vibrational quasicontinuum in the case of cubic anharmonicity. The shape of this curve may be determined by substituting the expression for the anharmonic frequency shift $\omega_{unh} = \partial H_{anh}/\partial I_1$, which follows from the record of the Hamiltonian of the anharmonic interaction between modes ν_1 and ν_5,

$$H_{anh} = \alpha_{155}(a_1^+ a_5 a_5 + a_1 a_5^+ a_5^+) \approx \alpha_{155} I_1^{1/2} I_5 \cos(\theta_1 - 2\theta_5) \qquad (17.6)$$

into (16.26). In the case of the CF_3I-molecule the distribution function f in (16.26) should be substituted by the Boltzmannian one.

The CF_3I-molecule is somewhat unique. Its excitation is practically completely free of the complications associated with the lower-level system, and the maximum in the spectral dependence of the kinetic coefficients corresponding to the excited states is close to the spectral maximum of the linear absorption.

Now let us turn to the SF_6-molecule. The energy of its dissociation is about forty quanta of CO_2-laser light. This molecule is also sensitive to excitation and may relatively easily dissociate in an intense IR-laser field when the irradiation energy flux is of the order of one joule per cm^2. However, the dissociation at such energy densities occurs only under the conditions of the two-frequency action in case the second field frequency is detuned from the main transition $0 \to 1$ toward longer waves. This is associated with the shift of the maxima of the kinetic-coefficient spectral dependencies toward low frequencies (red shift). Such a shift is typical of the modes which do not possess third-order anharmonicity. The triply degenerate mode ν_3 of the SF_6-molecule excited by the laser field is an example of such a mode.

Should the vibrational motion in the SF_6-molecule be completely stochastic as is the case for the CF_3I-molecule, we could compute the red shift value by means of expressions (16.31)–(16.33) on the basis of the approach to the vibrational spectrum of the triply degenerate mode which was developed in Lecture 15. However, we have no reason to suspect the stochasticity in the SF_6-molecule to be complete. The point is that the hybrid vibration $\nu_2 + \nu_6$ closest to mode ν_3, which is the preferable partner in the formation of the Fermi resonance, is detuned in frequency by more than twenty reciprocal centimeters.

Moreover, there is an experimentally observed difference between the absorption spectrum of a molecule excited by laser light and that of a molecule quasi-statically heated to the same energy. In addition, there is experimental evidence that the distribution function of a molecule in phase space is sensitive to the frequency of laser light. Everything said above manifests itself in the spectral dependence of kinetic coefficients which cannot, thus, be calculated using the canonic distribution. However, we can say as well that the assumption of the localization of the excitation energy in the resonant mode also leads to disagreement with experiment. If we substitute the distribution function corresponding to the localization of excitation into expression (16.26),

$$f = A\delta(n - I_1 - I_2 - I_3)\delta(\Delta - 2\alpha I_1 - \beta I_2 - \beta I_3), \qquad (17.7)$$

where A is the normalization factor, then the spectral dependence of the kinetic coefficients will be too wide.

The determination of the stochasticity region, which means defining function f, is the main problem encountered when describing the excitation dynamics of the high levels of small polyatomics. In some cases as, for example, for the molecule of silicon tetrafluorine, SiF_4, it is possible to achieve an acceptable consistency with the experimental data by taking a simple distribution function f in the form

$$f = A\delta(I_s - I_1 - I_2 - I_3), \qquad (17.8)$$

which assumes that the molecule, after it has absorbed a certain number of quanta I_s into the mode being excited, distributes all the additional quanta among all the other degrees of freedom (the leading-in-excitation model). The red shift of the dissociation yield maximum which has been experimentally observed to be about 80 cm^{-1}, corresponds to $I_s = 13$–14. For other molecules, say, for some halogenic hydrocarbons, the experimentally observed spectral dependence of the kinetic coefficients for the molecules excited with laser light is so complicated that it seems impossible to find a somewhat reasonable and simple distribution function.

The question of the distribution function for an SF_6-molecule still remains open.

Figure 17.5 shows the experimental dependencies which give an idea about the absorption line of some laser-excited small polyatomics.

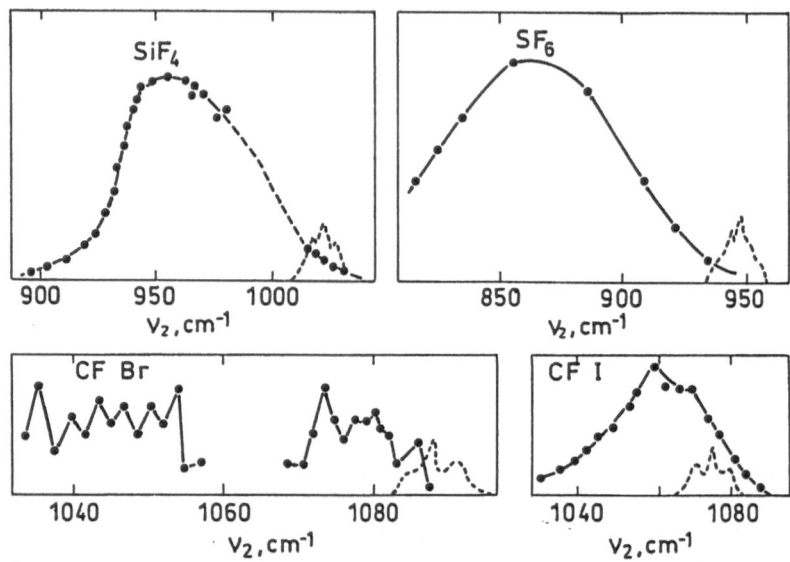

Fig. 17.5. Dependencies of the exit to the dissociation (SiF_4, SF_6, CF_3Br) and of the absorbed energy (CF_3I) on the second field frequency. Dotted line schematically shows the linear absorption spectra on transition $0 \rightarrow 1$. Ordinates are given in arbitrary units

Now let us consider the lower levels of a molecule. In contrast to the case of CF_3I-molecule the processes developing at the lower levels of an SF_6-molecule are typical of small polyatomics. As was mentioned at the beginning of this lecture, at a given frequency of excitation the molecule may get into the vibrational quasi-continuum not from every starting level (the bottleneck effect). In other words, the molecules get stuck (localized) in the system of lower levels; some of them absorb only two quanta; others, just one quantum; still others do not absorb any quanta at all. All this results in a value of the q-factor significantly different from unity, reaching, in the fields of moderate intensity (10^6–10^7 W/cm^2),the values of 10–30 %. The reason for this is the absence of a chain of successive resonances between the starting level and the vibrational quasicontinuum, i.e., the presence of a resonant structure of multiphoton transitions.

At room temperature the gas of SF_6-molecules possesses a rather low q-factor, however, the structure associated with the multiphoton resonances does not manifest itself in the absorption spectrum. This is caused by the superposition of a large number of nonlinear absorption spectra corresponding to the different starting states of the molecule. It is analogous to the formation of hot bands in the linear absorption spectra. In other words, if at some frequency only a small fraction of molecules is promoted into the vibrational quasicontinuum, a small shift of the exciting field frequency should not result in a change of the value of this fraction, because a relatively small number of some other molecules corresponding to other starting states will be excited.

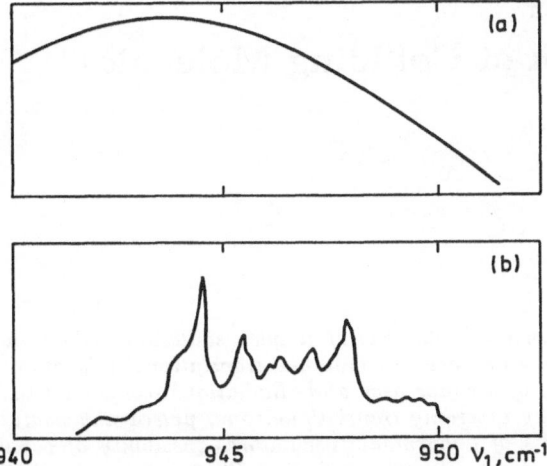

Fig. 17.6. Spectra of the absorption of the laser radiation energy in the system of lower levels of the SF_6-molecule at 300 K (**a**) and 150 K (**b**). Ordinates are given in arbitrary units

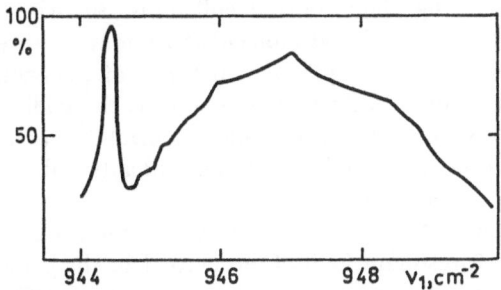

Fig. 17.7. Dependence of the dissociation probability of gas-dynamically cooledSF_6-molecule on the first field frequency. Frequency of second laser field is 933 cm^{-1}; energy density, 20 J/cm^2

   With the temperature decrease the number of possible starting states drastically diminishes. As a result, one may clearly observe the structure of the multiphoton resonances in the absorption spectrum. Figure 17.6 shows the absorption spectra of lower levels in an SF_6-molecule at temperatures of 300 and 150 K. A similar structure is observed in the frequency dependence of the q-factor for the gas of SF_6-molecules cooled in a gas-dynamic jet (Fig. 17.7).

   The excitation and dissociation of small polyatomics in intense IR laser fields has a number of important applications. We mention only laser isotope separation, the laser rectification of gaseous mixtures and IR laser photochemistry. The non-thermal action of radiation on matter required for these applications is provided by the spectral selectivity of the high vibrational state excitation which may be significantly increased, if necessary, by using the resonances in the system of lower levels.

Lecture Eighteen

Excitation of Colliding Molecules

Vibrational exchange in collisions of simple oscillators. Vibrational heating in the absence and in the presence of vibrational-translational relaxation. V-V'-exchange, preferable heating of a softer oscillator. Radiational-collisional cascade, vibrational temperature greatly exceeding translational one, pulsed and continuous irradiation regimes. Collisions of anharmonic oscillators, possibility of population inversion. Polyatomic molecules, bottleneck broadening. Photon-assisted collisions, the Rabi and Weisskopf frequencies.

So far we have been considering the excitation of molecules by intense resonant, radiation ignoring their mutual collisions. In other words, we have been dealing with a single isolated molecule rather than an ensemble of such molecules. This treatment is applicable to a gas of molecules at rather low pressures ($p\tau \leq 10^{-9}$ torr $\cdot$ s). At higher pressures the collisional processes of the V-V-, V-V'-, V-T-relaxations become essential (Lecture 12). The excitation of molecules under collisional conditions radically differs from the process of their collisionless excitation.

Let us proceed with the simplest case of a system of harmonic oscillators which is exposed to a resonant field of intensity I in the presence of a fast V-V-exchange processes. Our purpose is to consider the heating of an ensemble of simple colliding oscillators by the laser field. The absorption of the radiation energy by some chosen transition or even by their entity causes an instantaneous violation of the equilibrium conditions, which is rapidly thermalized due to the V-V-exchange processes. To analyze this situation it is necessary to use the kinetic coefficients. The dynamics of such a system is described by (12.36), which contains on its right-hand side the additional radiative terms given at the end of Lecture 12. The finite-difference form of this equation will be given later; now we make the following diffusion approximation:

$$\dot{\rho}(x) = W \left[N \frac{\partial}{\partial x} x \frac{\partial \rho}{\partial x} + \frac{\partial}{\partial x} x \rho \right] + \frac{\partial}{\partial x} \frac{I\sigma(x)}{\hbar\omega} \frac{\partial \rho}{\partial x}, \qquad (18.1)$$

where $N = \int x\rho(x)dx$ is the average number of quanta in the oscillator, and W, the probability of the V-V-exchange in the collision of an unexcited molecule, with the molecule promoted to the first vibrational level (see, for example, (12.6)). This equation is a direct consequence of the nonlinear Eq. (12.36), the nonlinearity being typical of collisional processes. The latter may be linearized for some specific types of collisional exchange probability.

One of such cases is that of the V–V-relaxation at weak collisions of harmonic oscillators. However, even in this extremely simple situation the nonlinearity of the basic process manifests itself. The point is that parameter N entering (18.1) is to be defined in a self-consistent way, i.e., from the solution of this equation itself. In other words, we have a nonlinear equation supplemented by the condition of, generally speaking, a nonlinear relation between the parameter and the solution of the equation. It is this circumstance that forbids us to write the stationary solution of (18.1) from the condition of population fluxes balance by taking $\dot{\rho} = 0$:

$$WNx\frac{\partial\rho}{\partial x} + Wx\rho + \frac{I\sigma(x)}{\hbar\omega}\frac{\partial\rho}{\partial x} = 0 \,,$$

which yields

$$\ln\rho = \text{const} + \int dx \frac{Wx}{WNx + I\sigma(x)/\hbar\omega} \,.$$

If we determine the constant value from the normalization condition and substitute the function ρ thus obtained into the expression $\int x\rho dx$, we get an equation for N which has no solution at positive $I\sigma/\hbar\omega$. Hence, we have to solve a time-dependent problem. Let us write down its solution, which is in many aspects typical of problems of physical kinetics. It leads to a rather evident answer (18.14) and (18.15), despite a somewhat lengthy procedure necessary to obtain it.

After the substitution

$$y = x/N \,, \tag{18.2}$$

(18.2) may be rewritten as

$$\dot{\rho}(t, y) = W\left[\frac{\partial}{\partial y}y\frac{\partial}{\partial y}y\right]\rho + \frac{1}{N^2}\frac{\partial}{\partial y}\frac{I\sigma(Ny)}{\hbar\omega}\frac{\partial}{\partial y}\rho \,. \tag{18.3}$$

At $I = 0$ the solution of (18.3) takes on the form

$$\rho(t, y) = \sum_m a_m e^{-mWt}e^{-y}L_m(y) \,, \tag{18.4}$$

where $L_m(y)$ is the Laguerre polynomial which, being multiplied by e^{-y}, is the eigenfunction of the first differential operator on the right-hand side of (18.3). In case $I \neq 0$, the solution of (18.3) should be tried in the form (18.4) assuming, however, the coefficients $a_m = a_m(t)$ to be time dependent. Naturally, in this case we should presume quantity N in (18.3) to be time dependent, too, $N = N(t)$. The normalization condition

$$1 = \int\rho(x, t)dx = \int N(t)\rho(t, y)dy$$

$$\sum_m a_m(t)e^{mWt}N(t)\int e^{-y}L_m(y)dy = a_0(t)N(t) \tag{18.5}$$

yields $a_0(t) = 1/N(t)$. In turn, the self-consistency condition,

$$N(t) = \int x\rho(x)dx = \int N^2(t)y\rho(y,t)dy$$

$$= \sum a_m(t)e^{-mWt}N^2(t)\int ye^{-y}L_m(y)dy$$

$$= N(t) - N^2(t)a_1(t)e^{-Wt}, \tag{18.6}$$

gives $a_1(t) \equiv 0$. Now let us substitute expression (18.4) with time-dependent coefficients into (18.3). The multiplication of both sides by $L_m(y)$ followed by the integration over dy gives

$$\dot{a}_n = \sum_m \left(\frac{1}{N^2(t)}\int L_n(y)\frac{\partial}{\partial y}\frac{I\sigma(Ny)}{\hbar\omega}\frac{\partial}{\partial y}e^{-y}L_m(y)dy\right)e^{(n-m)Wt}a_m(t), \tag{18.7}$$

which yields equations for a_0 and a_1:

$$\dot{a}_0 = \sum_m \frac{1}{N^2}\int \frac{\partial}{\partial y}\frac{I\sigma(Ny)}{\hbar\omega}\frac{\partial}{\partial y}e^{-y}L_m(y)dye^{-mWt}a_m(t), \tag{18.8}$$

$$\dot{a}_1 = \sum_m \frac{1}{N^2}\int (1-y)\frac{\partial}{\partial y}\frac{I\sigma(Ny)}{\hbar\omega}\frac{\partial}{\partial y}e^{-y}L_m(y)dye^{(1-m)Wt}a_m(t). \tag{18.9}$$

If we take into account the condition of self-consistence $a_1(t) \equiv 0$ and the normalization condition $a_0(t) = 1/N(t)$ and assume that $I\sigma$ is small with respect to W, $I\sigma/\hbar\omega \ll W$, it turns out that Eqs. (18.8) and (18.9) are sufficient to determine $N(t)$. Indeed, (18.8) along with (18.4) yields

$$\dot{a}_0 = \frac{1}{N^2}\frac{I\sigma(Ny)}{\hbar\omega}\frac{\partial}{\partial y}\rho(y)|_{y=0}. \tag{18.10}$$

This form of notation valid for continuous y does not permit us to draw complete analogies between the populations of discrete energy levels and the corresponding variable of the diffusion problem. For this analogy to be justified, from (18.9) along with (18.4), after separate integration in parts, we obtain

$$\int \frac{I\sigma(Ny)}{\hbar\omega}\frac{\partial}{\partial y}\rho dy = \frac{I\sigma(Ny)}{\hbar\omega}\frac{\partial\rho}{\partial y}|_{y=0}. \tag{18.11}$$

Hence,

$$\dot{a}_0 = \frac{1}{N^2}\int \frac{I\sigma(Ny)}{\hbar\omega}\frac{\partial}{\partial y}\rho dy. \tag{18.12}$$

The smallness of the parameter $I\sigma/\hbar\omega$ against W permits us to retain only the first term from the expansion of (18.4) in this equation, which is impossible in the very (18.10) due to the formal divergence of some derivatives approaching zero. Then,

$$\dot{N}(t) = \int \frac{I\sigma(Ny)}{N\hbar\omega}e^{-y}dy. \tag{18.13}$$

If we now use $\sigma(x) = \sigma_{01}x$, which holds for a simple oscillator, we finally get

$$\dot{N} = I\sigma_{01}/\hbar\omega \,, \tag{18.14}$$

$$\rho(x,t) = \frac{1}{N(t)} \exp\left[-\frac{x}{N(t)}\right]. \tag{18.15}$$

Therefore, in the case of the excitation of the entity of colliding oscillators by resonant radiation, when we pay attention only to the fast process of the resonant V–V-exchange, one can see that there is no stationary distribution. One may speak only about the quasi-stationary distribution $(a_0 \gg a_2, \ldots, a_m, \ldots)$, while the average number of quanta per molecule, i.e., its vibrational temperature, linearly increases with time.

The process of the energy build up by molecules, i.e., their heating, ceases when V–T-relaxation becomes essential. In this case we may think of a system achieveing the stationary state. We shall consider two aspects of this phenomenon: the general form of the stationary distribution function, and the particular way this function is established in one of the simplest cases, in which the rates of the V–T-exchange and radiative excitation (pumping) are small in comparison with the V–V-relaxation rate.

The stationary distribution function may be found by equating the left-hand sides of the balance equations

$$\begin{aligned}
\dot{\rho}_n &= W[(n+1)((N+1)\rho_{n+1} - N\rho_n) + n(N\rho_{n-1} - (N+1)\rho_n)] \\
&\quad + \frac{I}{\hbar\omega}(\sigma_n^{n+1}\rho_{n+1} + \sigma_n^{n-1}\rho_{n-1} - \sigma_{n+1}^n\rho_n\sigma_{n-1}^n\rho_n) \\
&\quad + R_n^{n+1}\rho_{n+1} - R_{n-1}^n\rho_n
\end{aligned} \tag{18.16}$$

to zero (see (18.1)). From this equation the following condition of detailed balance may be obtained:

$$\begin{aligned}
&W(n+1)(N+1)\rho_{n+1} - W(n+1)N\rho_n \\
&\quad + \frac{I}{\hbar\omega}\sigma_n^{n+1}\rho_{n+1} - \frac{I}{\hbar\omega}\sigma_{n+1}^n\rho_n + R_n^{n+1}\rho_{n+1} = 0\,. \tag{18.17}
\end{aligned}$$

This, in turn, yields the relation

$$\frac{\rho_{n+1}}{\rho_n} = \frac{W(n+1)N + I\sigma_{n+1}^n/\hbar\omega}{W(n+1)(N+1) + I\sigma_n^{n+1}/\hbar\omega + R_n^{n+1}}\,, \tag{18.18}$$

which allows us to determine all the populations, provided the value of ρ_0 is known. The value of ρ_0 may be found from the normalization conditions, while the value of N is determined from the conditions of self-consistency.

Now let us consider the following didactic example. Let $\sigma_0^1 = \sigma_1^0 = \sigma \neq 0$, $\sigma_n^{n+1} = 0$ at any $n \geq 1$, $I\sigma_0^1/\hbar\omega \gg W, R_n^{n+1}$, $R_n^{n+1} = (n+1)R$. In other words, we are going to discuss the stationary distribution in the following situation. Transition $0 \to 1$ is saturated by the radiation field and this is the only source of energy supply into the system. On higher levels populated in the course of the V–V-exchange, the energy dissipation takes place as a result of the V–T-relaxation. It follows from (18.18) that at $n \geq 1$,

$$\frac{\rho_{n+1}}{\rho_n} = \frac{WN}{W(N+1) + R} = C,$$

(18.19)

i.e., the Boltzmann distribution in energies is valid for high levels. Let us find the characteristic temperature of this distribution from the condition of pumping (heating) and the energy dissipation-rate balance. Let us use the generating function method for the purpose. If we denote

$$\Phi(\lambda) = \sum_n e^{\lambda n} \rho_n,$$

(18.20)

and bear in mind equality $\rho_0 = \rho_1$ (it is valid due to the assumption of transition $0 \to 1$) being saturated and relation (18.19), then, summing up the geometric progression, we obtain

$$\Phi(\lambda) = \rho_0 \left(1 + \frac{e^\lambda}{1 + Ce^\lambda}\right).$$

(18.21)

From the normalization condition $\Phi(0) = 1$ it follows that

$$\rho_0 = \frac{1 - C}{2 - C}.$$

(18.22)

In addition, the self-consistency condition entails

$$N = \frac{1}{(2 - C)(1 - C)}.$$

(18.23)

Inserting here the value of constant C from (18.19), we obtain the quadratic equation for N. Its solution is

$$N = \frac{W + R}{W} \left(1 + \sqrt{1 + W/R}\right),$$

(18.24)

which at $W \gg R$ leads to

$$N = \sqrt{W/R}.$$

(18.25)

Thus, the distribution function is of the Boltzmannian type (12.39).

Therefore, we see that the population distribution thus obtained arises from the flow of the energy which is absorbed by the $0 \to 1$ transition and then reaches the higher levels due to the V–V-exchange where it is dissipated due to the V–T-relaxation of higher levels. In this process the system is heated to the temperature $(\hbar\omega/k)\sqrt{W/R}$. The independence of the vibrational temperature on the light intensity is an evident consequence of the $0 \to 1$-transition saturation – the rate of the quanta absorption equals the rate of the V–V-exchange.

And now let us turn to the study of the process of the establishment of the distribution function in the case in which the V–V-exchange rate significantly exceeds the rates of both the V–T-relaxation and the radiation excitation, $W \gg R$, $I\sigma/\hbar\omega$. It is convenient to carry out such an analysis using the diffusion Eq. (18.1), where an additional term describing the V–T-relaxation is inserted into the right-hand side

$$\frac{\partial}{\partial x} R(x)\rho(x)$$

$$\dot{\rho}(x) = W \left(N \frac{\partial}{\partial x} x \frac{\partial}{\partial x} \rho + \frac{\partial}{\partial x} x\rho \right) + \frac{\partial}{\partial x} \frac{I\sigma(x)}{\hbar\omega} \frac{\partial}{\partial x} \rho + \frac{\partial}{\partial x} R\rho \,. \qquad (18.26)$$

If we repeat the whole procedure which led us from (18.1) to expression (18.13), we obtain

$$\dot{N} = - \int \frac{I\sigma(x)}{\hbar\omega} \frac{\partial}{\partial x} \rho(x)dx - \int R(x)\rho(x)dx \,. \qquad (18.27)$$

This equation, in fact, describes the dynamics of vibrational heating. Owing to the relatively slow rate of the radiative excitation and the V–T-relaxation we may substitute the Boltzmann distribution $\rho(x)$ (18.15) which is established due to the fast V–V-exchange. A similar substitution was made when solving (18.1). If, as previously, we presume that $\sigma(x) = \sigma_{01}x$ and $R(x) = R_0^1 x$, then we get the following equation:

$$\dot{N} = - \int \frac{I\sigma_{01}x}{N\hbar\omega} \frac{\partial}{\partial x} e^{-x/N} dx - \int \frac{R_0^1 x}{N} e^{-x/N} dx \,, \qquad (18.28)$$

whose solution has the form

$$N(t) = \frac{I\sigma_{01}}{R_0^1 \hbar\omega} (1 - \exp(-R_0^1 t)) \,. \qquad (18.29)$$

Therefore, the stationary distribution (18.15) is established in a time equal to the reciprocal rate of the V–T-relaxation and is characterized by a temperature equal to $(\hbar\omega/k)I\sigma_{01}/\hbar\omega R_0^1$. The physical meaning of the relations obtained is rather obvious and is as follows. In the case of a simple oscillator the rate of photon absorption does not depend on the type of distribution function, since the difference of the upward and downward radiative transition rates does not depend on the state number and comprises

$$(n+1)\frac{I\sigma_{01}}{\hbar\omega} - n\frac{I\sigma_{01}}{\hbar\omega} = \frac{I\sigma_{01}}{\hbar\omega} \,.$$

The probability of the V–T-relaxation in the model we are using increases with the level number. So, the excitation is effective up to a certain characteristic level n, for which the rates of the energy absorption and dissipation become equal to $nR_0^1 \approx I\sigma_{01}/\hbar\omega$. The number of this characteristic level above which the excitation is no longer effective is, in fact, by order of magnitude (and exactly in the case of a simple oscillator) the average number of the absorbed quanta (the vibrational temperature of the molecule).

We have often mentioned the existence of V–V'-processes in which vibrational exchange occurs between particles of different types with different energies of the vibrational quanta. In such systems the vibrational temperatures of the particles of different types may differ from each other and from the translational temperature common to them. Let us consider the V–V'-exchange in a system that contains simple oscillators of two types with fractional concentrations f and f', $f + f' = 1$, and the quantum energies differing by $\hbar\Delta$. Using (12.32) along with the harmonicity of the oscillators considered, we may write

$$
\begin{aligned}
\dot{\rho}_n &= W f(n+1)(N+1)\rho_{n+1} - W f(n+1)N\rho_n - W fn(N+1)\rho_n \\
&\quad + W fnN\rho_{n-1} + W'' f'(n+1)(N'+1)\rho_{n+1} \\
&\quad - W'' f' \exp(-\hbar\Delta/kT)(n+1)N'\rho_n - W'' f'n(N'+1)\rho_n \\
&\quad + W'' f' \exp(-\hbar\Delta/kT)nN'\rho_{n-1} , \\
\dot{\rho}'_n &= W' f'(n+1)(N'+1)\rho'_{n+1} - W' f'(n+1)N'\rho'_n \\
&\quad - W' f'n(N'+1)\rho'_n + W' f'nN'\rho'_{n-1} \\
&\quad + W'' f \exp(-\hbar\Delta/kT)(n+1)(N+1)\rho'_{n+1} - W'' f(n+1)N\rho'_n \\
&\quad - W'' f \exp(-\hbar\Delta/kT)n(N+1)\rho'_n + W'' fnN\rho'_{n-1} .
\end{aligned}
\tag{18.30}
$$

Here the nonprimed symbols correspond to an oscillator of the first type, and the primed ones, to another. Symbol W'' denotes the probability of the slowest V–V'-process, i.e., the probability of a quanta exchange between the soft and hard oscillators on the main transition. If we now take $\dot{\rho}_n = \dot{\rho}'_n = 0$ and according to the usual procedure equate the population fluxes to zero, then for the neighboring levels the following relations are obtained:

$$
\frac{\rho_{n+1}}{\rho_n} = \frac{W fN + W'' f' \exp(-\hbar\Delta/kT)N'}{W f(N+1) + W'' f'(N'+1)} = C ,
\tag{18.31}
$$

$$
\frac{\rho'_{n+1}}{\rho'_n} = \frac{W' f'N' + W'' fN}{W' f'(N'+1) + W'' f(N+1) \exp(-\hbar\omega/kT)} = C' .
\tag{18.32}
$$

Now, using the generating functions

$$
\Phi(\lambda) = \sum_n e^{n\lambda}\rho_n = \frac{1-C}{1-e^{\lambda}C} , \quad \Phi'(\lambda) = \sum_n e^{n\lambda}\rho'_n = \frac{1-C'}{1-e^{\lambda}C'} ,
$$

we write down the conditions of self-consistency

$$
N = C/(1-C) , \quad N' = C'/(1-C')
$$

and substitute the values C and C' from (18.31) and (18.32). Then we obtain the following equations for N and N':

$$
N[N'(1 - \exp(-\hbar\Delta/kT)) + 1] = N' \exp(-\hbar\Delta/kT) ,
\tag{18.33}
$$

$$
N'[N(\exp(0 - \hbar\Delta/kT) - 1) - \exp(-\hbar\Delta/kT)] = N .
\tag{18.34}
$$

These two equations are seen to coincide. This results from the fact that the total number of quanta in the course of the V–V'-process remains unchanged, being the free parameter which is determined by the given initial conditions; so it cannot be found from the condition for the process to be stationary. It follows from (18.33) and (18.34) that the average number of quanta for the harder molecules is less than that for the softer ones:

$$N = N' \frac{\exp(-\hbar\Delta/kT)}{N'(1 - \exp(-\hbar\Delta/kT)) + 1} .$$ (18.35)

Let us pay special attention to two extremes. At $N'\hbar\Delta/kT \ll 1$, from (18.35) we have

$$N \approx N' .$$ (18.36)

Hence, in this case the oscillators may be considered resonant if at the typical excitation level the difference in the energy of the oscillators is less than the translational temperature. In the opposite extreme, $N'\hbar\Delta/kT \gg 1$, we have

$$N = \frac{\exp(-\hbar\Delta/kT)}{1 - \exp(-\hbar\Delta/kT)} ,$$ (18.37)

i.e., the number of quanta in the harder oscillator turns out to be small, its levels being populated in accord with the Bose statistics with the energy quantum corresponding to the detuning. All the energy is concentrated in this case in the softer oscillator. If the detuning is small against the translational temperature $\hbar\Delta \ll kT$, the typical excitation extent of the hard oscillator comprises $N = kT/\hbar\Delta$.

The problems considered so far allow us to analyze the situations one comes across under real conditions of resonant laser action on a gas of colliding molecules. Of course, a harmonic oscillator model is too simple for a detailed treatment of these processes; however, under collisional conditions in which the duration of a single impact itself is so small, that its product by the typical detuning (rotational structure, anharmonicity) is much less than unity, the applicability of such a model is more justified than in the collisionless conditions.

Usually, the experiments under the collisional conditions are carried out in the pressure range of 1–100 torr. At such pressures it is relatively easy to realize the regime of the uniform illumination of rather large gas volumes using either the pulsed or the continuous-action lasers.

Let us proceed with the pulsed regimes assuming the pulse duration shorter than the V–T-relaxation time. Then from (18.14) we may determine the energy of the pulse required for the given heating of the molecular vibrational degrees of freedom,

$$N = \int \frac{I\sigma}{\hbar\omega} dt = \frac{F\sigma_{01}}{\hbar\omega} ,$$ (18.38)

where F is the irradiation energy density. At $N \approx 10$ which corresponds to the temperature of 15,000 K at $\hbar\omega \approx 2 \cdot 10^{-20}$ J and $\sigma_{01} \approx 10^{-18}$ cm^2, the value of F comprises 0.2 J/cm^2. We shall estimate the duration of such a pulse from the following plain considerations. For the model leading to the linear growth of the vibrational temperature (18.14) and to the nonstationary Boltzmann distribution (18.15) to be valid, the following inequalities should be satisfied:

$$p/\tau_{VV} = W \gg \frac{I\sigma_{01}}{\hbar\omega} = \frac{F\sigma_{01}}{\tau_{\text{pulse}}\hbar\omega} = \frac{N}{\tau_{\text{pulse}}} \gg R = \frac{p}{\tau_{VT}}, \qquad (18.39)$$

where p is the gas pressure, τ_{pulse} is the duration of the radiation pulse, while τ_{VV} and τ_{VT} denote the typical times of the corresponding relaxation processes at the pressure of 1 Tor. From this it may be easily seen that

$$\tau_{VT}/Np \gg \tau_{\text{pulse}} \gg \tau_{VV}/Np. \qquad (18.40)$$

Since the times of the V–T- and V–V-relaxations differ significantly, this implies that the pulse duration may have a rather wide range of values. For many molecules $\tau_{VT} \approx 10^{-3}$ torr $\cdot$ s, while $\tau_{VV} \approx 10^{-8}$ torr $\cdot$ s. Then at $p = 1$ torr and $N = 10$ the pulse duration may be found in the interval from 1 ns to 100 μs. Therefore, the heating of the vibrational degrees of freedom of the colliding molecules, the so-called radiative-collisional cascade process, may occur in laser fields of quite moderate intensity of the order of 1 kW/cm^2 and higher.

In the continuous irradiation regime we have to take into consideration the V–T-relaxation, which restricts an infinite growth of the molecular energy. The fact that the duration of a collision event in a gas is much shorter than the intercollisional period becomes essential here. So, if from the viewpoint of the collisional energy transfer the oscillators may be treated as resonant, from the viewpoint of the radiative excitation process which occurs during the intercollisional period, the anharmonicity of vibrations is, generally speaking, significant. However, if the collisional line broadening (i.e., $1/T_2$) that is not necessarily caused by the V–V-exchange only ensures the resonance of oscillators, then we may take advantage of formula (18.27) to estimate the extent of vibrational heating. At times exceeding $1/R_{01}$,

$$N = \frac{I\sigma_{01}}{\hbar\omega}\frac{\tau_{VT}}{p}. \qquad (18.41)$$

At $\tau_{VT} = 10^{-3}$ torr $\cdot$ s, $\sigma_{01} = 10^{-18}$ cm^2, $\hbar\omega = 2 \cdot 10^{-20}$ J and a pressure of 1 torr it turns out that an intensity of 2 W/cm^2 is sufficient to achieve $N = 10$.

In case the V–V-exchange rate W is not too large and there are no additional broadening interactions, falling out of resonance due to anharmonicity may lead to the fact that the radiation is absorbed only on the general transition. However, even in this case, because of the V–V-exchange, the vibrational temperature may reach rather high values. If the condition of the $0 \to 1$ transition saturation is satisfied, the number of the quanta stored according to (18.25) comprises

$$N = \sqrt{\tau_{VT}/\tau_{VV}}, \qquad (18.42)$$

which at typical values of the relaxation times, as a rule, exceeds the dissociation energy. The intensity required for this process should exceed some typical value,

$$I \gg (\hbar\omega/\sigma)p/\tau_{VV}. \qquad (18.43)$$

At $\hbar\omega = 2 \cdot 10^{-20}$ J, $\sigma = 10^{-18}$ cm^2, $\tau_{VV} = 100$ torr $\cdot$ ns and $p = 1$ torr this intensity should be above 200 kW/cm^2.

Therefore, we see which conditions should be fulfilled for the resonant irradiation of the colliding molecules to lead to strong vibrational heating. Since in the given consideration we presume the translational temperature to remain low, say, at the expense of heat conductivity, a gap appears between the vibrational and translational temperatures. Such nonequilibrium heating may entail consequences essentially different from those of equilibrium heating. In particular, laser radiation may, thus, initiate or maintain chemical reactions between molecules which remain cold. In principle, in this situation the formation of a compound unstable at high translational temperatures is possible. It is also evident that such high vibrational heating may, in fact, imply the selective dissociation of the specific molecular component of the gaseous medium resonant with respect to radiation. Other components in this process remain practically unexcited. Of course, everything said above is valid when the eigenfrequencies of other medium components significantly differ from the resonant frequency. Otherwise, the excitation of molecules in the process of the almost-resonant V–V'-exchange becomes essential.

The description of a system that is a mixture of two molecules with similar properties, coupled via the V–V'-exchange, is much more lengthy and includes the determination of the pumping and relaxation rates for the molecules of both types. However, if the rates of the V–V- and V–V'-processes are essentially higher than the rate of radiative pumping $I\sigma/\hbar\omega$ and the V–T-relaxation rate, the relation between the latter processes is determined only by the total number of the accumulated quanta. The distribution of these quanta among the molecules is given by formula (18.35). In other words, in this approximation the relation between the vibrational temperatures of hard and soft molecules does not depend upon the distribution of roles the molecules play in the process considered: which one is excited by radiation and which one serves as an outlet of energy flux in the course of the $V - T$-relaxation. A hard molecule is always less excited and this difference may be considerable at $N'\hbar\Delta/kT \gg 1$. Such schemes have been proposed many times for the laser separation of molecules with different isotopic compositions.

The process of the vibrational exchange, in which the transfer of a hard quantum to a soft quantum and some translational energy is preferable, takes place not only in the mixture of hard and soft harmonic molecules but in a gas of identical anharmonic molecules as well. It is this process that establishes the Rich-Treanor distribution discussed in Lecture 12. If the molecular anharmonicity is negative, i.e., if, as is usually the case, the energy of the vibrational quantum decreases with molecular excitation, the probability of a quantum transfer from the slightly excited molecule to the molecule excited to a considerably higher extent is greater that the probability of the inverse process. Therefore, in a gas of anharmonic molecules the V–V-exchange tries to create a situation in which a small number of "hot" molecules takes the energy from a large number of "cold" ones, thus cooling them still morer, while the "hot" molecules are heated further. In the limit of large times this process might concentrate all the vibrational energy in a single molecule heated practically infinitely high, the decrease in the entropy and in the energy of vibrational motion being accompanied by a corresponding increase in the entropy and by an additional heating of the translational degrees of freedom. The divergence present in the Rich-Treanor distribution at $\alpha < 0$ (see (12.47) and Fig. 12.5) corresponds to this particular situation. Of course, under real conditions this does not occur because the V–T-relaxation prevents high excitation of the molecular states. In this case, as mentioned in Lecture 12, the Gordiez-Osipov-Shelepin distribution is established. But this distribution is not stationary either, since it corresponds to the situation in which the number of quanta in the system of molecules is not conserved, which results in a cooling of their vibrational degrees of freedom down to the translational temperature.

It is evident that the distribution of molecules among the vibrational states in the presence of the V–T-relaxation may be stationary only if there is a source supplying the system with vibrational energy. This energy may be provided by the absorption of resonant laser radiation. A complete solution of the problem concerning the establishment of a stationary nonequilibrium distribution in this case, which would take into account an anharmonicity, the V–V- and V–T-relaxations as well as the pumping by the radiation field, is lengthy. However, under the condition that the V–V-relaxation rate essentially exceeds both the V–T-relaxation and pumping rates, $W \gg R, I\sigma/\hbar\omega$, the whole problem may be split into two simpler ones. The first is reduced to determining the molecular distribution function at a fixed number N of quanta; the second, to the determination of the radiation parameter values required to maintain the number of quanta at this level in a stationary way.

We shall not consider here the second problem but only mention that its solution should be obtained by equating the left-hand side of expression (18.27) to zero, where the distribution function found from the solution of the first problem is substituted. This function is to be found from the conditions of the detailed balance similar to (18.17), which in the case of an anharmonic oscillator due to (12.42) and the assumption of $W_{n,i}^{n+1,i-1} = (n+1)iW$ (at $n > i$), yield an expression analogous to (18.18):

$$\frac{\rho_{n+1}}{\rho_n} = \frac{W(n+1)(N+A_n) + I\sigma_{n+1}^n/\hbar\omega}{W(n+1)(N+1+B_n) + I\sigma_n^{n+1}/\hbar\omega + R_n^{n+1}}, \qquad (18.44)$$

where

$$A_n = \sum_{i \leq n} i \left[\exp \frac{E_i - E_{i-1} - (E_{n+1} - E_n)}{kT} - 1 \right] \rho_i \,,$$

$$B_n = \sum_{i > n} i \left[\exp \frac{E_{n+1} - E_n - (E_i - E_{i-1})}{kT} - 1 \right] \rho_{i-1} \,.$$

Quantities A_n and B_n characterize the difference between (18.44) and (18.18) that the anharmonicity accounts for. If now, as we did earlier, we take $R_n^{n+1} = (n+1)R$ and $\sigma_{n+1}^n = \sigma_n^{n+1} = (n+1)\sigma$, we get

$$\frac{\rho_{n+1}}{\rho_n} = \frac{N + A_n + I\sigma/W\hbar\omega}{N + 1 + B_n + I\sigma/W\hbar\omega + R/W}. \tag{18.45}$$

Using (18.45) we can, in principle, determine the distribution function at an arbitrary value of parameter N, which is to be found from the condition of the stationary character of the distribution. Without clarifying the concrete form of distribution (18.45) we shall study the question of what conditions in the system considered lead to a population inversion, which is of practical interest. We can see from (18.45) that for the inversion to arise between the $(n + 1)$th and n-th levels the following conditions should be satisfied:

$$A_n > 1 + B_n + R/W \approx 1 + B_n. \tag{18.46}$$

In other words, the difference between A_n and B_n should be above unity, $A_n - B_n > 1$. Let $E = n\hbar\omega - \alpha n(n + 1)$ and $\alpha/kT \ll 1$. Then the exponent in the expressions for A_n and B_n may be expanded into a series with only the first term retained. In this case,

$$A_n - B_n \approx \frac{2\alpha}{kT}\left[\sum_{i\leq n}i(i+1)\rho_i + \sum_{i>n}i(i+1)\rho_i - n\sum_{i\leq n}i\rho_i - n\sum_{i>n}(i+1)\rho_i\right]. \tag{18.47}$$

The bracketed value coincides with N^2 by order of magnitude. Hence, condition (18.46) may be fulfilled only if there is a sufficiently large number of quanta in the system

$$N \geq \sqrt{kT/2\alpha}, \tag{18.48}$$

which at $T = 300$ K and $\alpha = 1$ cm^{-1} comprises $N \geq 10$, corresponding at $\hbar\omega \approx 10^3$ cm^{-1} to a vibrational temperature of 15,000 K. However, we should not forget that it is more difficult to create a high vibrational temperature in an anharmonic oscillator, primarily because pumping is performed only on a single transition as opposed to the case of the harmonic oscillator. It should also be mentioned that the difference of the populations of the upper and lower levels of the pumping transition may not be large and may even be diminished with the vibrational heating, which may considerably complicate the process of heating itself. However, only under the conditions of sufficiently effective pumping is it possible to obtain the population inversion that follows from the Rich-Treanor distribution.

Everything we have said so far in this lecture refers, strictly speaking, to diatomic molecules only. In the case of polyatomics, where the collisional transfer of vibrational excitation from one mode to another is possible, the approaches developed here are inapplicable. It may looks like that the collisions between polyatomics may be considered as collisions of independent oscillators representing the vibrational modes of a molecule. But such an approach is valid in a rather restricted realm of situations. Indeed, for large molecules is not at all applicable, since their vibrational motion is thermal alones while the fact that the translational motion is involved in the process of intermolecular collisions does not result in any significant changes, because the heat capacity of translational degrees of freedom is relatively small. In small polyatomics at levels of excitation not exceeding the threshold of complete stochastization of the vibrational motion such an approach may be applied if under the action of collisions the rate of the resonant vibrational exchange between the same modes of different molecules is greater than that between the different modes of a single molecule. However, one should keep in mind that in the presence of the Fermi resonances the situation may be quite the opposite, i.e., the latter process may develop at a higher rate than the resonant V–V-exchange.

The consideration of the collisions of polyatomics is aggravated by the fact that the typical time of excitation transfer from one mode to another at the expense of the Fermi resonances, leading to a spectral broadening of the vibrational bands as the molecular excitation energy increases, may be comparable to or even shorter than the time of the molecular collision events. In such a situation the main contribution to the vibrational exchange process is rendered by the strong collisions, i.e., the collisions with small values of the impact parameter in which a large number of quanta are transferred during a single collision event. In this case the relaxation of the distribution function to its equilibrium shape occurs practically in a single collision and is described by the kinetic equation in the τ-approximation (12.40).

From everything, we have said above it follows that the resonant excitation of large molecules goes on in the presence of collisions on in the same way as in their absence, and thus needs no special treatment. The analysis of the excitation of small polyatomic molecules is essentially more complicated and should be performed separately for high and low excitation levels. In the case of the lower levels one may think of applying the isolated-mode and colliding-oscillator model, while for higher levels the τ-approximation is more appropriate. The description of the collisions of highly excited molecules with slightly excited ones (the collisions between molecules of "hot" and "cold" ensembles) comprises an important problem, which has not been completely solved yet.

However, it is evident that collisional exchange removes the complications arising in the system of lower levels of small polyatomics when they are excited by monochromatic laser radiation (the bottleneck effect). The essence of this phenomena is the following. Sharp resonances in the absorption spectrum of lower levels result in the fact that only a relatively small fraction of molecules (those which are in resonance with radiation) is promoted into highly excited states. But as we know, in these sufficiently high vibrational states the spectrum of eigenfrequencies of excited molecules becomes considerably broader; so they becomes resonant to those unexcited molecules, which comprise the "cold" ensemble, remaining out of resonance with the laser field. The vibrational exchange with them becomes resonant and cold molecules easily get into an excited state. Such a process is especially effective when the IR active vibrations are involved in it. Then the resonant V–V-exchange reduces, in fact, to the emission of quanta by an excited molecule followed by the absorption of quanta by an unexcited one. In other words, the excited molecules stand for the wide-band laser radiation that effectively excites the molecules of the cold ensemble, thus broadening the bottleneck.

In conclusion let us turn to the so-called photon-assist collisions. So far we have not considered the interactions of molecules with the radiation field during the time of collision event, assuming that in such a short time interval photon absorption is hardly probable. However, processes exist in which the absorption of a photon during the time of collision is significant. They are called radiative or photon-assist collisions and cannot be considered by the relaxation-matrix method (Lecture 4) but should be treated dynamically as the process of collision between two particles in a radiation field. From the variety of problems of photon-assist collision physics we mention only two that are of some practical importance.

The molecules (atoms) which in a collisionless situation do not reveal any dipole-active behavior at the frequency of the field may become dipole-active during collisions. This is an evident consequence of the change in the symmetry of the compound system temporally formed during the collision time. This implies that the transition dipole moment is a function of time and that is why the population for the resonant field comprises (Lecture 2)

$$\rho_{22} = \sin^2 \int \frac{\mathcal{E}_0(t)\mu(t)}{2\hbar} dt. \qquad (18.49)$$

This evidently leads to the absorption of radiation by a gas which is transparent at a given frequency at low pressures (in the absence of collisions). The line width of such an absorption is determined by the typical duration of the collision event where the $\mu(t)$ value is nonzero. If for the estimate we use the typical dimension of 10^{-8} cm and the typical particle velocity of 10^4 cm/s, the line width is about 300 cm^{-1}. If at the moment of closest approach the dipole moment reaches its maximum value of 0.3 D, the population during a single photon-assist collision can be changed by the value of the order of unity in the fields of intensity 10^{12} W/cm^2.

However, let us note that such intensities should significantly affect the very process of the collision of resonant molecules. Indeed, if the Rabi frequency becomes comparable or exceeds the so-called Weisskopf frequency, i.e., the reciprocal time of flight over the distance equal to the Weisskopf radius (12.14), i.e., if

$$\Omega_R = \frac{\mu\mathcal{E}_0}{\hbar} \geq \frac{\sqrt{\hbar v_T^3}}{\mu} = \Omega_W, \qquad (18.50)$$

then the collisional processes lead to a change in the populations of the quasi-energy levels of the pair of colliding particles (the compound system) rather than the energy levels of the particles themselves. It is the transitions between these levels that the relaxation matrix elements should be calculated for, according to the procedure given in Lecture 4. A large number of energy levels become resonant, thus essentially complicating the process of energy transfer. If the Weisskopf radius r_W noticeably exceeds the typical atomic radius of 10^{-8} cm (the situation is to be expected under conditions of strict resonance), the intensity estimate (18.50) leads to values of $10^7 - 10^9$ W/cm^2, noticeably smaller than the value obtained from (18.49).

From the viewpoint of applications it is essential that such effects manifest themselves as the dependence of the relaxation times (the line widths) upon the radiation intensity.

Lecture Nineteen

Excitation of Electronic Transitions in Molecules

Diatomic molecule. Franck-Condon principle. Polyatomic molecules, generalization of Franck-Condon principle. Term-to-term transition at vibrational motion. Repulsion of "intersecting" terms. Landau-Zener transition. Intramolecular conversion. Photochemistry. Single- and multiphoton excitation. IR–UV irradiation, possibility of selective breaking of bonds. Change of distribution functions of reactants under the action of radiation. Chemical-kinetic equation. IR-laser irradiation.

So far we have restricted ourselves to a consideration of the molecular vibrational excitation. However, as we know, vibrational motion is not the only kind possible for molecules – electronic excitation is also possible. The corresponding electronic energy levels may be excited either directly by the field, or in collisions with excited particles or electrons, or as a result of high vibrational excitation, etc.

The case of electronic transition in a diatomic molecule is the most evident one. Let us consider two possibilities of the upper electronic term population: the population caused by the radiation field at the transition frequency (Fig. 19.1 a) and that due to the high vibrational excitation of the lower electronic term (Fig. 19.1 b). Let us begin with the former case. The radiation wavelength needed for the electronic transition excitation is known to belong to the visible or UV ranges of the spectrum. If the vibrational motion of the nuclei constituting the molecule is treated classically, it becomes evident that the frequency of the electronic transition depends on the distance between the nuclei and varies over the vibration period. The frequency corresponding to the slowest phase of the motion of the nuclei, i.e., to the motion in the vicinity of the turning points, remains practically unchanged during the longest time interval. This implies that it is the turning points that will render the largest contribution to the net probability of the electronic transition.

Everything said above remains valid in a quantum consideration of the subject, which amounts to the following. The zero-approximation wavefunction is the product of the electronic wavefunction $\psi(r, R)$, i.e., the function of the coordinates of electrons r depending on the nuclear coordinates as on parameters, by the wavefunction $\psi(R)$, i.e., the function of the nuclear coordinates. The matrix element of the electronic transition dipole moment operator, therefore, has the form

$$\mu_{0,V}^{1,V'} = \int \psi_0(r, R)\psi_V(R)\mu(r, R)\psi_1^*(r, R)\psi_{V'}^*(R)d^{(n)}r d^{(m)}R. \qquad (19.1)$$

This means that the electronic transition owing to the rapidity of the electronic motion in comparison with that of the nuclei occurs without changing the nuclear coordinates, i.e., vertically in the scheme of the electronic terms. This statement has acquired the name of the Franck-Condon principle. The validity of this principle is ensured by the smallness of the Born-Oppenheimer parameter (Lecture 15).

If now we integrate over all the electronic variables in expression (19.1), it will take on the form

$$\mu_{0,V}^{1,V'} = \int \psi_V(R)\mu_0^1(R)\psi_{V'}^*(R)d^{(m)}R. \tag{19.2}$$

The physical meaning of quantity $\mu_0^1(R)$ is fairly obvious. It represents the matrix element of the dipole transition operator at fixed positions of nuclei. This expression shows that the typical scale of quantity $\mu_0^1(R)$ being, in fact, the size of the electronic cloud, this quantity should be considerably larger than the de Broglie wavelength of the nuclei (see (15.4)). So we may neglect the dependence of matrix element $\mu_0^1(R)$ on the internuclear distance R in the domains corresponding to the maximum contribution to integral (19.2). Then quantity $\mu_{0,V}^{1,V'}$ is maximum for those pairs of wavefunctions $\psi_V(R)$ and $\psi_{V'}(R)$ whose overlapping is maximum. The vibrational wavefunctions are known to have distinct maxima at the turning points. This is a straightforward consequence of the fact that the probability for a particle to be found is maximum in that domain in space in which its motion is slowest. This means that the main contribution to the overlapping integral will be rendered by the vicinities of the turning points, as pictured in Fig. 19.1 a.

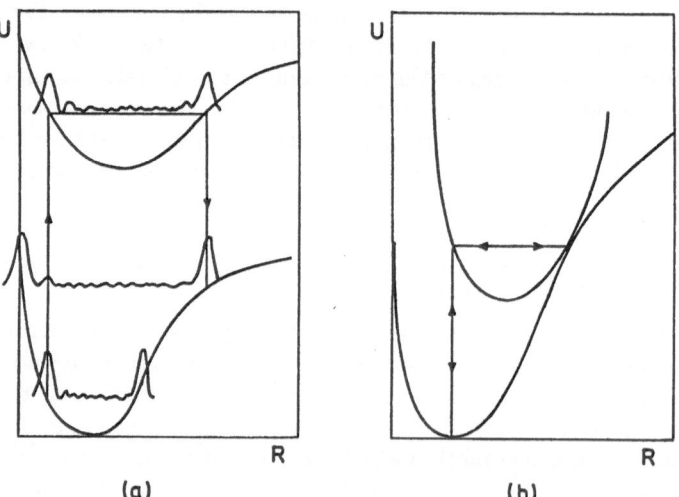

(a)   (b)

Fig. 19.1. Excitation of electronic term: (**a**) electronic transition induced by laser field, (**b**) excitation of electronic term due to nonadiabatic transition at laser excitation of vibrations

In accordance with the Franck-Condon principle the process of the excitation of electronic transitions in molecules under the action of a radiation field may be reduced to vertical transitions from the turning points. The emission of radiation by molecules (luminescence) also occurs due to the vertical transitions from the turning points, but this time due to downward ones. Owing to the asymmetry of the term shapes and the difference in the position of their minima, the transitions from the right and left turning points occur, as a rule, at different frequencies (Fig. 19.1 a). This implies that when exciting the electronic term by radiation at one of these frequencies (pumping), the inverse transition (luminescence) usually takes place at another frequency. If this frequency is less than that of pumping, the luminescence is referred to as Stokes luminescence; otherwise, anti-Stokes luminescence (Lecture 13) occurs.

Note that although the overlapping integral is maximum for the vibrational states with strictly coinciding turning points, it does not, however, vanish for the adjacent vibrational states. The measure of the transition force is the value of the overlapping integral whose modulus squared is referred to as the Franck-Condon factor. Therefore, the probability of the transition is the very probability of the electronic transition between the terms considered multiplied by the Franck-Condon factor. Of course, this approach applies in case the Franck-Condon factor does not vanish. Then the turning points are close to each other and the electronic wavefunctions may be considered uniform as we pass on from one turning point to another.

Everything said above refers to a diatomic molecule, i.e., to one-dimensional vibrational motion. For the polyatomics this motion is multidimensional. Considered classically, the trajectory of such multidimensional motion, evidently, has no turning points. Then the main contribution to the overlapping integral is rendered by the vicinities of the points, where the internuclear forces of the lower and upper terms coincide. While the molecule moves through the vicinities of these very points the frequency of the electronic transition depends only slightly on the coordinates of the nuclei. The point is as follows. The Franck-Condon principle states that the electronic transition may occur with a noticeable probability only if the coordinates and the momenta of the nuclei do not change as a result of the transition. Hence, for the electronic transition to have time to occur, the nuclei must reside for a certain time interval in a region of coordinate space in which their momenta in the lower and upper terms have had the possibility to coincide. This requirement is automatically met near the turning points. If there are no turning points, i.e., there are no points at which the nuclei come to rest in the course of vibrational motion, then, for the Franck-Condon principle to be satisfied, the trajectories with coinciding parts in the lower and upper terms should necessarily be present. In quantum terms the vicinities of such points represent the regions of the synchronism of the de Broglie waves corresponding to the vibrational motion.

Now let us consider the transition from one electronic term to another in the process of the vibrational motion. This is possible for terms possessing a specific structure: they should be nested in each other rather deeply and approach each other at some values of the internuclear distance (Fig. 19.1 b).

The excitation process is effective when the interterm distance becomes comparable to or even less than the vibrational quantum energy. It is clear that the adiabatic approximation does not apply in this case. Properly speaking, we mean the intersecting terms, i.e., the terms which, having different symmetries at fixed positions of the nuclei at some values of the internuclear distance, may possess the same energy. The nonadiabatic corrections, i.e., the allowance for the possibility of momentum transfer from nuclei to electrons or the transfer of angular momentum from the rotational motion of the molecule as a whole to vibrational angular momentum or to electronic spin motion leads to the interaction of different states corresponding to the terms considered. Note that when the terms are close to each other the interaction of their states arises in the presence of an external constant or varying fields as well.

Let us elucidate everything that has been said above with an example of the repulsion of electronic terms of different symmetry arising from the nonadiabatic interaction of molecular rotation with electronic angular momentum. Let $\widehat{\mathcal{R}}$ be the operator of the angular momentum of the rotation of a molecule as a whole, and $\widehat{L}$ be the operator of the electronic angular momentum. Let us pass on to the coordinate system rotating with the molecule. This system is noninertial and the Hamiltonian should be supplemented by the scalar product of $B\widehat{\mathcal{R}}/\hbar$ by $\widehat{L}\hbar$ (Lecture 15). If the matrix element of operator $\widehat{L}$ corresponding to the transitions between the states of different terms is nonzero, the insertion of product $B(\widehat{\mathcal{R}}\boldsymbol{L})$ into the Hamiltonian leads to the interaction of the terms corresponding to the states differing by unity in the value of the quantum number R. This implies that the rotation of the molecule as a whole may be slowed down at the expense of the acceleration of the rotation in the molecular electronic cloud in accordance with the law of conservation of angular momentum. If, in addition, the vibrational motion slightly interacts with both the electronic motion and the rotation, for the vibrational motion at the moment of crossing the region of intersecting terms, a situation quite similar to that previously discussed in Lecture 9 arises. This means that if we denote the amplitude of the probability for the molecule to be found on the lower electronic term by symbol $\psi_1(R,t)$ and that of its being found on the upper term by $\psi_2(R,t)$ and choose the time reference point at the moment of passing the point of intersection, then, assuming velocity v of the nuclear motion in the intersection region to be uniform, we obtain a system of Schrödinger equations for ψ_1 and ψ_2:

$$j\hbar\dot{\psi}_1 = F_1 vt\psi_1 + V\psi_2,\tag{19.3}$$

$$j\hbar\dot{\psi}_2 = F_2 vt\psi_2 + V\psi_1,\tag{19.4}$$

where F is the derivative of the potential energy over the coordinate taken at the point of intersection, $F_{1,2} = \partial U_{1,2}/\partial R$, while V is the interaction energy resulting from the additional term $B(\widehat{\mathcal{R}}\widehat{\boldsymbol{L}})$ in the Hamiltonian. System (19.3), (19.4) reduces to system (9.39), where $\alpha = v(F_2 - F_1)/\hbar$. From the solution of (9.39) it follows that the probability of the transition becomes of the order of unity when $V^2 \approx \hbar v(F_2 - F_1)$ (Lecture 9).

The discussion above refers to the transmission of rotational motion from nuclei to electrons. However, in exactly the same manner, one of the directions of the vibrational motion of the nuclei may play the role of the nonadiabatic interaction which ensures the Landau-Zener character of the transition between the terms for all the other vibrations. The same applies to an external field whose strength along with the dipole moment operator matrix element determine quantity V.

Note that the region of the intersection of electronic terms is rather sensitive to slight disturbances which, being taken into account, always lead to the effect discussed above, which is sometimes referred to as the repulsion of electronic terms. When the transition between the intersecting terms takes place due to the nonadiabaticity of the electron motion, one can speak of electron terms for all degrees of freedom except the one providing the nonadiabaticity of the transition. In addition, we can say that the case of the transition between the electronic terms under the action of an oscillating field may also be considered as a consequence of the terms repulsion effect, if the terms meant are the quasi-energetic, not energetic, ones.

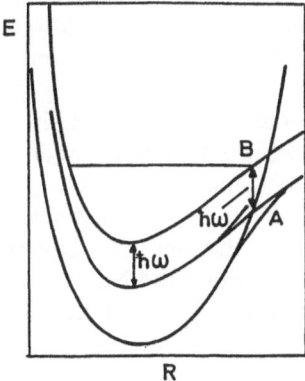

Fig. 19.2. Landau-Zener transition in the region of the intersection of terms

So, of special interest is the case of the electromagnetic field oscillating at a frequency close to that of transition. Figure 19.2 shows in bold solid curves two nested electronic terms of different symmetry which, therefore, may intersect. The thin line represents an excited electronic term corresponding to the field energy value diminished by one quantum or, if the field is considered classically, to the quasi-energy level with index minus one. Between this and the main term the Landau-Zener transition occurs due to the interaction with the field in the course of vibrational motion through the region of intersection. In other words, this process may be represented as the transition between the electronic terms from the turning point of term A into the turning point of term B under the action of the same radiation which causes the vibrational excitation.

The molecule being promoted to the upper electronic level may fall down to the lower level, starting from the left turning point, with the emission of a quantum of visible or UV radiation. The observation of the visible luminescence of small polyatomics caused by irradiation with resonant IR light has become the first experimental evidence of a strong vibrational excitation of these molecules under collisionless conditions.

In the physics of molecules there is a well-known process called intramolecular conversion. This process is, in fact, inverse to that considered above. The molecule, having absorbed a visible or UV radiation quantum in the transition between the left turning points, undergoes, upon arrival at the right turning point, the Landau-Zener transition induced by the nonadiabatic interaction in the region of term intersection (Fig. 19.1 b). As a result, we obtain a molecule which is highly excited vibrationally in the general electronic state, instead of a molecule excited electronically. Dealing with polyatomics one should, however, keep in mind that the points of intersection of the electronic terms in multidimensional configuration space (normal-coordinate space) are not passed by the molecule in each vibration but are achieved in the course of extensive wandering along the multidimensional potential surface. So, the time of intramolecular conversion in a polyatomic is longer than in the case of a diatomic molecule (compare with the process of dissociation, Lecture 17). The typical times of intramolecular conversion are to be found in the range of 10^{-12} s to 10^{-6} s.

The excitation of molecular electronic terms in single- and multi-quantum electronic transitions is being widely investigated and used in chemistry; it is the subject of so-called photochemistry. The absorption of a quantum (quanta) by a molecule considerably affects its chemical activity. Such a molecule, generally speaking, decays in a way distinct from, say, thermal heating (pyrolysis) with the formation of nonequilibrium radicals. Besides, the chemical reactions in which the electronically excited, but not fragmented, molecules may participate may also differ noticeably from those for molecules in the electronic ground state.

Note that if the change in the fragmentation character at electronic excitation is a purely molecular effect, the change in the chemical activity at excitation refers to atoms as well, where this phenomenon is more prominent. So, for instance, the excitation of atoms of noble gases causes their properties to become somewhat similar to those of the atoms of alkaline metals, rending them capable of participating in so-called harpoon reactions with halogen atoms. It is these reactions that are used for pumping the active medium of the excimer lasers.

The resonant absorption of photons and the distinction between the reactions thus obtained and thermal reactions are typical of photochemical processes. However, high radiation intensity is not essential for all of them. For the elementary process of classical photochemistry the energy of the quantum rather than the radiation intensity is of importance. When absorbed by a molecule, this energy should not thermalize until the molecule enters a chemical reaction which is distinct from a thermal reaction. The critical parameter here is the ratio between the rates of reaction and relaxation, which, being independent of the radiation intensity, is determined by the state the molecule is promoted to, by the cross section of the collisional relaxation, etc. However, the growth of the radiation intensity has two important consequences. First, the rate of the molecular excitation increases with intensity; second, the multiphoton processes become noticeable. In the latter case the irradiation brings about the population of high states corresponding, say, to the radiation of the vacuum UV range, which cannot be achieved using the traditional methods of classical photochemistry. The resonant excitation of such high states leads to an essential modification of the molecule's chemical properties. The rich spectrum of the electronic-vibrational states facilitates the multiphoton processes in molecules in comparison to atoms, due to a higher density of the intermediate resonances. In this way we may rather easily obtain molecular ions, including those with excited electronic states. This may be exemplified by the ionization of the CF_3I-molecule with the radiation of excimer KrF at a wavelength of 248 nm, which results in the formation of positive CF_3I^+- and CF^+-ions corresponding to the excitation of different electronic states of the molecule. Ion-yield dependencies on the radiation intensity are different for different types of ion; this points to the distinction of the photonity of the multiphoton process this ions are formed in.

The high excitation rate of the electronic-vibrational state of the molecule achieved at a high intensity of radiation resonant with respect to the corresponding transition is also important. The point is that in a strong laser field the excitation may, in principle, occur so fast that it becomes possible to neglect not only the thermal relaxation of electronic energy but the intramolecular conversion as well. Then the reactions between the molecules in the given electronic-vibrational states become possible. Of course, in this case the reactant pressures and the cross sections of the corresponding reactions are naturally assumed to be sufficiently large to ensure the rate of reactions exceeding the rate of the intramolecular conversion.

As we have mentioned many times, the electronic transitions in molecules are electronic-vibrational, and vibrational motion is essential. The point is that a change of the vibrational state causes a change in both the spectral dependence of the electronic-transition absorption probability and the consequences of these transitions. So, the combination of vibrational and electronic excitation (the IR–UV process) promises more diverse possibilities of acting on a medium. By an intense IR-laser field known to be capable of inducing a strong vibrational excitation of polyatomics and by the action of any factor causing an electronic transition (visible or UV radiation, electronic impact, the Pennings collision, etc.) a molecule may be promoted into a high excited vibrational state of one of the upper electronic terms. The ensuing destiny of the molecule, and how it is transformed depend on the specific vibrational state, on whether it corresponds to the molecular decay over a certain dissociation channel or to some type of internal conversion, and on whether this state is capable of persisting during an appreciable time period, decaying only via the radiative channel which corresponds to the transition to lower electronic terms accompanied by luminescence. Hence, by changing the initial state of the molecule we modify, generally speaking, the type of its further transformation. In other words, changing the intensity and the frequency of the IR-laser field, we may to some degree control the molecular transformation in the electronic transition. Effects of this kind have been observed in experiment.

As an example we may mention the CF_2Cl_2-molecule. This molecule has two vibrational frequencies falling into the operational range of the CO_2-laser. These frequencies correspond to the C–Cl and C–F bonds. By appropriately adjusting the radiation intensity on each of these frequencies we may excite the molecule to the same level of vibrational excitation on these two different frequencies. However, the vibrational wavefunctions corresponding to them will be essentially distinct. One of these functions corresponds to the localization of the excitation energy in the C–Cl bond while the other corresponds to that in the C–F bond. The probabilities of electronic transitions from these vibrationally excited states turn out to be different. This fact has been proved using the excitation of electronic transitions by electromagnetic radiation (KrF- and ArF-lasers) as well as electronic impact. Both electronic impact and the UV laser irradiation lead under these conditions to molecular decay. The fact that the distribution of fragments over the decay channels is dependent on the type of specific vibrations excited by the IR laser in the frame of the general electronic term was observed both in the form of the difference of the products of the chemical reactions these molecules may participate in, obtained in different cases and from direct observation with the use of the time-of-flight mass-spectrometer.

Let us emphasize that the combined action of IR and UV radiations permits us to expect at least in principle the selective fragmentation of molecules. Such a phenomenon is practically impossible in the case of vibrational excitation in the framework of a single electronic term due to the stochastization of the vibrational motion at energies close to the energy of the bond rupture (Lecture 17). In order to realize the selective break of the bond using the excitation to the highest electronic term it is necessary to swing the vibrations in the general term with the wavefunction overlapping in the best possible way with the wave package corresponding to the motion of the molecule into the dissociation valley. In this case the overlapping with the wave packages going into other dissociation valleys or directed onto the intramolecular conversion should be minimum. Finding such vibrational states for particular molecules presents a serious theoretical problem.

In the simplified form the selective IR–UV process being discussed may be visualized as follows. Let two normal vibrations directed along the x- and y-axes (Fig. 19.3) be possible in the potential well of the lower electronic term. In the upper electronic term corresponding to strong electronic excitation the motion along the x- and y-directions leads to decay over the differing channels. If the IR-laser action causes a strong swinging of the vibrations in the lower electronic term along, say, the x-axis, then, when the molecule is promoted to the upper electronic term, the momentum projection on the x-axis will

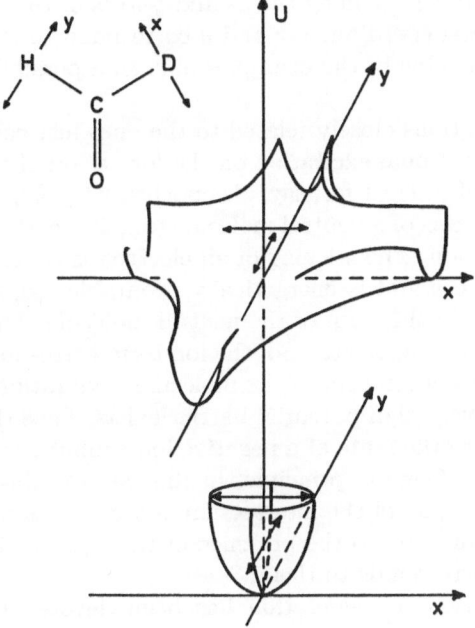

Fig. 19.3. The idea of selective breaking of the bonds in the IR–UV process

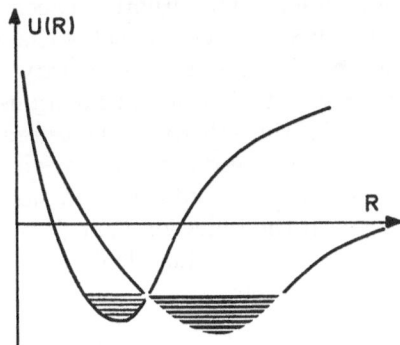

Fig. 19.4. General electronic terms of neutral molecule and free electron and of negative ion

be conserved in accord with the Franck-Condon principle. This determines the advantageous exit of the molecule into the dissociation valley oriented along the x-axis. In this case the highly excited electronic term corresponds to a rather loose molecule, the bottom of the potential well is fairly flat thus facilitating the consideration of the dissociation process.

The picture under consideration assumes the symmetry of the potential curves along the x- and y-directions along with the possibility of a separate excitation of the vibrations along the x- and y-axes of the main term. This is possible if the normal coordinates x and y correspond to the same vibrations of different isotopes. One of the examples of such a possibility is the HDCO-molecule.

Among the questions closely related to the ones just considered is that of the influence of vibrational excitation on the formation of negative molecular ions. This may be of interest for, say, plasmochemistry. Figure 19.4 shows the general electronic terms of a neutral molecule together with a free electron and that of its negative ion. The sticking of an electron is caused by the action of the polarization forces and is energetically profitable. So, the corresponding term is situated below the term of the neutral molecule. The electron usually gets onto the antibonding orbital. So, the ion term corresponds to large inter-nuclear distances at equilibrium. If the molecule is vibrationally unexcited, as a rule, a comparatively high potential barrier exists, since the transition from the neutral state into the state of a negative ion cannot occur because it does not obey the Franck-Condon principle. In the case of vibrational excitation, one may reach the region of the intersection of terms, where a Landau-Zener-type transition occurs due to the presence of the polarization forces and the electron sticking corresponds to them.

Our entire previous consideration has been devoted to the elementary processes of interaction between resonant laser radiation and atoms and molecules. However, in practice one rarely deals with the immediate results of elementary processes. One often studies them on the basis of the results of the ensuing chemical reactions. The situation is considerably complicated by the fact that the laser radiation may noticeably change the kinetics of chemical processes, not only by changing the elementary event of the reaction but at the expense of the reactant distribution functions which may turn out to be nonthermal or nonequilibrium-thermal, due to radiative-collisional cascade processes, for example. So, we need a consideration of chemical kinetics that would take into account the laser influence on both the elementary events and the reaction process in the radiation field. If we restrict ourselves to the consideration of binary reactions, a kinetic equation of the following form is usually sufficient for the purpose:

$$\dot{\mathcal{F}}_k(\Gamma_k) = \int \mathcal{L}_k(\Gamma_k, \Gamma_k')\mathcal{F}_k(\Gamma_k')d\Gamma_k' + \int I\sigma_k(\Gamma_k', \Gamma_k)\mathcal{F}_k(\Gamma_k')d\Gamma_k'$$

$$+ \sum_i \int W_{ik}(\Gamma_i, \Gamma_k, \Gamma_i', \Gamma_k')[\mathcal{F}_i(\Gamma_i')\mathcal{F}_k(\Gamma_k') - \mathcal{F}_i(\Gamma_i)\mathcal{F}_k(\Gamma_k)]$$

$$\times d\Gamma_i d\Gamma_k d\Gamma_i' d\Gamma_k' + \sum_{ij} \int W_{kij}(\Gamma_i, \Gamma_j, \Gamma_k)$$

$$\times [\mathcal{F}_i(\Gamma_i)\mathcal{F}_j(\Gamma_j) - \mathcal{F}_k(\Gamma_k)]d\Gamma_i d\Gamma_j , \tag{19.5}$$

where Γ_i is the set of phase variables of a classical motion and the quantum numbers of the quantized motions of the i-type particle.

The first term on the right-hand side of (19.5) represents the action of the Liouville operator $\mathcal{L}$ on the distribution function $\mathcal{F}$ and describes the internal molecular motion. If we exclude the nonadiabatic process of the intramolecular conversion from our consideration, the internal molecular motion turns out to be the fastest. It is this motion that forms the molecular spectrum. If we choose the quantum numbers of these adiabatic eigenenergy states and the integrals of classical motion as the quantum numbers and the phase variables Γ, operators $\mathcal{L}_k$ vanish up to the small nonadiabatic terms describing the intramolecular conversion.

The second term is accounted for by the interaction of the molecule with the radiation. It is written down assuming the incoherent character of this interaction. Together with the third term (compare with the Boltzmann collision integral and with (12.23)) the second one describes the radiative-collisional process of the change in the distribution function. The last term on the right-hand side of (19.5) corresponds to the chemical reaction in the course of which the binary collision of the particles of types i and j result in the formation of a particle of type k, and to the backward reaction. It is this term that leads to the fact that, in contrast to everything considered in the previous lectures, the distribution function $\mathcal{F}$ does not retain its normalization, being responsible for the change in the particle concentrations during the reactions. It is often convenient to represent the distribution function $\mathcal{F}$ in the form

$$\mathcal{F}_k(\Gamma_k) = c_k f_k(\Gamma_k) \,, \tag{19.6}$$

where c_k are the variable concentrations and f_k, the distribution functions of the particles over the phase volume normalized per unity.

In the case in which the rate of the change in concentration c_k is less than the rate of the relaxation of distribution f_k to its stationary or quasi-stationary form, the substitution of (19.6) into (19.5) leads to the well-known equations of chemical kinetics:

$$\dot{c}_k = \sum_{ij} K_{kij} c_i c_j - \sum_{ij} K_{ijk} c_k \,, \tag{19.7}$$

where the reaction rate constants

$$K_{kij} = \int W_{kij} f_i f_j d\Gamma_i d\Gamma_J d\Gamma_k \,, \tag{19.8}$$

$$K_{ijk} = \int W_{ijk} f_k d\Gamma_i d\Gamma_j d\Gamma_k \,, \tag{19.9}$$

are determined by the stationary (quasi-stationary) distribution functions obtained from the solution of (19.5).

In case the reactions are such that the rates of the establishment of concentration exceed the rates of the establishment of (quasi-)stationary distributions, as may be the case, for instance, for slow intramolecular conversion or for the slow V–T-relaxation cooling of the products of fast conversion, then the passage to the (19.7)-type equations by means of substituting (19.6) into (19.5) is impossible. This is the situation in the case of the classical photochemistry we have touched upon in this lecture.

Let us dwell on what we have said above in more detail. A fraction of molecules of a certain type transits into a higher energetic state as a result of the electronic and/or vibrational excitation. The density of the molecule–partners in the reaction under consideration and the cross section of this reaction are such that entering the reaction for the excited molecules is more preferable than relaxation. Then the situation occurs in which two ensembles of molecules – the highly and slightly excited ones – behave, with respect to the chemical reaction, as molecules of different types. It is convenient in this case to represent the distribution function as the sum of two terms:

$$\mathcal{F}_k = c'_k f'_k + c''_k f''_k \,. \tag{19.10}$$

The distribution functions of the excited f''_k- and unexcited f'_k-molecules enter this expression with concentrations c''_k and c'_k respectively. In this case it appears possible to obtain (19.7)-type equations for c'_k and c''_k. Instead of one concentration these equations contain two concentrations, c'_k and c''_k, changing slowly compared to the time of establishment of the corresponding distribution functions f'_k and f''_k. It is natural that in contrast to (19.7), equations for c'_k and c''_k should explicitly contain the kinetic coefficients describing the interaction with the field. The corresponding field terms of the chemical-kinetic equation may be obtained by the substitution of (19.10) into the second term on the right-hand side of (19.5).

If functions f'_k and f''_k may be considered Boltzmannian (19.5) gives the usual equations of chemical kinetics (19.7) with the only exception that each substance k is now characterized by two concentrations – those of its excited and unexcited components.

The latter remark refers to the case of photochemical reactions. The kinetic equations for these reactions contain the laser-radiation intensity explicitly. Nevertheless, the laser radiation may essentially affect the chemical reactions described by (19.7)-type equations as well. These equations do not contain the laser radiation intensity in explicit form. Such a situation is typical of the IR-laser action on the vibrational degrees of freedom of the molecule when laser irradiation noticeably changes the distribution functions of molecules and the reaction rates constants in accord with Eqs. (19.8) and (19.9). The distinction of the distribution function from the Boltzmannian one usually entails the violation of the detailed balance principle. This may qualitatively change the properties of the chemical-kinetic equations. It has the effect of shifting the chemical equilibrium state corresponding to the stationary solution of (19.7). As an example we may cite the reaction of the substitution of bromine by iodine in $(CF_3)_3CBr$ in the process of non-equilibrium IR-laser dissociation. This is practically impossible in the case of equilibrium heating due to the instability of the $(CF_3)_3C$ radical.

Laser Thermochemistry.
Processes on a Surface

Laser energy supply, its rate and resonant properties. Ambiguity, hysteresis, catastrophes. Back loops, periodic regimes. Stochastic attractors. Laser thermochemical pyrolysis and synthesis. Nontriviality of the products of laser thermochemistry. Thermochemistry on phase boundary surface, oxidation and combustion of metals in laser radiation field. Spatial instability. Cold wall thermochemistry. Physical adsorption. Van der Waals forces. Polarizational interaction of molecules in resonant laser field. Deepening the adsorption potential, hampering the diffusion through radiation-transparent structures with a developed surface.

In our consideration of the interaction of laser radiation with molecules, we have so far paid attention primarily to the ability of resonant radiation to input the energy immediately into the desired molecule and/or into the desired degree of freedom of the molecule leading to the appearance of long-living non-equilibrium distributions. However, in laser fields of reasonable intensity the relaxation processes do not always permit this nonequilibricity to develop fully. Nevertheless, even in this situation laser radiation may significantly affect the progress of chemical reactions. The point is that laser radiation provides the mixture of the reactants with an input of laser energy at a rate significantly exceeding that of chemical reactions. In the latter case the equations necessary to describe the processes under study may be obtained from the general Eq. (19.5) in the way that simpler Eqs. (18.14) and (18.15) were obtained from the more general Eq. (18.11). If we substitute the Boltzmann distribution with variable temperature and variable concentrations in (19.5) and bear in mind the self-consistency and normalization conditions, we obtain the following equations for temperature and concentrations:

$$c\dot{T} = \sum_i I(t)\sigma_i(\omega, T)n_i - \dot{Q},$$

$$\dot{n}_i = \sum_{j,k} K_{ijk}(T)n_j n_i - \sum_{j,k} K_{kij}(T)n_k,$$

(20.1)

where c is the heat capacity; $\dot{Q}$, the heat losses; and indices i, j and k specify the type of the molecule.

Before considering the kinetics of the chemical reactions described by system (20.1) we dwell upon some important features of the stationary solutions of this system. The second of Eqs. (20.1) at a fixed temperature has a unique stationary solution corresponding to chemical equilibrium. The values of the

stationary concentrations obtained therefrom determine the absorption coefficients $n_i\sigma_i$, which, in turn, determine the stationary temperature of the
system in accordance with the first of Eqs. (20.1). But this equation may, generally speaking, possess several stationary solutions at the same intensity $I(t)$.
So, for example, two stationary temperatures are possible if the cross section
increases with temperature in the case of laser radiation tuned in resonance
with the absorption line of the product of the pyrolysis. Low temperature corresponds to the case in which there is a small amount of absorbing substance,
and a small flux of absorbed energy is withdrawn from the system on account
of thermal conductivity with a small temperature gradient. High temperature
corresponds to the case when there is a great deal of absorbing substance and
a large flux of the absorbed energy is withdrawn at the expense of thermal
conductivity with a large temperature gradient.

Indeed, let us consider the first of Eqs. (20.1), say, under the assumption
of the Lorentz shape of the absorption line

$$\sigma_i(\omega, T) = \frac{\sigma_0 \kappa T}{(\omega - \omega_i)^2 + \kappa^2 T^2} \, .$$

The equilibrium concentration of the product of pyrolysis is determined by
the Boltzmann factor $n_i(T) = n_0 \exp(-E_i/kT)$. The heat flux depends on the
thermoconductivity λ and the cooler temperature T_r. In accord with Newton's
formula, $\dot{Q} = \lambda(T - T_r)$. Then the equilibrium temperature in the case of strict
resonance, $\omega = \omega_i$, is determined from the condition

$$\sigma_0 n_0 \frac{\exp(-E_i/kT)}{\kappa T} = \lambda \frac{T - T_r}{I} \, ,$$

the left- and right-hand sides of which are plotted in Fig. 20.1 versus temperature. At small I the slope in the dependence of the right-hand side is large
and the equation has only one root. At large I the slope is small and the
equation has three roots of which only two (1 and 3 in Fig. 20.1) correspond
to the stable equilibrium.

A set of parameters (intensity I and temperature T_r in our case) at which
the first of Eqs. (20.1) first reveals several stationary solutions is a peculiar
point which is called the "catastrophe critical point" or "catastrophe point".
This name originates from the fact that in this point is the beginning of the
region in which large variations of function may be caused by small changes
in arguments. The catastrophe points are not specific features peculiar to the
resonant interaction of radiation with matter; they have been known for a
long time. As an example, we may give the triple point of the phase transition
"vapor-liquid", where the parameters are known to be the temperature and
pressure.

In the case under consideration the set of quantities determining the catastrophe points includes the laser radiation intensity as an essential parameter.
This is a point of distinction between laser thermochemistry and traditional
thermochemistry, and is, in fact, a definition of the former. Note that for laser

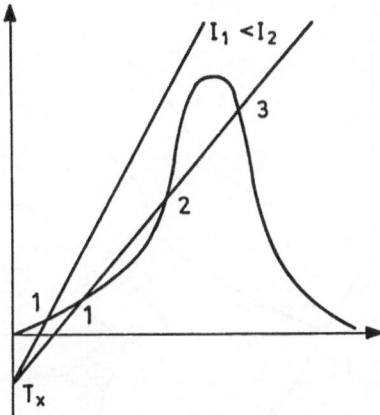

Fig. 20.1. Change in character of solution of equation for stationary temperatures under change in intensity of resonant irradiation

thermochemistry, as it follows from the previous discussion, not only a high rate of the energy supply but also its resonant character are essential.

Let us begin our consideration of thermochemical processes with the simplest case of the reversible decomposition of a substance according to the scheme

$$A \rightleftarrows B. \tag{20.2}$$

Examples of such a decomposition are the pyrolysis reactions

$$S_2O_6F_2 \rightleftarrows 2\,S_3F, \quad N_2O_4 \rightleftarrows 2\,NO_3. \tag{20.3}$$

It is convenient to show the qualitative peculiarities of reactions of this type using the standard ideas of the theory of catastrophes.

Let us consider the concentration of the products of pyrolysis as a function of the cooler temperature and the intensity of the radiation resonant with respect to the product (Fig. 20.2). In the absence of laser radiation ($I = 0$) the growth of the cooler temperature entails the growth of the reactants-mixture temperature which equals it. This, in turn, corresponds to an equilibrium increase in the concentrations of the product of pyrolysis. If the intensity of laser radiation is small, there are no qualitative changes in the picture when the reactant mixture is heated. One should bear in mind that the temperature of the reactant mixture turns out to be slightly greater than the cooler temperature. With a further increase of intensity the dependence of the product concentration on the cooler temperature changes qualitatively – a region of ambiguity appears in which there are two stationary solutions. One of them corresponds to the small concentration of the reaction product resonant with respect to the radiation, to the small radiation absorption and, hence, to the small difference between the temperatures of the reactant mixture and the cooler. Another solution gives the high concentration of the absorbing substance and, hence, the great difference between the temperatures of the mixture and the cooler.

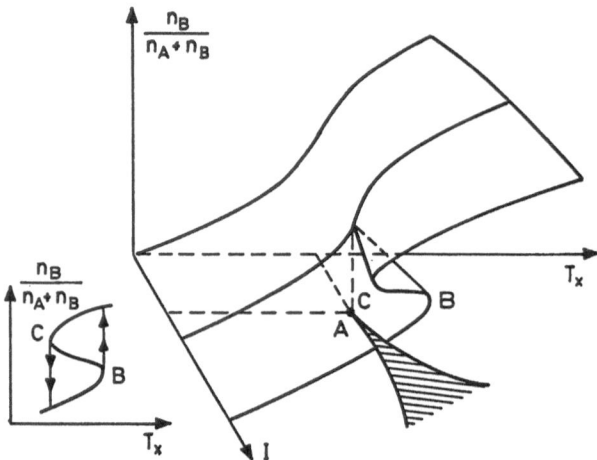

Fig. 20.2. Dependence of relative concentration of pyrolysis $A \rightleftarrows B$ product on cooler temperature T_c and laser intensity I. Peculiarity A is called critical point of frill-type catastrophe. Region of ambiguity of the concentration dependence on $I - T_c$ plane is hatched. Hysteresis in the dependence of $n_B/(n_A + n_B)$ on T_c is shown separately

The dependence shown in Fig. 20.2 allows us to trace the change in the reactant concentrations at a slow variation of the cooler temperature and the radiation intensity. The well-known hysteresis phenomena, usual in the case of ambiguous dependencies, are observed in this case. Indeed, if one fixes the radiation intensity above the catastrophe critical point and slowly increases the cooler temperature, the set of parameters determining the state of the mixture reaches point B, where the temperature derivative of the concentration becomes infinite and the point representing the change of the system parameters jumps to the upper sheet of the frill. This corresponds to the jump of both the concentration and the temperature of the reactant mixture. When the cooler temperature decreases, the inverse transition to the lower sheet of the frill takes place at point C. The process of the actual transition is essentially a nonstationary one. Its description should allow for the nonzero terms of the left-hand side of (20.1).

It should be noted that there are thermochemical reactions which reveal hysteresis behavior at a fixed cooler temperature under the change of the resonant laser irradiation intensity.

The subject of thermochemistry is far from being exhausted by a discussion of stationary problems. It further includes the consideration of the dynamics of the chemical reactions taking place under the action of time-variable and constant laser-radiation fields. In the simplest case a set of parameters determining the position of the catastrophe point should contain the irradiation duration along with its intensity. The resonant laser-thermochemical processes resemble in their main features well-known processes of metal hardening during thermal treatment. But, generally speaking, the situation is much more diverse, since the laser thermochemical system is not closed and dynamical behavior is its inherent fundamental property. This becomes essential when there are many reactants and reaction products.

The point is as follows. Conventional equations of the chemical kinetics for a closed system (the second of Eqs. (20.1)) always have a stable stationary solution corresponding to the chemical equilibrium. This property is unique for the large family of nonlinear differential first-order equations. In our case it is a consequence of the detailed balance condition and the relations between the rates of direct and backward reactions. Heating the reactant mixture with resonant laser radiation makes the system unclosed. Adding the thermal balance equation (the first of Eqs. (20.1)) to the system of chemical-kinetics equations may lead to the loss of stability of the stationary solutions. Then the nonstationary solutions become essential.

When the number of reactants is small one may observe the known periodic solutions which are represented by limiting cycles in the phase space. In other words, the periodic regime is established in the system irrespective of the initial conditions. In this regime the time dependence of the concentration is cyclic. Usually the energy absorption and the heating of the reactant mixture dominate during the first phase of the cycle, while in another phase the absorption is less than the energy losses, and the temperature of the mixture decreases. For illustration we may use the previously considered example of the pyrolysis reaction $A \rightleftarrows B$, with the only distinction that the laser radiation is absorbed by initial substance A instead of product B. In addition, let the heat capacity of the cooler be finite so that its temperature increases with the heat flow. Then Fig. 20.3 shows how the representing point motion "winds" onto the hysteresis curve of Fig. 20.2, thus testifying to the existence of a limiting cycle. The motion in the limiting cycle has an explicit phase of the system heating (the lower sheet of the frill) and the cooling phase (the upper sheet of the frill) when the amount of absorbing substance is small.

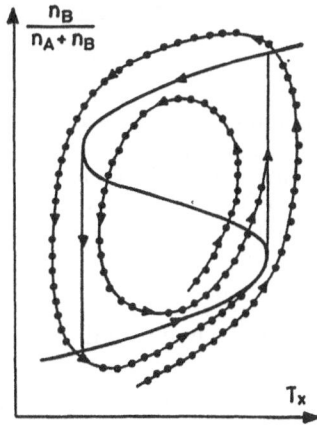

Fig. 20.3. Limiting cycle under the resonant energy input in the decomposing substance at laser pyrolysis with scheme $A \rightleftarrows B$

Along with limiting points (the stable stationary solutions) the limiting cycles are the simplest limiting sets toward which the phase trajectory of non-linear differential equations tends. In multidimensional problems, the topology of the limiting sets is much more involved. They cover appreciable domains of the phase space and possess a rather complex geometric structure. So, the phase trajectories tending toward such limiting sets are whimsical, irregular and, in fact, random. Such limiting sets are called stochastic attractors.

From the mathematical standpoint the most fully investigated attractor is the so-called Lorentz strange attractor. This is a three-dimensional stochastic attractor of three nonlinear coupled differential equations of the first order. The typical form of the phase trajectory tending toward the Lorentz strange attractor is given in Fig. 20.4. For more details we recommend "Introduction into the Theory of Vibrations and Waves" by M. I. Rabinovitch and D. I. Trubetskov.

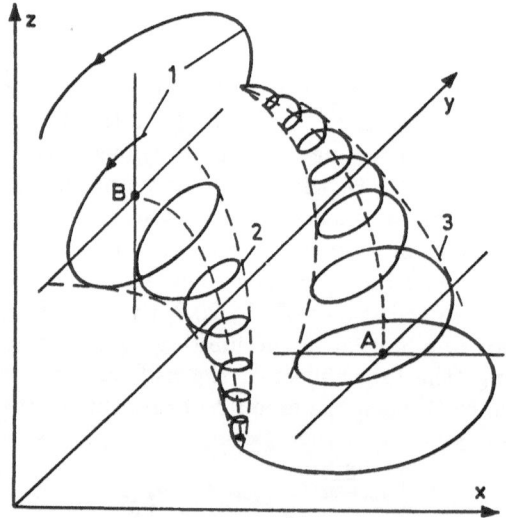

Fig. 20.4. The Lorentz stochastic attractor. Point A is stable in plane xy and unstable along the z-axis; point B is stable in plane yz and unstable along the x-direction. The trajectory of representing point 1 enters the region of attraction B, moves along the surface through funnel 2, passes onto plane xy, where, being trapped by the attraction region of point A, returns through funnel 3 onto plane yz but this time into another point of attraction-region B

Here we shall only point to its main qualitative features. The stochastic Lorentz attractor is the result of the interaction of two vortices in three-dimensional space. One vortex corresponds to the motion trajectories stable in the xy-plane and unstable in the z-direction; another vortex is stable in the yz-plane and unstable along the x-axis. Both vortices resemble the funnels formed when a liquid flows out of a vessel with the only distinction that the "stem" of one vortex runs into the "cap" of another and vice versa. Such an interaction of two vortices leads to stochastic instability of the phase flows.

In the case of strongly interacting systems of large dimensionality the stochastic attractors occupy practically all the volume of the phase space. So, if laser radiation acting on a complex chemical system removes it from the equilibrium position (from the stable stationary state), then the system, as a rule, enters the attraction region of the stochastic attractor. This means, that the concentrations of all substances, that can be obtained in the chemical reactions, turn out to have the same order-of-magnitude values irregularly changing in time. This is why laser thermochemical reactions allow the production of substances which cannot be formed in regular thermal reactions. An example is the production of ammonia from a mixture of oxygen, nitrogen and water vapor under irradiation with CO_2-laser light which is resonantly absorbed by the H_2O-molecules. This process has a certain threshold value of laser radiation intensity. The chemical scheme of such a process is extremely complex and includes about thirty equations. However, the main reactions are the first ones describing the bonding of atmospheric nitrogen:

$$H_2O + \hbar\omega_{CO_2} \rightarrow H + OH, \quad OH + OH \rightarrow H_2O_2,$$
$$N_2 + 2\,H_2O_2 \rightarrow N_2O_4 + 2H_2, \quad N_2O_4 \rightarrow 2\,NO_2, \text{ and others.} \tag{20.4}$$

In the previous consideration the back loops leading to a loss of stability of the stationary solutions of chemical-kinetic equations were introduced into thermochemistry via the thermal balance equation on account of the time (frequency, i.e., resonant over the absorption frequency) properties of the radiation which brings the energy into the system. The spatial structure of the field was ignored. However, it may also lead to a loss of stability.

This fact finds its most striking manifestation in the thermochemical processes on the surface when the thickness of the substance layer with the changed constants becomes comparable with the wavelength. In this process the resonant character of interaction manifests itself not via the correspondence of the radiation frequency to the frequency of some transition but via the correspondence of the radiation wavelength to the optical thickness of the film formed on the surface.

A typical example is the oxidation of metals, say, of copper or tungsten, in the air under the action of the CO_2-laser radiation normally incident on the interface. The heating of metals leads to the formation of the oxide film. The optical constants of the oxide differ from those of the air. So, the film, in fact, represents a Fabry-Perot resonator. The transparency of this resonator varies with the film thickness, reaching its maximum value at $d = \lambda/4$, where λ is the radiation wavelength in the substance. The kinetics of this process is rather complex. It is determined by the rates of the chemical reaction and the diffusion of the metal (or oxygen) atoms through the film substance, by the thermal conductivity, by the laser intensity, etc. In the experiment, depending on the parameters used, one may observe all the peculiarities of the behavior typical of laser thermochemistry – jumps of the stationary states (catastrophes), periodic regimes (limiting cycles), and chaos (stochastic attractors).

Indeed, with good heat conductivity and a high reaction rate the film growth stops at some fixed thickness at which the absorption becomes equal to the heat withdrawal. If the heat withdrawal is poor and the chemical reaction does not develops too fast, the film growth persists inertially after passing the absorption maximum that leads to a decrease of absorption and an increase of reflection. If the system has enough time to get cold before reaching the next turn of the absorption growth, the oxidation process may be replaced by reduction, leading to a decrease in film thickness. Reduction, in turn, is superseded by oxidation. As a result, the periodic regime is established. If cooling is not sufficient, the oxidation reaction after reaching maximum absorption develops explosively. Naturally, processes of this kind are very complicated if the formation of different oxides (protoxide, oxide, peroxide, etc.), including the metastable ones, is possible.

Let us note the evident importance of the above-described resonant thermochemical processes for the practical needs of laser technology.

Surface laser thermochemistry is only one of the possible examples of the resonant laser action on the interface between two media. The range of phenomena taking place under the action of the resonant laser radiation under essentially heterophase conditions, say, at the interface between a gas and a solid, is significantly wider.

The existence of a boundary between phases with different properties (heat capacity, thermoconductivity, density, structure, etc.) allows the non-equilibricity of the processes developing under intense resonant laser irradiation to manifest itself as spatial nonequilibricity. Such nonequilibricity may be realized in many ways, e.g. by temperature discontinuity on the phase boundary, nonequilibricity arising near the surface at the collisionless dissociation of the gas molecules in its vicinity, nonequlibricity obtained in the course of radiative-collisional cascade excitation of the gas near the surface, etc.

The sharply developed spatial nonequilibricity which leaves the wall cold permits the implementation of processes of a different chemical nature, chemical adsorption among them. This process makes it possible to cover the cold wall with thin layers, and to produce epitaxial and other crystalline or amorphous films. It also permits chemical etching of a surface by the products of laser thermo- or photochemical reactions or by the substances produced during laser synthesis of nonequilibrium distributions.

We may exemplify these applications by the growth of amorphous-silicon films in the process of the collisionless dissociation of SiF_4- and SiH_4-molecules under the action of the radiation of pulsed CO_2-lasers and by the etching of germanium by the products of the pyrolysis of CF_3I- and CF_3Br-molecules which are resonantly heated by the radiation of continuous-wave CO_2-lasers. In the former case the radiation is focused into the gaseous medium near the glass plate on which the amorphous silicon was precipitated. In the latter case the radiation is transmitted to the gaseous medium via an etched germanium sample, transparent in this range of wavelengths.

We have mentioned the chemical processes developing on the phase boundary under the action of resonant laser radiation. The existence of a wall for these processes is not principal. It only fixes the changes in the gaseous phase caused by laser irradiation. These changes were considered in detail in this and the previous lectures.

However, there exists an interaction of molecules (atoms) with the wall which does not involve any chemical changes. This is physical adsorption. The typical interaction energy of a molecule with the wall in this case is significantly lower than in the case of chemical adsorption. So, the elementary act of physical adsorption may turn out to be sensitive to laser light. Let us consider this in more detail.

Physical adsorption is caused by the Van der Waals forces. In classical physics these forces are explained by the redistribution of charges under the rapprochement of interacting molecules. But we need a quantum description since we are interested in the action of resonant laser light on physical adsorption. In the quantum-mechanical treatment of the problem in the simplest case of the Van der Waals interaction of two approaching two-level molecules we have to use the scheme of the levels of a compound system composed of these molecules (Fig. 12.1). However, in finding the Van der Waals forces, we should use the full expression for the dipole-dipole interaction and retain the second-order terms of the perturbation-theory series in contrast to the case of vibrational exchange, where it turns out that it is sufficient to take into account the dipole-dipole interaction retaining only the first term of the perturbation series; so, we may restrict ourselves to item (12.1). In fact, we should allow for the repulsion of levels (0,0) and (1,1) due to the interaction

$$\widehat{V}_r = (\mu_1^+ \mu_2^+ + \mu_1^- \mu_2^-)/R^3(t)\,, \tag{20.5}$$

which was not considered in Lecture 12. The value of this repulsion is

$$V_{vdW} = V_r^2/2\hbar\omega \propto \mu^4/R^6\,, \tag{20.6}$$

which corresponds to the well-known Van der Waals correction.

The quasi-energy method is a convenient tool for the description of the intermolecular interaction in the oscillating electromagnetic field. The scheme of the quasi-energy levels for two two-level molecules is given in Fig. 20.5, where we choose the radiation quantum $\hbar\omega$ close to the frequency of transition $0 \to 1$. This figure shows that along with the repulsion of levels $|1,1,0\rangle$ and $|0,0,0\rangle$, the interaction of the near-resonant levels $|0,0,0\rangle$, $|0,1,-1\rangle$, $1,0,-1\rangle$ and $1,1,-2\rangle$ appears. Besides, resonant levels $|0,1,-1\rangle$ and $|1,0,-1\rangle$ are coupled via the V–V-exchange (see (12.1) and Fig. 12.1). Due to the resonance, levels $|0,1,-1\rangle$, $|1,0,-1\rangle$, $|1,1,-2\rangle$ are closer to level $|0,0,0\rangle$ compared to level $|1,1,0\rangle$ and, so, they may displace it more appreciably. The dependence on distance in this case turns out to be the same as for the vibrational-vibrational interaction V_{VV}, i.e., proportional to $1/R^3$.

In fact, this is the nature of the influence of resonant laser irradiation on the Van der Waals interaction. Of course, to obtain an estimate we should perform multiple averaging that allows for the orientation and rotation of molecules, the experiment geometry, etc.

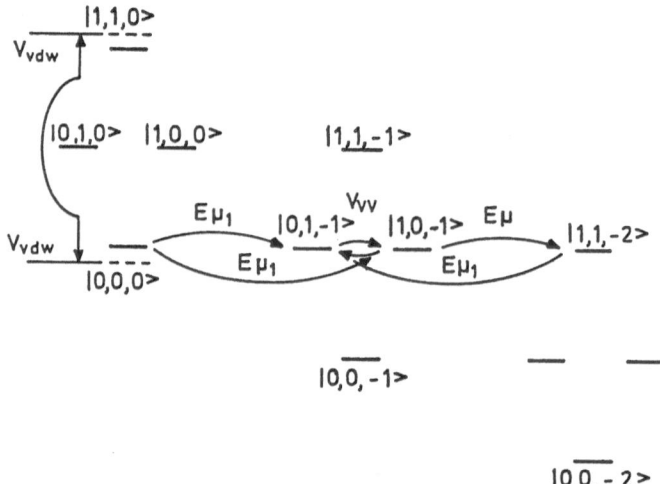

Fig. 20.5. Scheme of quasi-energy levels of two approaching two-level molecules in near-resonant external field: V_{VV}, vibrational interaction; V_{Vdw}, Van der Waals interaction, which results in displacement of levels to positions denoted by dotted line; $E\mu$, energy of interaction with radiation

As we can see, in the terms of the quasi-energy method the appearance of the intermolecular forces in the resonant field involves a V–V-exchange process which does not result in the dissipation of the radiation field energy. If we consider both the molecular system and the radiation field quantum-mechanically, the process of the modification of the intermolecular interaction by the field is as follows. Having absorbed the radiation quantum, one molecule of the interacting pair transits to the upper level and acquires the momentum of the absorbed photon. Then, in the process of the V–V-exchange, i.e., in the process of the virtual emission and absorption of another photon distinct from the laser photon, the excitation of the first molecule is transferred to another molecule with the corresponding change in the momenta of both molecules leading to their attraction (or repulsion). The second molecule returns the photon to the laser radiation field in the induced transition. Therefore, ideally, the change in the intermolecular interaction occurs without any loss of laser radiation quanta.

The procedure mentioned above of averaging in the isotropic space considerably masks the extent of the effect of the change in the intermolecular interaction potential caused by the resonant radiation field, leading in second-order perturbation theory to a slight change of the chemical potential: as a rule, much less than kT. Near the surface the situation is significantly different. The reason is the "image-force" phenomenon well-known in electrodynamics: a metal (or dielectric) surface creates the mirror-image of a dipole near the surface, which is responsible for the electrostatic interaction of the dipole with the surface. The same effect gives the layer of adsorbed, and thus spatially fixed by the surface, molecules of the same sort as in the gaseous phase. In such a situation the averaging does not eliminate the linear-over-perturbation effect.

Let us mention that if there is a layer of adsorbed molecules on the surface, still another mechanism of the change in the adsorption potential exists. The point is as follows. The molecules adsorbed on the surface and the molecules in the gaseous phase, placed in a resonant laser radiation field, may have a vibrational temperature considerably exceeding both the temperature of the translational motion along the surface and the adsorption potential. This implies that at the surface we are dealing with a difference between the vibrational and translational temperatures completely analogous to that considered in Lecture 18. The vibrational temperature exceeds the translational one. One of the reasons for this may be that the vibrational frequencies of the molecule significantly exceed those of the phonons in the condensed medium, and, hence, the relaxation of the molecular vibrational energy into the wall requires multiphonon transitions which have a small probability. At the same time the molecular translational motion effectively interacts with the phonon reservoir of the condensed medium and, hence, is effectively thermalized. The vibrational excitation loosening the molecule makes its dipole moment increase, thus increasing the usual Van der Waals interaction (see (20.6) and Lecture 15).

Experimental observation of the change in the adsorption potential under the action of resonant laser radiation may be conveniently performed in systems with a developed surface $(S^{1/2} \gg V^{1/3})$, where this relatively small effect may be detected owing to the accumulation phenomenon. The point is that the change in the adsorption potential considerably (exponentially, due to the Boltzmann dependence) changes the time the molecule spends on the surface. This, in turn, significantly affects the rate of gas diffusion in the system with the developed surface. The deepening of the potential causes a decrease in the diffusion coefficient, and a decrease of the adsorption potential makes it increase.

Resonant laser control of gas flows through systems with a developed surface by means of changing the adsorption potential is a process which takes place under conditions of local thermodynamic equilibrium. However, the radiation field may affect the kinetic properties of the molecules. This should manifest itself in thermodynamically nonequilibrium processes. If the excited and unexcited molecules interact differently with a wall that is either clean or covered with a layer of adsorbed molecules of the same type, we may expect phenomena analogous to the light-induced drift considered in Lecture 8. In principle, near the wall it is possible to create microscopic non-equilibricity exciting a certain Doppler component of the gas absorption line by the laser field. This will lead to the appearance of gas flows directed against or along the laser radiation propagation, depending on the sign of the radiation frequency detuning with respect to the center of the Doppler line and on the sign of the change in the corresponding transport kinetic coefficient under excitation.

Lecture Twenty-one

Self-Influence of Laser Radiation
in Intense Resonant Interactions

Polarization and susceptibility. Self-influence at collisionless multiphoton excitation of molecules. Shift of dispersion curve. Pulse shortening. Self-focusing, self-defocusing, waveguide modes. Wave front inversion in four-wave interaction. Inverting mirror. Dynamic holography. Delayed self-influence at thermochemical interactions.

Intense radiation is well known to act strongly on the resonant medium changing its population distribution. The change in the population distribution, in its turn, causes an appreciable change in the microscopic properties of the medium, the field-induced polarization among them. In the case of optically rather thick media such a change in polarization leads to the effects of radiation self-influence. The investigation of the properties of radiation in media with a pronounced self-influence is the subject of nonlinear optics. Let us note that nonlinear optics also deals with parametric effects like the generation of harmonics, the sum and difference frequencies, etc., whose quantum nature is considered in Lecture 13. Medium polarization is usually described in nonlinear optics without the microscopic consideration of the mechanisms of its formation. The description is, as a rule, based on phenomenological coefficient like the quadratic or cubic susceptibility, etc. However, in this lecture we shall particularly dwell on the mechanisms of the formation of nonlinear polarizability on the field frequency. The well-known effects of nonlinear optics related to such a consideration are self-focusing (or self-refraction, in a more general case) and restoration of the wave front under the four-wave interaction. They become naturally apparent in the resonant interaction of intense laser radiation with matter.

In Lecture 3 we considered one of the coherent self-influence processes – self-induced transparency, where the effect is determined by the area under the pulse rather than simply by the field amplitude. Such a situation is somewhat exotic and is not observed even for practically reversible systems (we mean the systems without irreversible damping), if they are complex enough. In a multilevel system with a complex spectrum the effect of coherent damping considerably simplifies the overall picture reducing it to the already-known linear dependence of the polarization on the field in case the field spectral width is much narrower than the spectrum of the transition frequencies. The proportionality factor is in this case determined by the distribution of populations among the levels. The latter, naturally, may depend on both the instantaneous field value and all the past field values (a system with memory). However, it is important that the instantaneous values of the polarization and the field strength are linearly coupled with each other:

$$P = 4\pi\chi E. \tag{21.1}$$

This physically rather obvious relation may be proved strictly with the help of the formalism developed in Lecture 11, when analyzing the excitation dynamics of spectrally complex multilevel systems. Then, summing up the diagram series for susceptibility χ we obtain

$$\chi \propto \sum_{k,n} \langle (\mu_n^k)^2 \rangle (x_n(\varepsilon) + x_n(\xi)) \rho_k. \tag{21.2}$$

This expression allows us to reduce the Kramers-Kronig relations (see (5.40)) which for a multilevel system with coherent damping, may be obtained without using the considerations associated with the irreversibility and the causality principle. Therefore, the problem is reduced to determining the instantaneous distribution of populations among the levels ρ_k.

Many processes may influence the formation of this distribution. Most of them have been considered in the previous lectures. Here one can think of three possibilities. The first is the processes of the pulsed action on the medium in which the population distribution at some moment in time turns out to be dependent on the flux of the radiation that has passed up to this moment through the medium. The processes of the collisionless excitation of polyatomics as well as the collisional excitation processes in the presence of the V–V-exchange in the absence of V–T- relaxation are examples of this case. The second possibility is the processes in which the quasi-stationary distribution of populations determined by the instantaneous intensity is established. This is the case when the rate of the distribution formation (in molecules this is usually the V–T-relaxation rate) is greater than the inverse typical time of the intensity variations. Such a situation is realized when resonant radiation passes through gases at rather high pressures. The third possibility is associated with the existence of the processes slow compared to the typical time of the intensity variation leading to the retardation of a system's response to laser radiation absorption. This is usually observed during laser stimulation of slow chemical reactions (laser thermochemistry (see Lecture 20)). In this situation the process may remain nonstationary even in a powerful laser field (limiting cycle, stochastization).

Let us dwell in more detail on the mechanism of nonlinearity formation and on the non-linear-optic consequences resulting from each of the three cases mentioned here. For a start, let us consider the collisionless multiphoton excitation of a small polyatomic. The spectrum of its vibrational states (Lecture 15) has a band structure. In this case the narrow vicinities of resonances become populated in the laser excitation process(see Fig. 21.1 and Lecture 11). In order to determine the real and imaginary parts of the susceptibility with the help of the Kramers-Kronig relations it is necessary to know the full spectral dependence of the probabilities of the quantum absorption and emission of photons by such a molecule. This spectral dependence is formed as follows. From each resonance vicinity at some definite frequency not equal to that of the excited field, transitions are possible, both up and down, to the unoccupied levels. The difference between the up- and down-transition probabilities determines the absorption cross section of the molecule located in a given zone and in a given narrow vicinity of the resonance. The averaging of this quantity over the starting states, i.e., over the resonance vicinities in different bands, gives the absorption cross section at a given frequency. The dependence of the cross section thus defined on the frequency is smooth and has no peculiarities near the frequency of the exciting (strong) field. In the classical limit it may be described by formula (15.20). Such an absorption line form ($\chi''(\omega) \propto \sigma(\omega)$) in accord with the Kramers-Kronig relations, unambiguously corresponds to the dispersion dependence $\chi'(\omega)$. In the process of molecular excitation the absorption line (as we already know from Lecture 16) is usually shifted to the red region of the spectrum. The dispersion curve is shifted accordingly. So, at a given frequency the absorption coefficient and the refraction index of the gas of such molecules changes with the extent of their excitation.

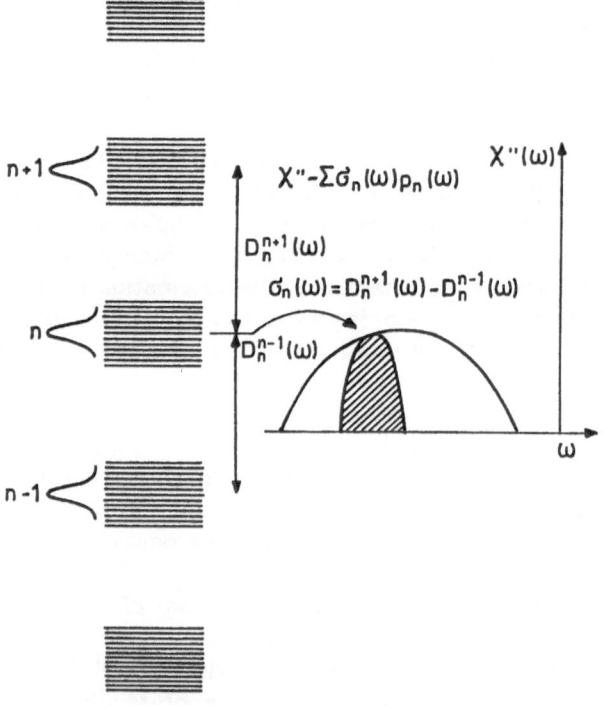

Fig. 21.1. Formation of absorption line of molecules excited by laser radiation. On plot of dependence $\chi''(\omega)$, hatched region corresponds to contribution of nth band in net absorption coefficient

Let us consider the following didactic example as an illustration. Let the molecular excitation be brought about by the radiation whose frequency lies in the region of longer waves with respect to the center of the unexcited molecular absorption line. As the excitation takes place, the absorption line is displaced to the red region of the spectrum and superposes its center on the frequency of the exciting field. Then, the absorption increases and the radiation does not pass through the gas-filled cell if, naturally, it may be considered optically thick. As a result, in the process of the pulsed action the shortening of the impulse of the radiation passing through the cell becomes possible. Of course, such a shortening of the pulse at the expense of the absorption of its tail is possible when its energy per unit area is not too large. A pulse of a larger energy causes a still greater displacement of the absorption line and brings its center below the exciting radiation frequency. As a result, the medium becomes transparent and the effects associated with the change of the refraction index may be observed.

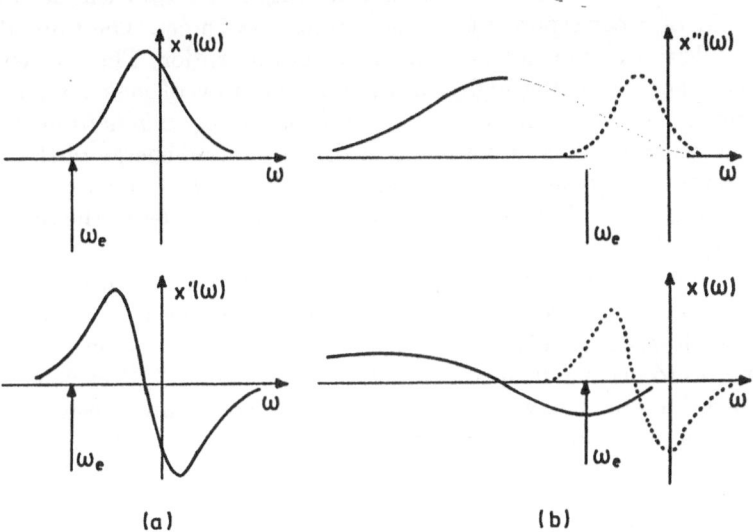

Fig. 21.2. Absorption and dispersion curves of slightly (a) and strongly (b) excited molecules. For comparison Fig. b shows Fig. a curves in dotted line

Figure 21.2 shows the change of the envelope of the refraction index $\chi'(\omega)$ resulting from the broadening and displacement of the absorption line. Depending on the position of the laser frequency ω_l with respect to the initial absorption line, both an increase and decrease in the refraction index may be observed with an increase of the irradiation energy density leading to self-focusing and self-defocusing, respectively.

Of course, there are more involved processes other than self-focusing and self-defocusing. If the radiation frequency is on the red side of the absorption line, then at a fixed point in space the growth of the refraction index may be superseded with time by its decrease in accordance with the increase of the past energy. In the case of laser beams with nonuniform transverse intensity distribution this may lead to the formation of complex profiles of the radiation distribution variable in time. Let us note the fact known from electrodynamics that specific waves refered to as waveguide modes may propagate along the extensive spatial nonuniformities of the refraction index. The radiation falling into the waveguide modes is trapped there and affects the substance refraction index in the place of its localization thus leading to the complex self-consistent time dependence of the spatial distribution of the electromagnetic field in the resonant medium.

The numerical estimate may be obtained in the following simple way. The typical polarizabilities χ' and χ'' comprise $\chi' \sim \chi'' \sim N\mu^2/\hbar\Delta\omega$, where N is the particle density and $\Delta\omega$, the typical width of the absorption line (the width of the region of the anomalous dispersion). At $\Delta\omega = 30$ cm^{-1}, $\mu = 0.3$ D and $N = 3 \cdot 10^{16}$ cm^{-3} (1 torr) the resonant addition to the refraction index comprises $4 \cdot 10^{-5}$. Then at a wavelength of $10\,\mu$m the accumulated phase shift of π corresponds to a distance of $z \approx 25$ cm. The same distance corresponds to the typical length of the linear absorption. The estimate given above corresponds to the high vibrational excitation of molecules when the spectrum of their linear absorption $\chi''(\omega)$ considerably differs from the initial one. This usually occurs at a density of the irradiation energy of the order of several tenths of a joule per square centimeter, i.e., when, in accord with Lecture 17, the molecule is excited to the energies of the order of the dissociation threshold.

With the increase of pressure the typical length at which the changes in the refraction index become apparent falls down reaching 1 cm at the pressure of 10 torr. However, in this case the collisional processes become essential. The first process to become involved is that of the resonant V–V-exchange, which does not bring about any principal changes in the picture considered above. The frequency dependencies $\chi'(\omega)$ and $\chi''(\omega)$ are, as previously, determined by the energy that passed through the system, though, perhaps, with some delay of the order of the V–V-exchange time. At large pressures (several dozen torr) the V–T-relaxation processes become essential at durations of laser pulses of several hundreds of nanoseconds. In this case at each time moment in each point of space the distribution of populations corresponding to the instantaneous value of the radiation intensity is established. Then both coefficients χ' and χ'' are proportional to the intensity, and adequate treatment of the situation corresponds to the usual problem formulation in nonlinear optics (Kerr uninertial nonlinearity). The estimate of the required intensity may be performed using formula (18.41) which, at a pressure of several dozens torr, produces the intensity value of several dozen watts per square centimeter.

The self-focusing and self-defocusing processes under conditions in which the refraction index depends on the radiation intensity are of common knowledge. Regardless of the nonlinearity formation mechanism they are considered in detail in courses on nonlinear optics and we shall not dwell on them here. We only note that in a gas that absorbs the radiation resonantly, both self-focusing and self-defocusing may be observed, depending on the position of the radiation frequency with respect to the center of the absorption line. There is no such dependence in the case of Kerr self-focusing.

Nonetheless, let us dwell at length on one of the many processes of nonlinear self-action, the so-called phenomenon of wave front inversion in the four-wave interaction in particular. Our interest in this problem is to be explained by the fact that this is a realization of the wave front inversion phenomenon which is the subject of extensive studies in quantum electronics, since it allows compensation for the phase inhomogeneities of active media.

The simplest account of this process may be given by the preset-field approximation. Let us assume that two coherent waves fall onto the nonlinear medium. One of them is the plane wave,

$$\mathcal{E}_1(x, t) = \mathcal{E}_1 \cos(\omega t - \boldsymbol{kx}),$$

while the other is the set of spatial harmonics with spectral amplitudes $\mathcal{E}_2(\boldsymbol{k'})$,

$$\mathcal{E}_2(x, t) = \int \mathcal{E}_2(\boldsymbol{k'}) \cos(\omega t - \boldsymbol{k'x}) dk'.$$

We are interested in the refraction index at frequency ω which is known to be determined by the population distribution. The latter, in turn, is determined not by the instantaneous intensity of the resulting field $\mathcal{E} = \mathcal{E}_1 + \mathcal{E}_2$ but by the intensity $\langle \mathcal{E}^2 \rangle_{TVT}$ averaged over the time period of the order of the V-T-relaxation time τ_{VT}. If the refraction index is proportional to the intensity thus defined (this is the simplest case), then after averaging we obtain

$$n \propto \mathcal{E}_1 \int \mathcal{E}_2(\boldsymbol{k'}) \cos(\boldsymbol{k'} - \boldsymbol{k}) \boldsymbol{x} d^3 k'. \tag{21.3}$$

In other words, the refraction index turns out to be spatially modulated in accord with the interference pattern of fields $\mathcal{E}_1(x, t)$ and $\mathcal{E}_2(x, t)$.

If we illuminate the medium thus prepared with the nonuniform refraction index by the weak plane wave of the same frequency ω with the wavevector opposite to that of the first wave, $E_3 = \mathcal{E}_3 \cos(\omega t + \boldsymbol{kx})$, then the running wave of the polarization $\mathcal{P}(x, t)$ arising under the action of the running field wave $\mathcal{E}_3(x, t)$ has not a single spatial harmonic, as in the case of the uniform medium but represents the set of the spatial harmonics,

$$n(x)\mathcal{E}_3(\boldsymbol{x}, t) = \mathcal{P}(\boldsymbol{x}, t) = \int \mathcal{P}(\boldsymbol{k''}) \cos(\omega t - \boldsymbol{k''x}) d^3 k'', \tag{21.4}$$

where, evidently,

$$\mathcal{P}(\boldsymbol{k''}) = \int \mathcal{P}(\boldsymbol{x}) \cos(\omega t - \boldsymbol{k''x}) d^3 x. \tag{21.5}$$

Note now that the amplitude of the polarization harmonic $\mathcal{P}(k)$ is related to the field harmonic $\mathcal{E}_2(k)$ by the following expression:

$$\mathcal{P}(\boldsymbol{k}'') \propto \mathcal{E}_1 \mathcal{E}_3 \int \mathcal{E}_2(\boldsymbol{k}') \cos(\boldsymbol{k}'' + \boldsymbol{k})\boldsymbol{x} \cos(\boldsymbol{k}' - \boldsymbol{k})\boldsymbol{x} d^3k' d^3x. \qquad (21.6)$$

The integration over x yields δ-functions which, in turn, after integration over k' give

$$\mathcal{P}(\boldsymbol{k}'') \propto \mathcal{E}_1 \mathcal{E}_3 \mathcal{E}_2(-\boldsymbol{k}''). \qquad (21.7)$$

Each harmonic of field $\mathcal{E}_2(x, t)$ is seen to correspond to the inphase polarization harmonic which propagates in the strictly opposite direction. If the synchronism conditions are fulfilled, which is assumed, each oscillating polarization harmonic characterized by a certain frequency and wave vector causes the emission of a plane electromagnetic wave at the same frequency with the same wavevector. Therefore, the plane wave $\mathcal{E}_3(x, t)$ causes the appearance of the spatial wave $\mathcal{E}_4(x, t)$ with the configuration of the field exactly coinciding with that of wave $\mathcal{E}_2(x, t)$ but with the wavevector strictly opposite to the wave vector of wave $\mathcal{E}_2(x, t)$ This is the essence of the wave front inversion (conjugation) at the interaction of four waves. Two oppositely aligned intense fields in the region of their mutual action create in the medium the mirror acting on any other field of the same frequency falling onto the region at an arbitrary angle. This mirror is somewhat specific, since it does not reflect the wave but inverts it, i.e., the "reflected" wave propagates in the direction strictly opposite to the direction of incident light.

The four-wave inversion of the wave front is an example of dynamic holography, where a holographic image is formed by the base field ($\mathcal{E}_1$) and the subject field ($\mathcal{E}_2$). This image is not fixed anywhere but is read out in the real-time scale (in the time scale of V–T-relaxation) by the restoring field ($\mathcal{E}_3$).

Note that the inversion of the wave front may be observed not only owing to the change of χ' but at the expense of the change in χ'' as well, i.e., due to the spatial modulation by the external fields of both the refraction index and the absorption coefficient.

The inversion of the wave front has been observed experimentally at the four-wave interaction of intense laser beams in the resonant medium. The pulsed CO_2-lasers have been employed as sources of laser beams while the resonantly absorbing medium was a gas of SF_6-molecules (the medium with an anomalous resonant dispersion) at the pressure of several dozen torr. The reflection coefficient reached the value of 30 %.

Now let us turn to considering the self-action in the systems in which the laser thermochemical processes take place. We have seen in Lecture 20 how back loops arise due to the nonlinear character of radiation absorption (nonlinear dependence of the temperature of the system on the irradiation intensity). These back loops lead to self-maintaining nonstationary processes (limiting cycles, chaotization). However, the back loops necessary for the appearance of nonstationary processes at stationary excitation may be realized at the expense of the nonlinear optical effects of radiation propagation in the medium. This is the essence of the third possibility mentioned at the beginning of this lecture.

This phenomenon can be summarized as follows. Laser radiation with a nonuniform transverse distribution of intensity propagating in a resonantly absorbing medium leads to a gradual heating of the illuminated region as well as to a heating of the region not exposed directly to irradiation, due to the thermoconductivity. Let the pyrolysis of the type described by formula (20.2) take place in the medium, the refraction indices at the radiation frequency of the initial substance (A) and the product of pyrolysis (B) being considerably different. So, the nonuniform temperature field established in the substance at some time moment causes a spatial inhomogeneity of the refraction index leading to a corresponding spatial distribution of the relative concentrations of the initial substance and the product of pyrolysis. This, in turn, leads to a redistribution of the radiation in space that changes the temperature field with time, etc. (i.e., self-action of the radiation is observed). The relatively slow rate of the heating and cooling processes and the presence of the hysteresis phenomena in the chemical reactions explains the considerable delay of the refraction index profile response to the change in the spatial intensity distribution.

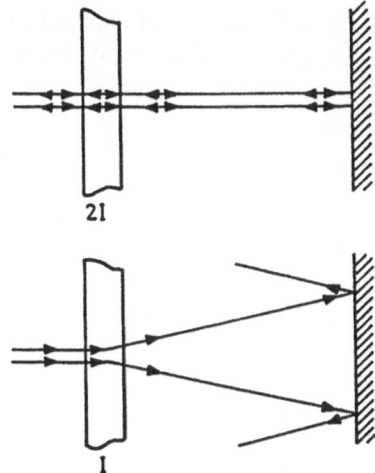

Fig. 21.3. Self-action of laser radiation propagating in a laser-thermochemical medium

Let us illustrate the mechanism of the phenomenon under consideration with a rather simple example. Let a narrow laser beam fall onto an optically thin nonlinear medium – gas $A+B$ (see 20.2). This beam, being reflected from a mirror situated at the distance greater than the radiation coherence length, returns and passes through the medium at the same place (Fig. 21.3). If the laser radiation intensity equals I, the total intensity of $2I$ passes through the medium (if we neglect the radiation divergence). Let the value I be such that the intensity of I corresponds to the temperature at which the representing point is below the catastrophe point, while the intensity of $2I$ corresponds to the representing point position above that point (Fig. 20.2). In such a situation there is no stationary regime in the system. Indeed, as the system is heated in the radiation field of intensity $2I$, the medium reaches point B (Fig. 20.2) and performs the transition to the upper frill sheet. The refraction index strongly changes, which brings about strong scattering of the radiation. After passing the medium this radiation, being reflected by the mirror, does not fall now into the region under consideration. Because of the decrease of intensity the medium gets cold and after reaching point C transits to the lower frill sheet. The entire process than start again.

It is evident that in continuous systems, where a change of the refraction index in some region leads to a change of the extent of heating in another domain, the dynamics of the process may be considerably more complex. The spatial variable-in-time goffering of the laser beam in the resonantly absorbing medium has been observed experimentally. The experiment was performed for CF_3I and SF_6 at pressures of the order of 100 torr using the continuous-wave CO_2-laser with a power of up to 10 W. The cell, about 20 cm long and containing a gas, was situated between the positive lens and its focal point. At this point the radiation distribution that passed through the cuvette was recorded using the IR imager. Almost-regular pulsations of this radiation divergence were observed. The period of these pulsations was between 10 and 100 s.

Afterword

The questions discussed in this book are organized and set forth proceeding from didactic considerations. Some of the topics are relatively new while others are well known and developed in detail. Though we believe that this book may be read without any additional sources, it is worthwhile mentioning the literature both reviewing subjects and treating some questions more fully. We give a list of literature, pointing out that our references do not reflect a priority of authors.

The questions of the coherent interaction of radiation with substance, photon echo, and self-induced transparency are discussed in considerable detail in reviews [12, 13] (see also [14]).

The formalism used when describing the relaxation processes is given in greater detail in [15, 16].

To become better acquainted with the questions concerning the saturation of homogeneously and inhomogeneously broadened lines, intra-Doppler spectroscopy, and dispersion relations, one may use books [17–19].

The relatively new phenomenon of light pressure exerted upon resonant absorbing particles is elucidated in review [20] (see also [A1–A3]). For those who want a more detailed knowledge of the important question in laser gaskinetics of light-induced drift, we mention papers [21–24].

There is a voluminous literature on the question of multiphoton processes in atoms and simple molecules. We cite only the fundamental works [25, 26], a monograph [27], and a book of collected articles [28] that contains a representative bibliography (see also [A4–A6]).

For a detailed acquaintance with the statistical approach to spectrally complex systems we recommend monograph [29] (see also [A7–A9]). The coherent processes in complex multilevel systems are considered in publications [30–32] (see also [28] and [A10]).

Papers [33] and monographs [34,35] are devoted to a detailed consideration of the physics of collisional processes (see also [A11–A12]).

Considering parametric interactions, we have paid attention mainly to their quantum nature. For an introduction to the traditional approach to these phenomena we recommend monographs [36–38] and [A13].

There are many papers devoted to the action of laser light on atoms and molecules. Reviews [39, 40], monographs [27, 41], and books [28, 42] are in best agreement with the aims of this book (see also [A14–A15]). The problems of the excitation of highly energetic states of atoms and molecules turn out to be closely related to the problems of the nonlinear theory of vibrations, and dynamic stochasticity. For a detailed understanding of these theories we recommend books [43–45] and review [46] (see also [A16–A20]). The peculiarities of the spectrum of small symmetric molecules important to understanding the processes of the excitation of such molecules are given in [47].

Concerning photochemical processes we recommend consultation of books [42–49]. Laser thermochemical processes are described in reviews [50–52]. On the question of the influence of laser radiation on adsorption processes in systems with a developed surface we recommend reference [53]. Wave front inversion using the mechanism of intense resonant interactions is considered in [54] (see also [A21]). The photochemical burning out of the gap and its applications in the low-temperature spectroscopy of solid solutions of organic molecules is discussed in the thorough paper [55].

Emphasizing once more that this list of literature is far from being complete, we would like to mention in conclusion that the rapid progress of quantum electronics promises further development of already-existing branches and the appearance of new, interesting studies of intense resonant interactions of laser light with matter. For practical applications one should expect the most significant results in the realm of chemical transformations caused by the action of laser light.

Appendix

1. Dicke's Cooperative Effect

An ensemble of particles placed in a field interacts with it, generally speaking, in a different way than a single particle does. Emitting and absorbing the radiation quanta, the particles interact with each other, phasing their Rabi oscillations. If such an interaction involves a large number of particles, the ensemble behaves cooperatively – the inphaseness brings about macroscopic polarization and increases the rate of spontaneous decay. The cooperative behavior, as a rule, becomes apparent when the distance between the particles is much less than the wavelength. Namely, in this case, the phases of all pairwise interactions turn out to be practically the same (close to zero).

Let us consider this phenomenon in more detail. Let the small volume $a \times a \times a$, $a \ll \lambda$, contain many two-level particles. For the convenience of calculations we assume that the lower level has the energy of $-\hbar\omega/2$, and the upper one, $+\hbar\omega/2$. Then, using the Pauli matrices, the system Hamiltonian may be written in the form

$$\widehat{H}_{\text{sys}} = \frac{1}{2}\hbar\omega \sum_{n=1}^{N} \widehat{\sigma}_{z,n} \,.$$

The Hamiltonian of the electromagnetic field is written as

$$\widehat{H}_f = \hbar\omega \sum_k A_k^+ A_k \,,$$

where operators A_k^+ and A_k correspond to the creation and the annihilation of photons of type k. In this notation the energy of the interaction of the photons with the particles may be represented as

$$\widehat{H}_{\text{int}} = V \sum_{k,n} (A_k^+ + A_k)\widehat{\sigma}_{x,n} + V \sum_{k,n} (A_k^+ - A_k)\widehat{\sigma}_{y,n}$$

$$= V \sum_k (A_k^+ + A_k) \sum_n \widehat{\sigma}_{x,n} + V \sum_k (A_k^+ - A_k) \sum_n \widehat{\sigma}_{y,n} \,,$$

where the interaction constant V is $V = \mu(2\pi\hbar\omega/v)^{1/2}$; and v, the volume. Let us note that the Hamiltonian of the system of particles and their interaction with radiation contains only one combination of the Pauli matrices, namely, the sum of these matrices over all the particles:

$$\sum_{n=1}^{N} \widehat{\sigma}_{z,n} = S_z , \quad \sum_{n=1}^{N} \widehat{\sigma}_{y,n} = S_y , \quad \sum_{n=1}^{N} \widehat{\sigma}_{x,n} = S_x .$$

The resulting spectrum of the ensemble of N particles is a set of N equidistant levels multiply degenerated, in general. Level k has the degeneracy multiplicity C_N^k since this is the number of ways to distribute k excitations among N particles. This spectrum is fairly simple but the lattice of allowed transitions corresponding to it is complex. Indeed, for each state of the kth level corresponding to a certain distribution of k excitations among N particles there are $N - k$ transitions with the additional excitation of one of the $N - k$ still unexcited particles. But the degeneracy of levels allows us to pass on to a new basis of states in which the dipole-moment-operator matrix turns out to be diagonal. In this basis each state of kth energy level corresponds to a single state of the $k + 1$th level to which the transition is allowed. These states are called the Dicke states. Due to the summation rule the force of the transition between the Dicke states is considerably greater than that for the transitions of a single particle.

The physical meaning of the Dicke states becomes more clear if one takes advantage of the well-known analogy between a two-level particle and a particle with spin $\frac{1}{2}$ placed in a magnetic field. It was emphasized above that the system Hamiltonian including the interaction terms contains the sums of the Pauli matrices over all the particles. In other words, the set of N particles is equivalent to the particle with a large value of spin S, $0 \leq S < N/2$. The structure of the interaction Hamiltonian,

$$(\widehat{A}^+ + \widehat{A})\widehat{S}_x + (\widehat{A}^+ - \widehat{A})\widehat{S}_y ,$$

is such that it leaves the total spin S unchanged, causing transitions between neighboring magnetic sublevels. So, the complete spectrum of Dicke states may be described by two quantum numbers: the total spin S and the magnetic quantum number M which obey the well-known selection rule $\Delta S = 0$ and $\Delta M = \pm 1$. The amplitude of the transition probability may be found in this case using the rules of determining the matrix elements of the vectors in the moment representation

$$V_{S,M}^{S,M+1} = V_0^1 (S - M)^{1/2}(S + M + 1)^{1/2} ,$$

where V_0^1 is the matrix element of the transition for a single particle. Sometimes notations R and R_z are used for quantum numbers S and M. For each value of S there are

$$\frac{2S + 1}{(N/2 + S + 1)} \frac{N!}{(N/2 - S)!(N/2 + S)!}$$

ways to obtain the total spin of S adding algebraically N spins of $1/2$. The analysis of this expression shows that the fractional number of states with small spin decreases at $S \to 0$ according to a power law while the fraction of states with large spin decreases as S increases in accord with Gauss law, reaching its maximum – unity – at $S = N/2$. The largest number of states have spin $S \sim N/2$. This may be easily understood in the classical limit. The procedure of adding N vectors of length $1/2$ is in outline analogous to the process of random walks with the step of $1/2$. The most probable result of such a summation is known to be the vector with the length of $N^{1/2}$. Therefore, the states with $S \sim N^{1/2}$ have the largest statistical weight. But these states are characterized by the high excitation energies of $\hbar\omega(N/2 + N^{1/2})$ and, so, at $\hbar\omega \ll kT$ they are practically void. The Boltzmann distribution preferably populates the states with lower energies, i.e. those with large spin:

$$S \le \frac{N}{2} \frac{1 - \exp(-\hbar\omega/kT)}{1 + \exp(-\hbar\omega/kT)} \ .$$

It is this circumstance that allows us to observe the cooperative phenomena. At $S \sim N/2$ the typical matrix element of transition is $\sim N$ times greater than that for a particle. Hence, $\sim N$ magnetic sublevels are passed approximately in the time the particle undergoes a Rabi oscillation. In fact, we deal with the precession of a large spin in a constant magnetic field which rotates with the same period as a spin $1/2$. In this case the medium polarization is proportional to the number of particles.

At high temperatures, when the statistically more profitable states with $S \sim N/2$ are populated, the medium polarization precees at the same Rabi frequency; but it turns out to be $N^{1/2}$ times less than in the previous case.

However, the most essential differences between systems with small and large values of S become apparent in the process of spontaneous decay. Indeed, the matrix element of the transition for the system with large S is N times greater than for a single particle. Hence, the rate of spontaneous transition $S, M \to S, M-1$ is N^2 times greater than for a single isolated particle. In other words, N particles in the Dicke state with large spin value emit N quanta N times faster than a particle emits a single quantum.

In the statistically more favorable states where $S \sim N^{1/2}$, the rate of spontaneous decay corresponds to the decay rate of a single particle, though in this case only $N^{1/2}$ quanta may be emitted. The other quanta are captured and may be rescued only if nonradiative interactions are present.

In conclusion, let us note that the possibility of the laser generation is also associated with the existence of the Dicke states with large $S \sim N^{1/2}$, since it is these states most strongly interacting with the radiation that turn out to be populated in the systems with large inversion.

2. Dipole-Dipole Interaction as a Process in Multilevel Systems

The dipole-dipole interaction presents one of the processes in multilevel systems. This is the two-quantum transition via the set of a large number of intermediate levels which appear if one adds the spectrum of one-photon excitations of the electromagnetic field to the compound spectrum of two particles given in fig.12.1. It is sufficient to pay attention to those field states which are added to the system state $(0,0)$ corresponding to the absence of excitation in both particles. In other words, of interest are such intermediate states in which both particles are unexcited and the photon with the definite wavevector k is emitted. The composite matrix element of transition $(1,0) \rightarrow (0,1)$ will be determined by the detuning of the intermediate state from the resonance $\hbar\omega - \hbar kc$ and by the matrix elements of the initial and final states with intermediate μE and μE, where $|E| = (4\pi kc\hbar/v)^{1/2}$ is the amplitude of the zero-point vibrations of the electromagnetic field in volume v corresponding to the field oscillator with wavevector k. Let us determine the cosine of the angle between μ and E in the simplest case. Let the particles be at distance $r_1 - r_2 = x$ from each other and the vectors of their dipole moments be parallel and directed along the y-axis normal to the x-axis. Then for the photon with wavevector $k = \{k_x, k_y, k_z\}$ the cosines being sought are the same for both particles and equal $(1 - k_y^2/k^2)^{1/2}$. Substitution into the expression for the composite matrix element (9.23), and integration over all intermediate states, i.e., over the volume and over the wavevector k space, with allowance for the phase shift $jk_x x$ and for two possible polarizations of a photon, yield in the polar system of coordinates $k_x = k\cos\theta$, $k_y = \sin\theta\cos\phi$, $k_z = k\sin\theta\sin\phi$:

$$
V_{01}^{10} = 2 \int \frac{8\pi\hbar kc}{v} \mu_1\mu_2 (1 - \cos^2\phi\sin^2\theta) \frac{e^{jkx\cos\theta}}{\hbar kc - \hbar\omega} \frac{k^2\sin\theta\,d\theta\,d\phi\,dk\,du}{(2\pi)^3}
$$

$$
= \frac{1}{\pi} \int k^2 \left[\frac{\sin kx}{kx} - \frac{\sin kx}{(kx)^3} \right] \frac{kc}{kc - \omega} \mu_1\mu_2\,dk .
$$

The first bracketed term corresponds to the dipole field in the far region and we ignore it. The integration of the second term yields

$$
-\frac{\mu_1\mu_2}{x^3} \frac{2}{\pi} \left\{ \cos\frac{\omega x}{c} \left[Si\frac{\omega x}{c} + \pi \right] - \sin\frac{\omega x}{c} Ci\frac{\omega x}{c} \right\},
$$

which at distances less than the wavelength produces $V_{01}^{10} = -\mu_1\mu_2/x^3$, i.e., expression (12.1). Let us note that the main contribution to the integral over k given above is rendered by the nonresonant transition of the photons with a wavelength of the order of the interparticle distance.

Bibliography

[1] Landau, L. D., Lifshitz, E. M.: Quantum Mechanics. Moscow, Fizmatgiz, 1974
[2] Lifshitz, E. M., Pitaevsky, L. P.: Physical Kinetics. Moscow, Nauka, 1979
[3] Silin, V. P.: Introduction to Kinetic Theory of Gases. Moscow, Nauka, 1971
[4] Rumer, Yu. B., Ryvkin, M. Sh.: Thermodynamics, Statistical Physics and Kinetics. Nauka, Moscow, 1972
[5] Andronov, A. A., Witt, A. A., Haikin, S. E.: Theory of Vibrations. Nauka, Moscow, 1981
[6] Sobelman, I. I.: Introduction to Atomic Spectra Theory. Nauka, Moscow, 1977
[7] Elyashevitch, M. A.: Atomic and Molecular Spectroscopy. Fizmatgiz, Moscow, 1962
[8] Hertzberg, G.: Infrared and Raman Spectra of Polyatomic Molecules. New York, 1945
[9] Mathews, J., Walker, R. L.: Mathematical Methods of Physics. Benjamin, Menlo Park, CA, 1970
[10] Karlov, N. V.: Lectures on Quantum Electronics. Nauka, Moscow, 1983
[11] Svelto, O.: Principles of Lasers, Plenum Press, New York London, 1982
[12] McCall, S. L., Hahn, E. L.: Self-Induced Transparency. Phys. Rev. **180** (1969) 457–485
[13] Poluektov, I. A., Popov, Yu. M., Roitberg, V. S.: Coherent Effects of Ultrashort Light Pulse Propagation in Resonant Media. Quantum Electronics (Russ.) **1**, no. 4 (1974) 1309 1974, vol. 1, 4, p. 757; 6, p. 1309
[14] Basov, N. G. (ed.): Coherent Cooperative Phenomena. FIAN Proceedings (Russ.) Nauka, Moscow, 1975, vol. 87
[15] Fain, V. M., Hanin, Ya. I.: Quantum Radiophysics. Sov. Radio, Moscow, 1972, vol. 1; 1975, vol. 2
[16] Zwanzig, R. W.: Lectures in Theoretical Physics. NBS Publications, Boulder, Colorado, USA, 1960; J. Chem. Phys. **33** (1960) 1338
[17] Yariv, A.: Quantum Electronics. Wiley, New York, 1988
[18] Letokhov, V. S., Chebotaev, V. P.: Principles of Non-Linear Laser Spectroscopy. Nauka, Moscow, 1975
[19] Rautian, S. G., Smirnov, G. I., Shalagin, A. M.: Non-Linear Resonances in Spectra of Atoms and Molecules. Nauka, Novosibirsk, 1979
[20] Kasantsev, A. P., Ryabenko, G. A., Surdutovitch, G. I., Iakovlev, V. P.: Scattering of Atoms by Light. Phys. Reports **129** (1985) 2
[21] Gelmukhanov, F. H., Shalagin, A. M.: Light-Induced Gas Diffusion. JETP Letters **29** (1979) 773–776
[22] Dykhne, A. M., Starostin, A. N.: Theory of Drift Motion of Molecules in the Field of Resonant IR Radiation. JETP **75** (1980) 1211–1226
[23] Kalyazin, A. L., Sazonov, V. N.: Anisotropy of Inelastic Scattering and Selective Diffusion of Gas Mixture Components under the Action of Laser Light. Quantum Electronics (Russ.) **6** (1979) 1620–1625
[24] Brzhazovsky, Yu. V., Vasilenko, L. S., Rubtsov, N. N.: SF_6 Diffusion under the Action of CO_2 Laser Radiation. JETP Letters **35** (1982) 527–529

[25] Keldysh, L. V.: Ionization in the Field of Strong Electromagnetic Wave. JETP **74** (1964) 1945–1952

[26] Zeldovitch, Ya. B.: Quasi-Energy of a Quantum System under Periodic Action. JETP **51** (1966) 1492–1509

[27] Delone, N. B., Krainov, V. P.: Atom in Strong Light Field. Energoatomizdat, Moscow, 1984

[28] Basov, N. G.(ed.): Multiphoton Proceses in Molecules. FIAN Proceedings. Naukla, Moscow, 1984, vol. 146

[29] Dyson, J. F.: Statistical Theory of the Energy Levels of Complex Systems. J. Math. Phys **3** no. 1 (1962) 140, 157, 166, parts I–III

[30] Akulin, V. M., Dykhne, A. M.: Dynamics of the Excitation of Zone-Type Multilevel Systems by Laser Field. JETP **73** (1977) 2098–2106

[31] Makarov, A. A., Platonenko, V. G., Tyaht, V. V.: Interaction of "Level-Zone" Quantum System with Quasi-Resonant Monochromatic Field. JETP **75** (1978) 2075–2092

[32] Fedorov, M. V.: Resonant Multiphoton Ionization of Atoms in the Field of Intensive Electromagnetic Wave. Izvestiya AN SSSR, Fizika **41** (1977) 2569–2576

[33] Treanor, C. E., Rich, J. W., Rehm, R. G.: Vibrational Relaxation of Anharmonic Oscillators with Energy Dominated Collision. J. Chem. Phys. **48** (1968) 1798

[34] Gordiez, B. F., Osipov, A. I., Shelepin, L. A.: Kinetic Processes in Gases and Molecular Lasers. Nauka, Moscow, 1980

[35] Yakovlenko, S. I.: Radiative Collisional Effects. Energoatomizdat, Moscow, 1984

[36] Bloembergen, N.: Non-Linear Optics Benjamin, New York, 1965

[37] Ahmanov, A. S., Khokhlov, R. V.: Problems of Non-Linear Optics. VINITI, Moscow, 1964

[38] Ahmanov, S. A., Koroteev, N. I.: Methods of Non-Linear Optics in Light Diffusion Spectroscopy. Nauka, Moscow, 1981

[39] Karlov, N. V., Prohorov, A. M.: Laser Isotope Separation Soviet Science Uspekhi **118** (1976) 583–609

[40] Karlov, N. V., Krynetsky, B. B., Mishin, V. A., Prohorov, A. M.: Selective Photoionization of Atoms and Its Application. Soviet Science Uspekhi **127** (1979) 593

[41] Letokhov, V. S.: Non-Linear Selective Photoprocesses in Atoms and Molecules. Nauka, Moscow, 1983

[42] Steinfeld, J. J.(ed.): Laser-Induced Chemical Processes. Plenum, New York, 1981

[43] Bogolyubov, N. N., Mitropolsky, Yu. A.: Asymptotic Methods in Non-Linear Oscillations Theory. Gostehizdat, Moscow, 1955

[44] Rabinovitch, M. I., Trubetskov, D. I.: Introduction to Theory of Vibrations and Waves. Nauka, Moscow, 1974

[45] Lichtenberg, A. J., Lieberman, M. A.: Regular and Stochastic Motion Springer, Berlin Heidelberg New York, 1983

[46] Zaslavsky, G. M., Chirikov, B. V.: Stochastic Instability of Non-Linear Oscillations. Soviet Physics Uspekhi **105** (1971) 7–13

[47] Hecht, K. T.: The Vibrational-Rotational Energy of the Terahedral XY_4 Molecule. J. Mol. Spectroscopy, **5** (1960) 355 part I, 355, part II

[48] Okabe, H.: Photochemistry of Small Molecules. Wiley, New York, 1978

[49] Molin, Yu. N., Panfilov, V. N., Petrov, A. K.: IR Photochemistry. Nauka, Novosibirsk, 1985

[50] Bunkin, F. V., Kirichenko, N. A., Lukjantchuk, B. S.: Thermochemical Action of Laser Radiation. Soviet Physics Uspekhi **138** (1982) 45–94

[51] Bunkin, F. V., Kirichenko, N. A., Lukjantchuk, B. S.: Laser Thermochemistry. Soviet Physics Izvestia AN SSSR, Fizika, **46** (1982) 1150–1169

[52] Bunkin, V. F., Kirichenko, N. A., Lukjantchuk, B. S.: Thermochemical and Thermokinetic Processes in the Field of Continuous Laser Radiation. Soviet Physics Izvestia AN SSSR, Fizika, **47** (1983) 2000–2016

[53] Karlov, N. V., Orlov, A. N., Petrov, Yu. N., Prokhorov, A. M.: Laser Control of Diffusive Gas Outflow into Vacuum through Obstacles with Developed Surface. Soviet Physics Izvestia AN SSSR, Fizika, **49** (1985) 500–505

[54] Basov, N. G., Kovalev, V. I., Faizulov, F. S.: Wave Front Inversion in Moderate IR Range of Wavelengths. in Wave Front Inversion of Radiation in Non-Linear Media. ed. by Bespalov, V. I. IPF Press AS USSR, Gorgy, 1982, pp. 18–39

[55] Rebane, L. Q., Grokhovsky, A. A., Kikas, J. V.: Low-Temperature Spectroscopy of Organic Molecules in Solids by Photochemical Hole Burning. Appl. Phys **B29** (1982) 235

Bibliography to the English Edition

[A1] Arimondo, E., Bambini, A., Stenholm, S.: Phys. Rev. **A24** N2 (1980) 898–909
[A2] Stenholm, S.: Appl. Phys., **15** (1978) 287
[A3] Martin, P. J., Gould, P. L., Oldaker, B. J., Miklich, A. H., Pritchard, D. E.: Phys. Rev. **A36**, N5 (1984) 2495–2498
[A4] Stebbing, R. E., Dunning, F. B. (eds.): Rydberg States of Atoms and Molecules. University Press, Cambridge, 1983
[A5] Bagfield, J. E., Stevens, K. W. H., Sanders, J. H., Stenholm, S., Moss, T. S. (eds.): Microwawe Multiphoton Processes in Highly Excited Atoms. Progress in Quantum Electronics. Pergamon, Oxford, 1980
[A6] Chin, S. L., Lambropolus, P. (ed.): Multiphoton Ionization of Atoms. Academic, New York, 1984
[A7] Israilev, F.: J. Phys A. Math. Gen. **22** (1989) 865–878
[A8] Friedrich, H., Wintgen, D.: Phys. Reports **183** N2 (1989) 37–79
[A9] Menta, M. L.: Random Matrices. Academic, New York, 1990
[A10] Akulin, V. M., Dykhne, A. M.: Soviet Phys. JETP **67** N4 (1988) 856-865
[A11] Rebentrost, F.: Nonadiabatic Molecular Collisions. Theoretical Chemistry Advances and Perspectives, vol. 65 Academic New York 1981.
[A12] Capitelli, M. (ed.): Nonequilibrium Vibrational Kinetics. Springer, Berlin Heidelberg, 1986
[A13] Meystre, P.: Elements of Ouantum Optics. Springer, Berlin Heidelberg, 1990
[A14] Stenholm, S. (ed.): Laser in Applied and Fundamental Research. Hilger, Bristol, 1985
[A15] Lin, S. H., Fujimura, Y., Neusser, H. J., Schlang, E. S., Orlando, F. L.: Multiphoton Spectroscopy of Molecules. Academic Press 1984
[A16] Zaslavsky, G. M.: Chaos in Dynamical Systems. Chur: Hardwood, 1985
[A17] Abraham, N. B., Arecchi, F. T., Lugiato, L. A. (eds.): Instabilities and Chaos in Quantum Optics. Plenum, New York, 1987
[A18] Moss, F., Lugiato, L. A., Schleich, W. (eds.): Noise and Chaos in Nonlinear Dynamical Systems. Cambrige University Press, 1989
[A19] Casatti, G., Guarnrri, I., Shepeliansky, D. L.: Physica **A163** (1990) 205–214
[A20] Berry, M. V., Rerciva, I. C., Weiss, N. C.: Dynamical Chaos. Princeton University Press, 1989
[A21] Zeldovich, B. Ya., Pilipetsky, N. F., Shkunov, V. V.: Principles of Phase Conjugation. Springer, Berlin Heidelberg, 1985

Subject Index

Adiabatic approximation 74, 104, 188
Adsorption 277
Almost equidistant levels 105
Anharmonicity
  constant 191
  tensor 193
Anti-Stokes process 164, 259
Area theorem 14
Artificial autoionization 180
Assorting by velocity 59
Autoionization 180
Automodel solution 25

Balance equation (kinetic equation)
  128
Band-band system 115, 117, 126, 129
Bloch equations 33, 74
Bogoljubov method 34, 68
Boltzmann equation 67, 81
Born approximation 149
Born-Oppenheimer parameter 189, 258
Bottleneck effect 231

Carplus-Schwinger expression 49
CARS 167
Cascade processes (radiative-collisional
  cascade) 90, 130
Catastrophe 270, 271
Clusters of lines 199
"Coarse-grain" characteristics 108, 181
"Coherent damping" 14, 15, 17,
  123, 129
Coherent effects 3, 10, 21, 144, 152,
  170, 184, 280
Collision
  operator 68
  integral 80
Collisional exchange 146, 175
Collisional ionization 184
Collisional relaxation 68, 140, 174
Composite matrix element 96, 98, 100,
  117, 161, 169
Compound system 147, 158, 162, 184

Coordinational compounds 215
Coriolis
  interaction 195
  tensor 196
Counter-wave method 60
Cubic non-linearity 160, 191, 219
Cubic polarizability 163

Decay channel switching 137
Density matrix 31, 34, 67, 138, 141
Detailed balance principle 268
Dicke effect, states 292
Diffusion equation 132, 135, 139
Diffusion in momenta 77
Diffusive energy build up 211
Dipole-dipole interaction 277, 294
Dispersion 46
Dissociation 234
Doppler broadening 54, 174

Electron terms 189
Ensemble average 109, 112, 114,
  118, 123
EPR 34

Fano problem (*see* "level-band")
Fermi
  golden rule 111
  resonance 203, 210, 254
Field broadening 11
Fine spectrum 199
Fokker-Planck diffusion 81
Franck-Condon principle 258

Gap in line 57
Gas-kinetic collisions 68, 80
Generalized function method 94
Generating function method 133
Gordiez-Osipov-Shelepin distribution
  156, 252

Harmonics generation method 162
Hierarchy of interactions 188